HUMAN ANATOMY AND PHYSIOLOGY

ENDOTHELIUM AND EPITHELIUM: COMPOSITION, FUNCTIONS AND PATHOLOGY

HUMAN ANATOMY AND PHYSIOLOGY

Additional books in this series can be found on Nova's website
under the Series tab.

Additional E-books in this series can be found on Nova's website
under the E-book tab.

CELL BIOLOGY RESEARCH PROGRESS

Additional books in this series can be found on Nova's website
under the Series tab.

Additional E-books in this series can be found on Nova's website
under the E-book tab.

HUMAN ANATOMY AND PHYSIOLOGY

ENDOTHELIUM AND EPITHELIUM: COMPOSITION, FUNCTIONS AND PATHOLOGY

JOÃO CARRASCO AND MATHEUS MOTA
EDITORS

Nova Science Publishers, Inc.
New York

For permission to use material from this book please contact us:
Telephone 631-231-7269; Fax 631-231-8175
Web Site: http://www.novapublishers.com

NOTICE TO THE READER

Additional color graphics may be available in the e-book version of this book.

LIBRARY OF CONGRESS CATALOGING-IN-PUBLICATION DATA
Endothelium and epithelium : composition, functions, and pathology / editors, Joco Carrasco and Matheus Mota.
p. ; cm.
Includes bibliographical references and index.
ISBN 978-1-61470-874-2 (hardcover)
1. Endothelium--Physiopathology. 2. Epithelium. I. Carrasco, Joco. II. Mota, Matheus.
[DNLM: 1. Endothelium--anatomy & histology. 2. Endothelium--physiopathology. 3. Epithelium--anatomy & histology. 4. Epithelium--physiopathology. QS 532.5.E7]
QP88.45E542 2011
611'.0187--dc23

2011027342

Published by Nova Science Publishers, Inc. + New York

Contents

Preface

This book presents topical research in the study of the composition, functions and pathology of endothelium and epithelium. Topics discussed include the pathophysiology of endothelium; the role of epithelial innate immunity in inflammatory diseases; morphofunctional characteristic features of statis endothelium and dynamic epithelial cell populations; changes of respiratory epithelium in bronchial asthma; interleukin-6 (IL6) signaling and its role in the development of cardiovascular disease; endothelium-dependent effects of heat shock on vasomotor tone; the Gasdermin Superfamily functioning in epithelial cells and tight junctions in upper airway epithelium.

Chapter 1 - The oral mucosa comprises a stratified keratinized or non-keratinized epithelium composed by keratinocytes overlaying the lamina propria, which contains glands that secrete mucus and serous secretions and other structures like nerves, blood and lymphatic vessels. The continuous epithelial renewing process is what maintains the structural integrity of the oral mucosa. The cells from the deepest layer of the epithelium constantly suffer mitotic divisions, subsequently migrating upwards and replacing the squamous cells from the upper layers, which in turn lose their nucleus and cytoplasmic organelles, have their metabolism ceased and then undergo a programmed death as they become fully keratinized. Two functional cell populations compose the epithelium: the progenitor population that provides new cells, termed proliferative keratinocytes or epithelial stem cells; and the maturing population, which cells undergo differentiation depending on the extracellular amount of calcium, forbe esters, retinoic acid and D3 vitamin.

The aim of this chapter is to succinctly explain the main characteristics of the oral squamous epithelium cells by presenting their most expressed proteins and morphological architecture as well as to demonstrate the differences between them and squamous cell carcinoma cells, passing through the dysplastic stages generally detected. The oral squamous cell carcinoma (OSCC) is certainly related to the malignant transformation of the oral epithelial cells stimulated by genetic and/or carcinogenic factors, such as tobacco, alcohol and ultraviolet radiation. In general, the dysplastic epithelial cells show some distinct morphological characteristics including nucleus: cytoplasm ratio increase, nuclear and cellular pleomorphism and atypical mitoses. The OSCC cells, on the other hand, tend to lose their epithelial markers expression and begin to express mesenchymal markers, for instance vimentin, through the so-called epithelial-mesenchymal transition (EMT). Normal epithelial cells are interconnected by integrins and cadherins via intercellular junctions (gap or tight) or via desmosomes. Accordingly, SCC cells demonstrate lower expression of adhesion molecules E-cadherin, catenins and integrins, which helps them to migrate and begin the invasion process. One of the most studied signaling pathways related to cancer development is the PI3K/Akt, presenting the tumor suppressor gene PTEN as a regulator. In cancer, PTEN is often down-regulated or mutated and, subsequently NF-kB, cyclin D1 and mTOR proteins show enhanced expression. Alternatively, the cells from normal epithelium present basal activity of the PI3K-Akt signaling pathway and its related proteins once they also participate in the proliferation mechanisms.

Chapter 2 - The endothelial cells (ECs) are squamous cells that form a simple epithelium and line the interior surface of the circulatory system. Different from the first idea of endothelium as an inert tissue, it has emerged as a dynamic tissue with metabolic and synthetic functions, exerting significant autocrine, paracrine and endocrine functions. Vasoconstrictor (endothelin-1) and vasodilator factors (nitric oxide, prostacyclin and endothelium-derived hyperpolarizing factor) are synthesized and released by ECs. Vasodilators acts primarily by lowering cytosolic calcium concentration in the adjacent vascular smooth musccle cells, whereas endothelin-1 has the opposite effect. Besides regulating vasomotor tone, these vasoactive substances as well as others produced by ECs have functions such as regulation of the recruitment and activity of inflammatory cells and proliferation of smooth muscle cells. Importantly, ECs also produce antithrombogenic agents such as plasminogen activator factor, thrombomodulin and prostacyclin, preventing thromboembolic diseases by keeping blood fluidity. However, in pathologic conditions ECs also produce thrombogenic factors such as platelets activator factor (PAF) and von Willebrand factor. The capillary endothelial cells act as a semi-permeable barrier, regulating the transfer of small and large molecules between blood and tissue. Ultrastructurally, these capillaries can be classified in continuous, fenestrated and sinusoidal, according to the continuity of the endothelial layer and its basal lamina. The distribution of these capillary types among different tissues is related to its functions. The transport through the endothelium occurs by specific mechanisms (i.e. facilitative glucose transporters and Na + dependent and independent amino acid transporters), and it is also mediated by transcellular and paracellular pathways that regulate the passage of ions, water and macromolecules, as well as cell polarity. In the heart, the endothelial cells play an obligatory role in regulating and maintaining cardiac function, in particular, at the endocardium and in the myocardium capillaries where endothelial cells directly interact with adjacent cardiomyocytes. Cardiac endothelial cells, like vascular endothelium, also express and release a variety of auto- and paracrine agents which directly influence heart vasculature density, as well as heart metabolism, contractile performance, rhythmicity, and hypertrophy. A blood-heart barrier has been proposed. At the early stages of cardiac failure, all major cardiovascular risk factors may cause cardiac endothelial activation as an adaptive response often followed by cardiac endothelial dysfunction. In addition, it will be presented the specific features of the brain capillary endothelial cells that form the blood-brain barrier. The blood-brain barrier constitutes the physical and metabolic interface between blood and brain. It is responsible for providing a suitable environment for neuronal function, regardless fluctuations in blood composition. The transcellular transport through the endothelial cells is very selective while the paracellular passage is restricted. The specific phenotype of the brain endothelium is induced and maintained by factors released mainly by astrocytic end-feet that surround the abluminal surface of the capillaries besides the influence of pericytes, neurons and microglias that together form the neurovascular unit.

Chapter 3 - Epithelium is the first line of defense. A long-recognized property of epithelial cells is their physical barrier function against potentially harmful substances and microbial pathogens. Increasing new evidence indicates an important role of epithelial cells as both initiators and regulators of innate and adaptive immune responses, as well as the transition from innate immunity to adaptive immunity. Epithelial cells have been found to functionally express various pattern-recognition receptors, including Toll-like receptors (TLRs), which recognize conserved microbial patterns that trigger inducible activation of innate immunity.

Recent studies have identified two novel epithelium-derived pro-allergic cytokines, thymic stromal lymphopoietin (TSLP) and interleukin (IL) 33, as key initiators of the Th2-dominant responses that cause allergic inflammatory disease such as atopic dermatitis, asthma and ocular allergy. TSLP could activate myeloid dendritic cells to produce OX40 ligand (OX40L) that triggers Th2 response. IL-33 has been recently identified as a functional ligand to ST2 receptor on Th2 cells that mediate allergic inflammation. In ocular surface, TSLP/OX40L/OX40 and IL-33/ST2 signaling has been also found to be significantly activated in atopic patients with allergic conjunctivitis, and in a murine model of experimental allergic conjunctivitis induced by short ragweed pollen, a ubiquitous allergen. Interestingly, TSLP and IL-33 have been observed to be largely induced by specific microbial pathogens, the TLR ligands, such as viral double-stranded RNA (dsRNA) and bacterial component flagellin, via TLR/MyD88, TLR/TRIF and NF-κB pathways. These discoveries suggest that epithelium-derived TSLP and IL-33 serve as a bridge that links innate and adaptive immune responses. TSLP and IL-33 might be the potential biomarkers and novel therapeutic targets for allergic diseases.

Discovery of T helper (Th) 17 cells as a new T helper cell subset distinct from Th1 and Th2 cells is a recent breakthrough in T cell biology. The differentiation of Th17 cells from naïve CD4+ T cells is initiated by transforming growth factor-β (TGF- β) and IL-6; and the further expansion of Th17 cells is induced by cytokines IL-23, IL-1β and IL-21. Interestingly, Th17 cells promote inflammation, autoimmunity and allergy frequently at mucosal epithelial sites that are rich in TGF-β, IL-6, IL-23, and IL-1β, the major inducing cytokines for Th17 lineage. Th17 cells produce IL-17A and IL-17F, as well as IL-22 and chemokine attractant ligand CCL20, the pro-inflammatory cytokines presenting an intriguing new bridge between adaptive and innate immunity. In ocular surface, Th17 differentiation has been shown to be enhanced by cytokines produced by corneal epithelium that are exposed to hyperosmotic stress, inflammatory stimuli and microbial pathogens. Th17 has been identified as a major T cell disorder involved in pathogenesis of human dry eye syndrome, an inflammatory autoimmune ocular surface disease. The increased levels of IL-17-producing Th17 cells and associated cytokines have been observed in human dry eye and murine experimental dry eye model. The desiccating stress has been demonstrated to promote the production of Th17 inducing cytokines by corneal and conjunctival epithelia, and creates an environment priming Th17 differentiation through a dendritic cell-mediated pathway. Th17 signaling molecules become novel therapeutic targets for treating dry eye conditions.

Chapter 4 - In biology and medicine, the big attention to structure features, functional reactance, regeneration and other properties of epithelial tissue is paid. The special interest causes the skin epidermis. Recently, there was considerable data about its histostructure complexity (Letuchaya and Ketlinskii, 1980; Savost'anov, 2001; Potten and.Booth, 2002). Thus, in the overwhelming majority of works, morphofunctional rearrangement of epitheliocytes only in vertical direction was investigated.

Besides, available data on quantitative basement cells morphology (quantitative correlations between various cells groups in it, cells distribution according to their area and form, correlations between progenitor patterns and their various derivatives) is obviously insufficient.

Essentially, the objective information on a basal epitheliocytes state represents the special interest from positions of modern view about the organization and reaction of epithelial layer in various situations.

Skin epidermis is intensively proliferating tissue. Therefore, it is difficult to give quantitative and quality value of structural and population basement cells rearrangements at change of vital organism conditions.

Considering this circumstance, we have carried out relative analysis of two types of epithelial tissue: the corneal endothelium, which has lost proliferation ability during a postnatal ontogenesis and the skin's multi-layered epithelium – actively regenerating in the specified period. It has enabled us to analyze more deeply the structural and functional organization of epithelial tissue in norm and at influence of different factors on epithelial histic structures.

In this article, we summarize our own material related to spatial organization of different epithelia (the posterior corneal epithelium - endothelium and ear skin epidermis), reviewed in 350 mature male BALB/cj mice. We also examined cornea and skin obtained from 60 adult nominally healthy men between 20 and 90 years old who died from cardiovascular disease or from accidental traumas (control to the experimental group –60 persons died from chronic alcohol intoxication); we also examined tumor tissue of two patients groups with breast cancer or fibroadenoma; each group consisted of 15 patients 40 to 60 years old.

Chapter 5 - Airway epithelium lines the upper part of the respiratory tract and functions to maintain airway integrity by providing both a physical barrier to inhaled noxious agents and a mechanism for their clearance *via* the mucociliary escalator. Ciliated columnar cells and goblet cells are the most important cell types in the normal respiratory epithelium. The other types of cells are represented by basal cells, brush cells and Kulchitsky cells (small granule cells). The normal architecture and function of airway epithelium is disturbed in chronic airway inflammatory diseases such as bronchial asthma. Airway remodeling is the general description for the thickening and restructuring of the airways seen in asthma patients. The characteristics of airway remodeling include inflammatory cell infiltration in *lamina propria*, subepithelial fibrosis, muscular and glandular hypertrophy, thickening of basement membrane and increased vascularity. Damage to the airway epithelium is one prominent feature of bronchial asthma. Epithelium damage includes gap openings, partial or total denudation, goblet cell metaplasia. The increased production of mucus can lead to the formation of plugs that contribute to airway obstruction. Mucin glycoproteins (MUC) are the major macromolecular component of mucus in normal airway epithelium. An increased expression of MUC2, MUC4 and MUC5B has been reported in asthmatic bronchial biopsies. At the epithelium level, in patients with bronchial asthma, appears an increased expression of Ki-67, p21 (WAF1) and decreased expression of FasL, Bcl-2 or Bax, too. Apoptosis of the airway epithelium is increasingly recognized as a potential mechanism by which damage may occur.

More work is now needed to comprehend all aspects of this disease, including minimizing the ongoing damage to airway epithelium and consequent airway remodeling.

Chapter 6 - The testes of mammals are paired organs that essentially perform two functions: steroid biosynthesis and production of spermatozoa. They consist in two compartments encased by the tunica albuginea: the seminiferous tubules, the functional units of the testes, and the intervening interstitial space. Within the seminiferous tubules, the seminiferous epithelium is compartmentalized by adjacent Sertoli cells(SCs) via specialized junctions creating the blood-testis barrier (BTB) where post-meiotic germ cell development takes place in a specialized microenvironment. The fully functional BTB consists in three components: an anatomical/physical barrier to restrict entry of molecules into the adluminal compartment of the seminiferous tubules; an immunological barrier (inside and outside the tubules), that limits the movement of immune cells and regulates the level of cytokines in the seminiferous epithelium; a physiological barrier (transporters and channels at the basolateral and apical membranes), that are highly dynamic to encounter the needs of germ and SCs. Together, these components are essential to the function of BTB in testes and the special microenvironment created is responsible for proper development of the germ cells into fully functional sperm.

Developing germ cells are unable to use glucose for their energy metabolism and preferentially use lactate, used by spermatocytes and spermatids, as a substrate for ATP production. Sertoli cells on the seminiferous epithelium produce the lactate utilized by the developing germ cells. Carbohydrate metabolism of SCs has been shown to present some unique features. These cells can metabolize various substrates (e.g.: glucose, fatty acids, ketone bodies), but preferentially metabolize glucose being the majority of it converted to lactate and not oxidized via the citric acid cycle.The reasons why SCs preferentially export lactate for germ cells are not entirely understood. However, there are evidences of an anti-apoptotic effect of lactate on germ cells suggesting that it plays a crucial role in spermatogenesis.

Several biochemical mechanisms might contribute to the modulation of lactate secretion by SCs, among which (a) the transport of glucose through the plasma membrane, mediated by a set of homologous glycoprotein molecules (GLUTs); (b) the interconversion of pyruvate to lactate by lactate dehydrogenase (LDH); and (c) the release of lactate from SCs, mediated by the action of the monocarboxylate transporters family (MCTs).

The understanding of the functioning and regulation of these processes are crucial steps to identify key mechanisms associated with seminiferous epithelium (dys)function, and their influence over male (in)fertility.

Chapter 7 - The pleiotropic cytokine interleukin-6 (IL6) is elaborated by many cells involved in the development of atherosclerosis, including smooth muscle and endothelial cells as well as monocytes and T cells. In turn, IL6 controls several aspects of vascular and haematopoietic cell function that become dysregulated during the development of the chronic inflammatory phenotype that drives atherogenesis. However many studies have linked IL6 to a range of often contradictory effects on cardiovascular function. In this chapter, we will examine how the unique position of IL6 at the junction of the adaptive and innate immune responses, and its ability to signal via two biochemically distinct IL6-IL6 receptor complexes, may explain the effects ascribed to it in various models of chronic inflammatory disease. We will also assess how an increased understanding of IL6 signalling and its inhibitory regulation, particularly by suppressor of cytokine signalling (SOCS) proteins, might be useful in suggesting new therapeutic strategies for treatment of specific cardiovascular diseases such as atherosclerosis.

Chapter 8 - Heat shock-mediated delayed protection against myocardial and cerebral ischemia has been attributed to induction of heat shock proteins (HSP), especially HSP70, in myocytesand neurons. However, heat shock preconditioning only induces HSP70 in endothelium. The purpose of this study was to examine the effect of heat shock on endothelium-dependent vasomotor tone.

Isolated mouse aortas were heated to 43.5°C for 15 min and returned to 37°C for 18 h.They were then mounted in a wire myograph to assess vasomotor responses to vasoconstrictors and vasodilatators. Intact heat shocked aortas showed reduction in phenylephrine- or 5-hydroxytryptamine-induced vasoconstriction,while no difference in vascular reactivity was detected in heat shocked endothelium-denuded aortas. Reduced vasoconstriction in heat shocked aortas was restored to control level by incubation with L-NAME, a nitric oxide synthase (NOS) inhibitor, but not by geldanamycin, a HSP90 inhibitor.Furthermore, the effect of heat shock was not observed in aortas fromeNOSnull mice, suggesting that vasomotor responsiveness reduced by heat shock was related to increased basal eNOS activity. Immunoblot of lysates from heat shocked aortas did not show upregulation of eNOS or other NOS isoforms, while showingrobust expression of HSP70; immunoprecipitation using eNOS polyclonal or monoclonal antibodies demonstrated HSP70 interaction witheNOSafter heat shock. In enzymatic assay (conversion of L-arginine to L-citrulline), HSP70 increased enzymatic activity of recombinanteNOS.

We conclude that heat shock decreases vasomotor responses to vasoconstrictors by enhancing eNOS basal activity, through interaction with HSP70, that is robustly induced. Such an increase of eNOS might be related to protective effect of heat shock against ischemia reported *in vivo*. Finally, interaction between HSP70 and eNOS opens the intriguing possibility that HSP70 could have vasoregulatory properties in addition to its cytoprotective functions.

Chapter 9 - The Gasdermin Superfamily is a novel gene group consisting of Gasdermin Family genes (*GSDMA*, *GSDMB*, *GSDMC* and *GSDMD* in humans) and Gasdermin-related genes (*DFNA5* and *DFNB59* in humans). The members of the Superfamily are expressed in the epithelial cells of a variable type of tissue and may have a differentiation-status specific role; however, their actual functions remain unknown. In this chapter, the current status of studies on the Superfamily genes is overviewed and their biological significance is discussed.

In the esophageal and gastric epithelia, each member of the Gasdermin Family shows a distinct expression pattern according to the differentiation status of the epithelial cells; *GSDMA* is expressed in differentiated cells in both the epithelia and, in the gastric mucosa, it induces apoptosis of the surface pit cells. On the contrary, *GSDMB* expression can be observed in the region harboring rapidly amplifying epithelial cells. According to this expression pattern and function in the normal epithelium, *GSDMA* is frequently suppressed in esophageal and gastric cancers but *GSDMB* is even over-expressed in them. The expression of Gasdermin-related genes is also observed in several organs including the nervous system. *DFNA5* is a causative gene of autosomal dominant deafness, whose expression was detected in the cochlea. *DFNB59* is responsible for autosomal recessive deafness, expressed in the inner ear and in neurons of the auditory pathway.

The Gasdermin Superfamily genes are expressed in many tissues in a manner of tissue-type and differentiation-status specificities, which implies their important function in cell differentiation. Therefore, phylogenetic conservation of the genes in the animals other than human and mouse is also expected; however, that is yet to be unveiled. Since Placozoa (a primitive form of the invertebrate) as well as amphioxus has at least one sequence coding hypothetical protein similar to human DFNB59, it is very likely that the superfamily genes diverged from a common ancestor gene and now harbor in the genome of other lower animals. A phylogenetic approach may disclose the biological significance of the superfamily.

Chapter 10 - The mucosal barrier of the upper respiratory tract including the nasal cavity, which is the first site of exposure to inhaled antigens, plays an important role in host defense in terms of innate immunity and is regulated in large part by tight junctions of epithelial cells. Tight junction molecules are expressed in both M cells (membranous or microfold cells) and dendritic cells (DCs) as well as epithelial cells of upper airway. Various antigens may be sampled, transported, and released to lymphocytes through the cells while they maintain the integrity of the airway epithelial barrier. On the other hand, epithelial-derived thymic stromal lymphopoietin (TSLP) triggers DC-mediated Th2-type inflammatory responses and is a master switch for allergic inflammatory diseases including allergic rhinitis. TSLP enhances the tight junctional barrier function of human nasal epithelial cells in vitro. Furthermore, we established a model of respiratory syncytial virus (RSV)-infection using human nasal epithelial cells in vitro, showing that tight junction molecules were markedly increased together with proinflammatory cytokine release. This review summarizes the recent progress in our understanding of the regulation of tight junctions in the upper airway epithelium under normal, allergic and RSV-infected condition.

Chapter 11 - We have recently reported that excess fibroblast growth factor-7 (FGF-7) in *Krt12*$^{rtTA/rtTA}$*/tetO-FGF-7* double transgenic mice induced by doxcycline (Dox) resulted in a corneal epithelial tumor resembling human ocular surface squamous neoplasia (OSSN) (Chikama et al. Am. J. Pathol. 172:638-649).

Here, we further investigate the role of β-catenin in mediating FGF-7 induced corneal tumorigenesis. Pannus surgically removed from OSSN patients all had nuclear β-catenin and exhibited excess FGF-7. *In vitro*, FGF-7 induces stabilization and nuclear translocation of β-catenin in human corneal epithelial cells.

In vivo, conditional knock-out of β-catenin gene (*Ctnnb1*) did not have significant effects on corneal morphogenesis or homeostasis, but abolished OSSN formation in Dox-treated *Krt12*$^{rtTA/rtTA}$*/tetO-FGF-7/tetO-Cre/Ctnnb1*$^{lox(E2-6)/lox(E2-6)}$ mice. These results suggest that β-catenin plays a pivotal role in mediating the formation of OSSN. Thus, the administration of inhibitor(s) that prevents activation of β-catenin may be beneficial for treating OSSN.

Chapter 12 - Lactation is under homeostatic control, and sustained milk stasis or weaning leads to mammary gland involution. This transition is under the control of systemic and local factors. Of all the phases of mammary gland development, involution is the least well understood. Recent epidemiological observations suggest that involution events promote breast diseases, including breast cancer. This has renewed interest in understanding involution, and its regulation. In this chapter we discuss the events occurring during mammary gland involution and the role of various systemic and local factors in regulating these events. We also briefly review about how these processes can influence the progression of breast diseases.

In: Endothelium and Epithelium
Editors: J. Carrasco and M. Mota, pp. 1-33
ISBN 978-1-61470-874-2

Chapter I

Epithelium: Composition, Functions and Pathology

M^a. Fátima, G. Klingbeil, Fernanda S. Giudice, Felipe F. Sperandio and Décio S. Pinto Jr.

Abstract

The oral mucosa comprises a stratified keratinized or non-keratinized epithelium composed by keratinocytes overlaying the lamina propria, which contains glands that secrete mucus and serous secretions and other structures like nerves, blood and lymphatic vessels. The continuous epithelial renewing process is what maintains the structural integrity of the oral mucosa. The cells from the deepest layer of the epithelium constantly suffer mitotic divisions, subsequently migrating upwards and replacing the squamous cells from the upper layers, which in turn lose their nucleus and cytoplasmic organelles, have their metabolism ceased and then undergo a programmed death as they become fully keratinized. Two functional cell populations compose the epithelium: the progenitor population that provides new cells, termed proliferative keratinocytes or epithelial stem cells; and the maturing population, which cells undergo differentiation depending on the extracellular amount of calcium, forbe esters, retinoic acid and D3 vitamin.

The aim of this chapter is to succinctly explain the main characteristics of the oral squamous epithelium cells by presenting their most expressed proteins and morphological architecture as well as to demonstrate the differences between them and squamous cell carcinoma cells, passing through the dysplastic stages generally detected. The oral squamous cell carcinoma (OSCC) is certainly related to the malignant transformation of the oral epithelial cells stimulated by genetic and/or carcinogenic factors, such as tobacco, alcohol and ultraviolet radiation. In general, the dysplastic epithelial cells show some distinct morphological characteristics including nucleus: cytoplasm ratio increase, nuclear and cellular pleomorphism and atypical mitoses. The OSCC cells, on the other hand, tend to lose their epithelial markers expression and begin to express mesenchymal markers, for instance vimentin, through the so-called epithelial-mesenchymal transition (EMT). Normal epithelial cells are interconnected by integrins and cadherins via intercellular junctions (gap or tight) or via desmosomes. Accordingly, SCC cells demonstrate lower expression of adhesion molecules E-cadherin, catenins and integrins, which helps them to migrate and begin the invasion process. One of the most studied signaling pathways related to cancer development is the PI3K/Akt, presenting the tumor suppressor gene PTEN as a regulator. In cancer, PTEN is often down-regulated or mutated and, subsequently NF-kB, cyclin D1 and mTOR proteins show enhanced expression. Alternatively, the cells from normal epithelium present basal activity of the PI3K-Akt signaling pathway and its related proteins once they also participate in the proliferation mechanisms.

Chapter

No other cavity in the human body shares a relationship to the external environment like the oral cavity [1]. One of the main constituents of our mouth, usually referred by dentists as "the soft tissue surrounding the tooth", the oral mucosa is recovered by a thin epithelium and has the capacity to produce mucous [2].

Basically, the oral mucosa serves as a "wrapper" covering and protecting various tissues and structures like glands, nerves, blood and lymphatic vessels. And indeed, the oral mucous epithelium is continuously renewed and thus shows very well adaptation to the function of protecting the underlying structures from aggressive chemical and mechanical stimuli [3,4]. Actually, the oral squamous epithelium, characterized by its well defined layers, shows specific alterations when exposed to mechanical, physical or chemical irritations. Such irritations may be represented by smoking, eating and drinking, for example, and all of them can alter the normal appearance of that epithelium [1].

Besides the aforementioned protection the epithelium allows the intermingling of the ducts of the salivary glands in the several oral regions maintaining the mucosa's moisture [5]. The wet surface and the relative lack of appendages are the hallmarks that distinguish the oral and esophageal mucosa from the skin. While the skin contains numerous follicles, sebaceous and sweat glands, the glandular component of the oral mucosa is represented primarily by the minor salivary glands. These glands are concentrated in the submucosa, and their secretions reach the mucosal surface via minor salivary ducts [3].

In the oral cavity, there are three different types of mucosa: 1- Specialized Mucosa, also called sensorial mucosa: covers the dorsum of the tongue and the vermilion of the lips. 2- Masticatory Mucosa: Gingiva and hard palate. 3- Lining Mucosa: covers the other portions of the tongue and the rest of the mouth. [3,6,7,8,9,].

As mentioned, the oral squamous epithelium is stratified, which means it has more than one layer of cells. In addition, a superficial layer comprised by keratin forms a top protective layer depending on the site of the mouth [4]. In that way, the oral epithelium possesses two differentiation patterns: stratified keratinized, that belongs to the masticatory mucosa; and stratified non-keratinized, for the lining mucosa. The non-keratinized epithelium may also show some type of incomplete keratinization that is denominated paraqueratinization (gingival and buccal mucosa, for example), on the contrary of the orthokeratinized epithelium that characterizes the actual keratinized epithelium (hard palate, for example) [10]. The non-keratinized epithelium is the most epithelial pattern found in the oral cavity, more than 60% [3,6,7] (Figure 1). The basal and spinous layers though, are present in both forms. Interestingly, the oral epithelium is very well adapted to its different functions at each part of the oral cavity. The mechanisms vary in order to develop each of those functions. For example, the masticatory mucosa is suiting and connected to the underlying connective tissue and its collagen fibers. On the other hand, the non-keratinized epithelium of the lining mucosa is more flexible and flaccid. The main function of all types of the oral epithelium is: mechanical and chemical protection, secretion and maintaining the humid, thus preventing the loss of water. The specialized mucosa is responsible for the thermal, gustative (tongue surface) and sensorial sensibility [3,6,7,8,9]. Cells located at the basal layer of the oral epithelium are organized in palisade and show a cubical shape. The basal layer composed by these cells covers the basement membrane that is localized above the lamina propria. In 1963 Pierce and cols discovered through the immunohistochemical technique the presence of some antigens from the basal membrane in the epithelial cells. The presence of antibodies based on ferritin indicated that the components from the basal lamina are produced by the endoplasmic reticulum from the basal cells [11]. In the same year, Hay and Revel performed autoradiographic testes and concluded that H^3-prolina was firstly observed in the epithelial cells and afterwards in the region of the basement membrane [12]. The basal cell layer synthesizes and secretes several substances, thus producing the basal lamina from the basement membrane [6,7]. These studies show that some components of the basement membrane possess epithelial origin [5].

Above the basal cell layer there is the spinous or prikle cell layer which is responsible for the epithelial thickness. The elliptical cell shape and their spines projections are responsible for the layer's name. This layer is composed of large overlapped polyhedral shaped cells once they have a higher amount of fluid in the cytoplasm. When compared to the basal cells they are wider and with rounder appearance [10] (Figure 1).

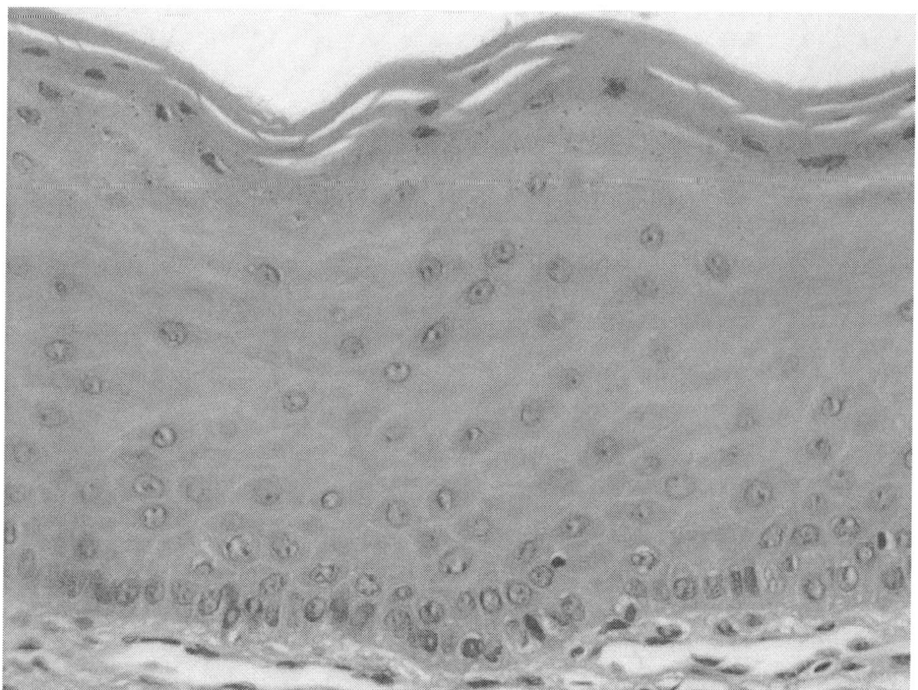

Figure 1. Oral non-keratinized epithelium: basal cells, prikle cell layer (spinous layer), gralunar layer absent, superficial layer, with nucleated cells.

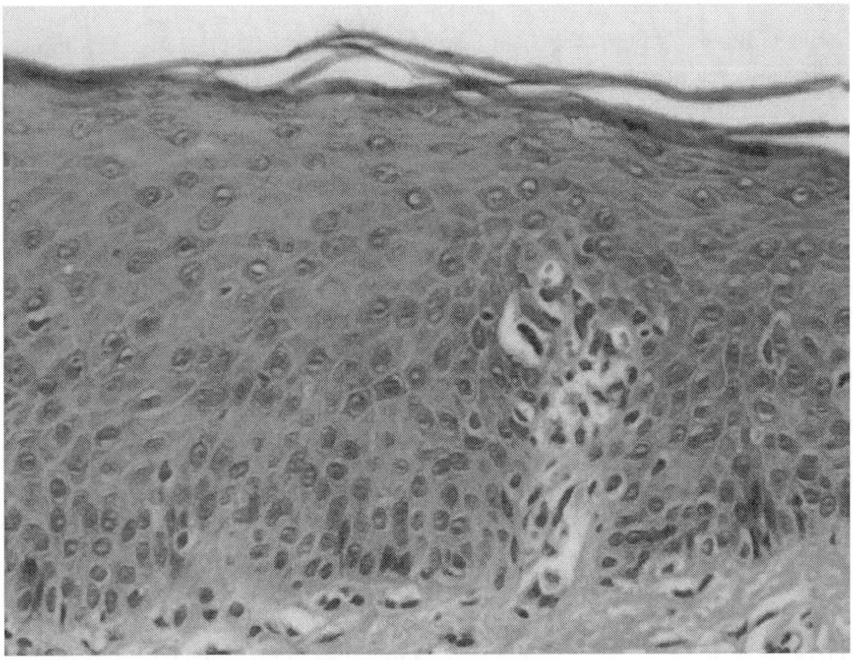

Figure 2. Oral parakeratinized epithelium: basal layer, spinous cell layer, granular layer, keratinized layer containing nucleated cells (incomplete keratinization).

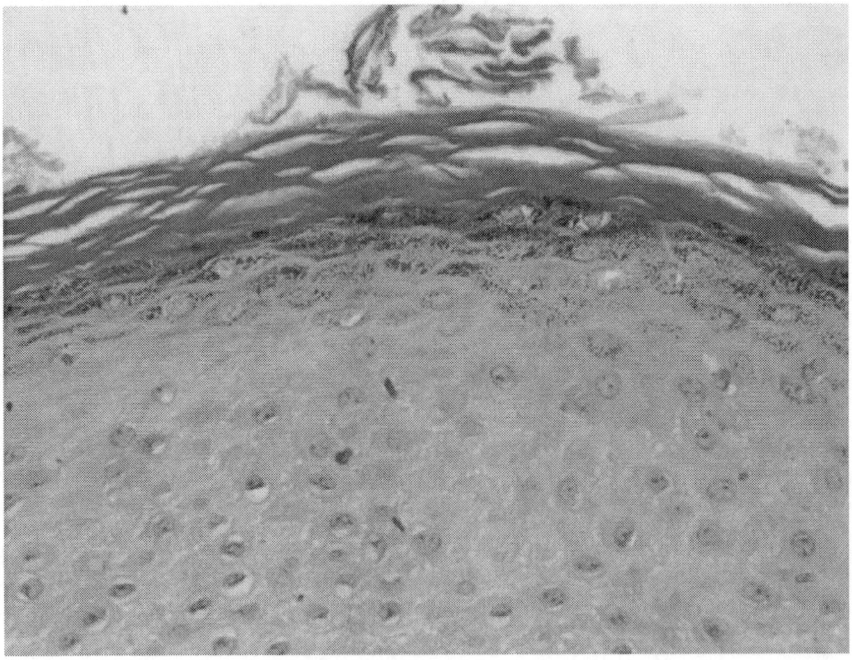

Figure 3. Oral orthokeratinized epithelium: granular layer, keratinized layer containing cells without nuclei (complete keratinization).

Both parakeratinized (Figure 2) and orthokeratinized (Figure 3) epithelia show two other superficial layers: the granular layer and the keratinized layer. The epithelial differentiation can be recognized by their granular and keratinized layer and by the expression of various proteins, like keratins and filaggrin by the cells that belong to these layers [13,14]. The filaggrin is present in the keratohyaline granules (KHG) in the granular and keratinized layers. It is a protein matrix that promotes the aggregation and disulfide bonds of the keratin filaments. It also participates in the regulation of the epithelium osmolarity and flexibility [15]. The cytoskeleton from the mammalian cells is composed by three filaments system: microfilaments (6nm), intermediate filaments (IFs) (10 to 12 nm) and microtubules (25nm). The IFs are divided in six different types (I to VI). One of the most typical structural proteins (IFs) of the epithelial cells is the keratin, which belong to IFs type I (acid) and IFs type II (basic to neutral) [16]. The IFs are chemically stable structures present in the cytoskeletal system of most eukaryotic cells [12,17,18,19,20]. The cells in the superficial layers of the epithelium that shows no keratinization are extended, crowded and bigger, with polyhedral shape with planned and flattened exterior. The renewal tissue process is responsible for the cells peeling and flaking as they age and dye. The epithelial cells maturation is therefore observed by the cell size increasing [2] as they senescence and migrate to the tissue surface [6,7,21]. The most appearing cell changes come from the synthesis of structural proteins, secretion of intercellular material and the emerging of the organelles [10]. Many of the structural proteins produced by oral squamous epithelium are essential for tissue integrity [20,22]. The cell keratinization in the keratinized epithelium is showed in the superficial layers reflecting the cell physiology and feature changes [6]. The difference between the orthokeratinized and parakeratinized epithelia emerges at the superficial layers. The orthokeratinized epithelium is the most common epithelium found in the masticatory mucosa. Its most superficial cells are flat and devoid of nuclei and their cytoplasm, filled with keratin, show a greater flattening, lifting off easily, while they are not viable [6]. In the parakeratinized epithelium these superficial cells, probably not viable for a long time, still possess their nuclei thus justifying their scaling [6]. Considered a pathological sign in the skin, the parakeratinization in the mouth is associated with a histological health feature. The flattened nucleated cells in the granular layer from both keratinized epithelium contain KHG in their cytoplasm, which are usually observed as spots or dark spots in the cytoplasm [6]. In the keratinized layer this KHG, micrograpules and other organelles are absent. In some regions of orthokeratinized epithelium, most cells of the keratinized layer are filled with an electro-dense homogeneous material, whereas in the parakeratinized epithelium the cytoplasmic filamentous structure are still recognizable in most cells and as aforementioned their nuclei still persist [22].

Therefore, the oral epithelium is mainly composed by keratinocytes, which are also called epithelial cells and are divided in the already mentioned layers. In smaller quantities, other cells like melanocytes, Langerhans, Merkel [6,8] and Granstein cells (6) can also be distinguished. The keratinocytes are responsible for the constant renewal process and for the structural integrity maintenance of the oral mucosa. The regeneration process happens in the deepest compartment, the basal layer, adjacent to the basement membrane. This process is achieved by a hierarchy of proliferative cells. The cells called basal, or proliferative keratinocytes are produced by mitotic divisions, migrate to the surface, and replace the older squamous cells [23,24]. It could be taken in consideration that epithelial cells consist of two different functional populations: a progenitor population (which function is to divide and provide new cells) and a maturing population or transient amplifying cells (cells that undergo the process of maturation and differentiation) [9,25]. The cells comprising the progenitor population receive the designation of epithelial stem cells [26].

Each stem cell located in the basal layer duplicates and generates two other cells. Accordingly, the progenitor compartment is not homogeneous but consists of two sub-populations of functionally distinct cells. A small population of progenitor cells has a very slow reproductive cycle and is considered representative of stem cells. The function of the stem cells is to retain the proliferative potential of the tissue, and once these cells do not divide frequently, they are important in preserving the genetic information of the tissue, since DNA is very vulnerable and subject to changes during the mitotic phase [26]. Each cell generated after the duplication returns to the cell cycle or follows toward differentiation. This decision depends on several factors such as the available amount of phorbol esters, retinoic acid and vitamin D3 [3].

In the oral epithelium the turnover rate is approximately about twice or thrice comparing with that of the epidermis [27], and a proliferation index of 58.8 has been reported in native oral mucosa [28]. The connective tissue plays also an important role in the epithelial differentiation, which is showed by some previous studies [29,30,31].

The basal cells containing integrins are connected by focal adhesion. The differentiation process involves migration, which is concomitant with the loss of integrin expression and increasing of the cadherins, which are adhesion mediators by intercellular junction type gap or tight or by desmossomes [3].

The keratinocytes self-renewal program (proliferation and differentiation) requires a balanced involvement of different tightly controlled signaling pathways that ensure tissue homeostasis. This keratinocyte proliferation is limited only to the basal layer of the epithelium. An aberrant suprabasal proliferation is associated with pathological states such as malignant transformation, indicating that the keratinocytes commitment to regulation fails to exit the cell cycle and to differentiate, and thus characterizing a neoplastic growth.

Keratins

The Keratins (Ks) are part of the predominant family of the cytoskeleton proteins present in epithelia [32]. They represent about 30 to 50% of all components of the cytoskeleton in the oral epithelium [33] and they are a major biochemical constituent of the tonofilaments [20]. The human keratin gene family is composed of 54 functional genes of the total of the 70 human intermediate filament genes [34]. The keratin bundles entwine the nucleus, goes and cross the cytoplasm and connect to the epithelial cell junctions, in this case the desmossomes and hemidesmossomes [34,35,36]. The expression of genes that encode the Ks can be modified by the presence of certain substances such as retinoids, calcium and vitamin D. Retinoids bind to nuclear receptors and interfere with the transcription process by interacting with response elements present in the genes of Ks [16]. Probably the main function of the Ks is the maintenance of the integrity and the mechanical stability of the keratinocytes and epithelial tissue. The keratins are divided into two sub-groups: α and β keratin. The α keratin is present in the mammalian cells and both sub-groups (α and β) are present in reptiles and birds cells [5]. The α keratin expression is constructed by two heterodimers organized in pairs: type I (acid), which includes the K9 to K20 and type II (basic to neutral), which includes the K1 to K8. They differ in its amino acid sequence and in the biochemical properties [17,18,19,32,34,36]. Some epithelial cells can express more than one pair of these heterodimers. The cell's differentiation program requires a specific keratin pair. Of the 28 type I keratin genes, 17 are epithelial keratins and 11 are hair keratins; and of the 26 type II keratin genes, 20 are epithelial keratins and 6 are hair keratins. Almost half of all human keratins are expressed in the hair follicle [34,37]. In the absence of the appropriated heterodimer pair these molecules become instable and susceptible to degradation by proteases action [32]. The keratin molecular weight ranges from about 40 kDa to 76 kDa. In the simple (one-layered epithelium) and glandular epithelium, these heterodimers molecules of low molecular weight (40 kDa) are represented by K8 and K18 [10,19,37,38,39,40] and sometimes the K7 [38,39,40]. K8 and K18 are focal expressed in the basal cell layer together with the K19 [37,41]. The basal keratinocytes of normal human stratified epithelia express characteristic keratins and adhesion molecules of the integrin family [43,44]. After migration from the basal layer, keratinocytes cease to divide, increase in size, assume a polygonal shape, and begin to synthesize new keratins. These new Ks, as well as new adhesion molecules, are normally present in the upper layers of stratified epithelia [13,45,46]. Assembly of involucrin in the cornified envelope and a gradual increase in fatty acids and ceramides in the upper layers of stratified epithelia are other biochemical changes associated with the keratinocytes' differentiation process in vivo [47,48]. As in the basal simple layered epithelium [41,37] the K19 (the smallest keratin) [49] is also expressed in the basal cells of the non-keratinized stratified oral squamous epithelium [34]. Some experiments showed that in the absence of its expression there is a functional compensation by the K18 [50]. The expression of K19 seems to decrease when the keratinocytes begin to produce involucrin [51]. K14 is a keratin normally found in the basal layer of stratified epithelia, where it pairs with K5 [52]. In oral tissues, it has been described in keratinizing palatal epithelia, sulcular epithelia, and in gingival keratinocyte cultures [13,53,54,55] K14 does not appear to be affected by the differentiation process, as does K19 [21]. The keratin filaments K5 and K14 are attached to the desmosomes and hemidesmosomes also expressed through all the layers of the stratified follicular outer root sheath [34]. The K15 can be also detected in the basal layer and is a partner of the K5 [38,39,40]. Their expression is very strong in the cells from the undifferentiating suprabasal cell layers and in the basal cells compartment, comprising the stem cells [56]. At the same time they are down-regulated in the cells from the suprabasal cell layers as the differentiation process takes place [56]. Different keratins emerge above the basal layer along with the cell differentiation process. K10 and involucrin are markers found in the upper layers of stratified epithelial tissues in vivo and in vitro [13,45,57] and as previously reported, the K10 and K1 keratins are related to the differentiation process [34]. Thus, as the major keratins they are also expressed in the ortokeratinized epithelium of the hard palate and in the parakeratinized epithelium of the gingiva [10,58]. They can be found in the non-keratinized epithelium as well but here in small amounts [10,59]. The keratins K13 and K4 are expressed in the non-keratinized [10] epithelium, in the suprabasal compartment. The K6 and K16 are expressed in non-keratinized epithelium [38], and in the more differentiated layers of the masticatory epithelium of the hard palate and gingiva. There are three isoforms of K6: K6a, K6b and K6c [37]. The junctional epithelium adjacent to the tooth surface also produces a different set of keratins, such as type K2, formerly referred as K2e, which is also present in the epidermis and K76, formerly called K2p [37]. The K76 is expressed in the supra basal layers like the K6 and K16, in the slightly orthokeratinized stratified squamous epithelium lining the gingiva, the masticatory epithelium and hard palate [19,60,61]. K76 differs from the K2 by the amino acid position in the rod domain and is limited homology, only 71%.

Even though both are expressed in the suprabasal cell layer, up third or fourth cell layer, the K76 is more closely related to the corneal K3, in 86% [61]. The presences of the K78 and K80 in the tongue were detected in some preliminary studies suggesting that their expression is related to a particular cellular differentiation pattern of a special epithelium or probably a transition phase [62]. The role of the keratin is not restricted to mechanical support of the cytosqueleton. In addition to the previous cited and observed mentioned roles it represents a constant functional and physiological structure [63,64] capable to assist in the epithelial polarity and membrane traffic [64]. The K20 is not expressed in stratified epithelium, but it is a characteristic marker of Merkel cells and of the human lingual taste buds [65,66]. In malignant transformation it can alter its distribution. It is a marker of adenocarcinoma, in special of gut origin, making from this property a sign of metastatic tumour from gut origin [65].

Kerathohialine Granules

Kerathohialine granules (KHG) are electropaque cellular inclusions which are found in cells of the granular cell layer of orthokeratinized and parakeratinized epithelium, where this layer can be very poor. This layer is absent in the non-keratinized epithelium and consequently KHG are not present [7]. Some authors consider that KHG are precursors of the matrix that surrounds the keratin layer. In oral epithelium, two different types of KHG can be found, although they are not always present in all regions. The first one is the same type present at the epidermal granular layer. The irregular KHG are associated with the long and thick tonofibrils, and are often surrounded by ribosomes. They appear limited to the complete orthokeratinized regions of the oral epithelium, such as granular layer of the hard palate and gum.

Eventually they can be present in a lesser extended area of parakeratinized epithelium at the same region [5,67,68]. The second type of KHG is peculiar to the oral epithelium and even less frequent in less keratinized areas. They are larger and regular compared with the epidermal type. They are surrounded by ribosomes, but lack the association with prominent bundles of tonofilaments. These typical oral granules are present at the most part of the epithelium covering mobile structures. Their features are associated with their large electropaque structure which is found between the pale and randomly disperse tonofilamens [5,67,68]. KHG contain profilaggrin which is the precursor of filaggrin [16,69]. KHG release profilaggrin when they undergo dephosphorylation and proteolysis turning in this way into filaggrin. When filaggrin binds to keratin filaments, it forms bundles that promote cellular collapse. This cellular collapse results in cells loss of original shape and they become flattened. Other structural proteins such as involucrin, loricrin, and tricohialina "small proline rich proteins" (SPRs) are synthesized and interact, due the action of transglutaminases, forming the cell envelope, located beneath the plasma membrane [69].

Tonofilaments

Tonofilaments belong to the IFs family. They are long filaments of about 8 nm of diameter of fibrous proteins which are synthesized by the ribosomes [9]. Tonofilaments of stratified squamous epithelium represent one of the main structural features of this tissue, the most abundant and largest structural element in the cytoplasm. All stratified epithelia contain IFs of similar composition and dimensions (7.5-10 nm diameters) [70]. Tonofilaments are present in every cell of both, keratinized and non-keratinized oral epithelium [5]. The concentration and organization of such filaments indicate the epithelial maturation grade [70]. With respect to the concentration and aggregation of these cytoplasmatic filaments in the epithelial basal cells, they are arranged in tonofibriles beams [70], and occupy about 13% of the volume of this tissue layer [8]. During the transition from basal to granular cell layer, in non-keratinized epithelium, the cells possess a greater amount of tonofilaments and an increase pronounced size of them, compared to cells of the keratinized epithelium. Yet in relation to the weight of non-keratinized epithelium, cells show a slight increase in volume, whereas at the keratinized epithelium, this volume expansion is less proportional compared to its weight [67,71]. This is partly due to the space before occupied by the nucleus and organelles and in part by the more compact packing system. This results in a difference in the filaments concentration at this level, which is more visibly pronounced. The space among tonofilaments at the keratinized layer of orthokeratinized epithelium is correlated to the transition from granular to keratin layer, when much water is lost, leaving a far greater dry material concentration. In the non-keratinized epithelium, equal amount of water and dry material are dispersed at the cytoplasm [71].

Filaggrin and Profilaggrin

The filaggrin ("filament aggregatin protein") is a cationic protein of 36 kDa, derived from a phosphorylated protein precursor of about 500 kDa called profilaggrin, which is expressed as KHG in the granular cell layer. Each polifilaggrin molecule has 10 to 12 filaggrin units [16]. Filaggrin is a late marker of keratinization that is usually not expressed in the oral non-keratinized epithelium [72]. When the cells from the granular layer reach the keratinized layer, the profilaggrin is converted to filaggrin [73]. The filaggrin acts as a matrix which binds to the IFs forming bands which are aggregated in parallel bundles through the keratin disulfide bonds [74] in the keratinized layer of the epidermis and other stratified epithelia, including the tongue and gingiva [52,75]. While profilaggrin is not able to bind keratins [76]. This protein is considered an important marker of the final stages of keratinocytes differentiation [16,77].

The filaggrin shows three main functions in cell differentiation: the maintenance of the epithelial texture, the IFs alignment and the controlling of the changes in the cell shape [69]. The loss of nuclei and cellular organelles are directly related to the transient expression of the filaggrin in the granular layer, while the same expression of profilaggrin conducts to the expression of a protein less toxic as the filaggrin. Studies from Dale and cols (1997) have demonstrated that cell contraction, nuclear membrane breakdown and nuclear condensation, related to terminal differentiation, are apoptosis phenotypes related to the expression of the protein filaggrin [78]. Fails on the expression of filaggrin are related to alteration on the terminal differentiation. This evidence strengthens the hypothesis of the correlation of their expression with endogenous apoptotic machinery [72]. Another correlation to this hypothesis is the compromised of the filaggrin expression with cell contacts, since the cell-cell adhesion has demonstrated to be also a regulator of apoptosis event [79]. An unpublished observation by Presland, Kuechle, Lewis, Fleckman, and Dale is the decrease in the levels of cell-cell adhesion, which is associated with changes in the distribution of desmosomal components, specially the desmoplakin, what is showed by cells expressing filaggrin. It seems that there is a correlation between the integrity of the keratin filaments web and the desmosome consistency [80]. Crash in keratin and vimentin filament bonds forming perinuclear aggregates and coils in epithelial cells can be also related to the overexpression of filaggrin [76,78].

Loricrin, Small Proline Rich Proteins and Involucrin

The expression of the protein loricrin, a glycine-serine and cysteine-rich protein, occurs in the terminal differentiation in the upper granular layer during the keratinization differentiation process. This expression is equal to the sulphur-rich component of the KHG, together with the profilaggrins. Loricrin is incorporated into the cornified cell envelope as a major component because it becomes cross-linked by the formation of N epsilon (gamma-glutamyl) lysine isopeptide bonds, justifying its important role in the structural shape of the cell envelope. *In vivo* tests showed that the humid tissues: oral epithelium and newborn epidermis have the highest levels of loricrin expression, however all mammalian stratified epithelium also expressed loricrin in a less manner [69,81]. The loricrin comprises 70-85% of the protein mass present in the keratinized layer and acts as a substrate for transglutaminases. The loricrins are grouped together through isopeptides bridges and in small quantity by the small proline rich proteins (SPRs) [69].

The small proline rich proteins (SPRs) normally located at the cell periphery, are classified in three family members: SPR1, SPR2 and SPR3, and they are restricted to terminally differentiating squamous cells. The mechanical properties of the cell envelope and of the cells in the keratinized layer can differ, depending on type and amount of the SPRs expressed on the tissue. The variants of the SPRs expression allow their adaptation to these different functions. Consequently, the epithelia mechanical properties can be altered in some pathology, as well as in wound repair, aging and other environmental stimuli [69]. SPRs can be cross-linked by transglutaminase that catalyses inter-protein bond formation [82]. Hohl and cols in 1995 have demonstrated in some in vitro studies that the detected levels of SPR2 and SPR3 were strongly increased after the addition of a transglutaminase inhibitor (LTB-2) [83]. These results have suggested that these SPRs are precursor proteins of the cornified cell envelope. These studies reveled that while SPR1 was restricted to the epidermal appendages areas, it was expressed in the sublingual and tongue epithelium. The SPR2 was restricted to lingual papillae and its expression in the epidermis was coherent. SPR3 was abundant in the oral and esophageal and totally absent in the epidermis [83]. Some SPRs are also expressed in non stratified epithelium, suggesting that this tissue may have a tendency to undergo squamous metaplasia, or can be connected to apoptotic process [84].

Involucrin

Involucrin, is a protein element, precursor of the cornified envelope, which is present in the inner face of the keratinocytes during its terminal differentiantion [80]. It is considered an earlier marker of the terminal differentiation, synthesized at the suprabasal epithelial compartment, profusely in the cytoplasm of the keratinocytes [85]. After involucrin expression, other proteins are incorporated to the cell envelop. It was observed by *in vitro* models that during the cell envelope formation, there is initially the incorporation of involucrin, followed by the aggregation of SPRs and loricrin [86].

Cell Junctions

Cells of some tissues are hold together by cell junctions [6]. Architecture and function of epithelial tissues are maintained by the cohesion among keratinocytes [87]. These mechanical joints promote adhesion between cells and junctions surfaces, allowing cell membranes to be closely attached, however without being completely adhered. These junctions are dense bodies that can be observed only by high-resolution microscopes. Each type of junction is a complex union mechanism, shared between the membranes of the involved cells [6]. Cells of the oral epithelium are interconnected by four different types of junctional structures: 1- tight junction (also called zonula occludens or macula) [6,7,32,70,88], 2- gap [89] or nexus junction [6,7,32,70], 3- adherens junctions [90] and desmosome [91] (also called macula adherens) [6,7,32,70] that are types of anchoring junctions responsible for the mechanical adhesion among adjacent cells and 4- focal adhesions [92,93]. Tight junction seems to possess a smaller role as a mechanical union and more like a sealant capable to seal the intercellular spaces into real separated compartments [10]. At oral epithelium it is quite rare and limited to upper layers [70], where it operates as a barrier and maintaining cell polarity [87]. It is not observed at the keratinized layer, especially when compared to desmosomes, thus justifying its rare frenquency [70]. It is found at glandular and epithelium lining the gut [87].

2- Gap junctions are quite frequent and observed in the basal but never in the keratinized layer. They are often close to the desmosomes. As the desmosomes, they are frequently displayed parallel to the plasma membranes of adjacent cells. There is not properly an intercellular space between them but they can vary this distance [70]. In this type of junction the outer membrane surface is separated by a slit of 2-5 nm [94] and it seems that its role is rather to facilitate the intercellular communication than providing cell adhesion [70]. 3- Desmosomes are the most peculiar and frequent junction type of the squamous stratified cell tissue [70] occupying about 50% of intercellular space of the keratinized layer [3,95].

The diameter of these discoid junctions is about 0.2–0.5 μm and interdesmosomal regions are frequently dilated with an intercellular split of 24–30 nm [96]. Its composition presents two electron-dense plaques in each involved cell. Desmosomes represent the most intense mechanical adhesion among the epithelial cells [7,70]. They are present in a great number of tissues subjected to mechanical forces as stratified squamous epithelia of the skin, oral epithelium, mucous membranes, myocardium and also other specialized tissues [96,97,98]. They are cell adhesion complexes (plaques) that promote epithelial cell binding or fix keratin filaments to the cell surface [32]. They possess two dense and thick intercellular regions that are associated to the cytoplasmic surface of the cell membrane of the opposite cell. Tonofilaments' bounds are inserted in these adhesion plaques [7]. This is well observed by seeing the elaborate plates and filaments system and consequently their resistance in separating these cells [70]. Desmosomes are responsible for the histological aspect of the spinous cell layer and its subsequent naming [7].

The desmosomes consist of two main protein groups: 1- the desmosomal cadherins, desmoglein and desmocollins (which reach the plasma membrane) [32]. These two desmosome transmembrane glycoproteins form a cell–cell adhesive structure prominent in epithelial tissues [99]. 2- A large group of proteins is present at the cytoplasmic surface of desmosomes, including the known structure "the desmosomal plaque", which works binding proteins, as cadherins, to the cytoplasmic keratin filaments. The cytoplasmic proteins, associated with this cytoplasmic plaque include: placoglobins, desmoplakins, plakofilins, envoplakins and periplakins. It is believed that the desmosomal cadherins, such as classical cadherins, present at junctional adhesion, are responsible for mediating cell adhesion through calcium dependent relation between the extracellular proteins domains that bind each cell to the other cells.

In addition to desmosomes, epithelial cells have junctional adhesions, which are composed by proteins on the plasma membrane, such as: E-cadherin and P-cadherins. These proteins are linked to a series of cytoskeletal cytoplasmic actin proteins [32]. In some types of epithelium where the cell turnover is slower, epithelial cells leave the basal layer, migrate and reach the epithelium surface in less than one month. This is possible through the independent cell-cell movement. To allow this path, there must be a constant mechanism of breaking and attaching this union. This phenomenon can be confirmed by observing cell division when cells possess fewer desmosomes compared to interphase cells. When these cells reach the surface, desmosomes tend to disintegrate. The observation of epithelial cells in culture showed no evidence of desmosomes on their surface when they were separated; however, when cells were placed in contact they were able to re-form these structures within 90 minutes [70]. Some evidence has proved the existence of different types or different groups of desmosomes. Even though they look similar in their structures, they may respond differently to various chemical substances. Thus, Borysenko and Revel [100] identified two types of desmosomes. One type can be dissociated at low concentrations of trypsin, and the other one does not respond to the same enzyme, but they can be broken by the ethylenediamine tetra-acetic acid (EDTA) [4].

When epithelial cells cover a non-cellular membrane, called "basal lamina", they possess structural features, which are similar to desmosomes, however called now as hemidesmosomes. These cells have one half of the structures and do not have the other half that belonged to other epithelial cell. There is a thickening of a limited area of basal cell cytoplasm, adjacent to the basement membrane, under which converge cytoplasmic fibrils. At this point, basement membrane shows a slight thickening and discrete aspect, which contains very thin filaments [4].

Considerations about Cell Cycle

To maintain the epithelial homeostasis in the oral cavity like other epithelia as aforementioned, epithelial cells (keratinocytes) are in a constant renewal process due to cell elimination by desquamation and apoptosis. Under normal conditions, cells produced by cell division at basal layer, which contains the population of highly proliferative cells (stem cells) and quiescent cells, are sufficient to replace those lost through desquamation process [101]. It is not easy to precise the time for the whole process of cell renewal in the epithelium, especially taking into consideration different types of oral epithelium [102,103]. The estimated time for the epidermis is around 12 to 75 days with an average of, approximately, 25 days. The time calculated for the oral epithelium varies from 20 to 40 days [102].

There are some differences according to the various locations of the oral cavity, like unique histological structures, difference in the number of cellular layers and thickness [104]. Another aspect is that cell renewal time in non-keratinized epithelium occurs more rapidly, while in orthokeratinized and parakeratinized epithelium (gingival and hard palate), this time is longer [102,103].

Control of cell division appears to be determined by local production of tissue hormones, called chalone. These hormones seem to be produced by post-mitotic cells, and possess the dual property of stimulating cell differentiation and inhibition of mitosis. Therefore the cell division rate is increased in regions where the concentration of chalones is low. Another systemic component, adrenaline, seems to be involved somehow in the action of the chalone [102,103]. Other factors such as diurnal rhythm, age, inflammation and stress seem to influence the mitotic activity. Interestingly, when the epithelium is faced with an inflammation, a small amount of inflammatory infiltrate may stimulate mitotic activity, whereas a larger amount of this infiltration can suppress the same activity [102,103].

Age Related to Development of the Oral Mucosa

In general all tissues suffer cumulative gradual changes with age, associated with slower cellular activities [70]. Among the epithelial tissues, the skin is the one in which cellular modifications that occur with age could be documented in detail, by outward signs, such as thinning, wrinkles (result from the changes of collagen fibers), hair color, decrease or loss. It must be noted that in both function and structure a large part of such changes is related to the exposure to the sunlight. But regarding to the oral epithelium, its location is protected from this type of environmental attacks [102,103]. Studies have shown a reduced mitotic activity of oral mucosa by the advancing of age [70]. However, some controversies persist about it, thus making some statements not conclusive [70].

When we observe the oral mucosa of an elderly person, we notice a pronounced decrease in thickness, looseness and dryness (xerostomia) and symptoms such as burning sensation, itching, grittiness, lack of sensitivity to certain tastes, especially in in post-menopausal women. In studies of Kano et al 2007, it was observed that on the laryngeal surfaces of epiglottis, the mean density of upper and middle levels of taste buds was significant decreased with ageing [105]. It becomes difficult to determine whether the extent of these specific changes are inevitable consequences of senescence programmed by its own tissues, or secondarily due to local changes, such as corrosion and abrasion following attrition, or even secondary to systemic factors such as: hormonal changes, nutritional or cardiovascular factors.

For years, differences in the epithelial structure and its relationship with underlying microcirculation and density have not been very good elucidated. Recent researches have shown that conditions for successful tissue repair must present adequate microvascular blood flow and tissue oxygen tension. Such conditions decrease with age increasing [102,106,107] and recent studies have demonstrated that the microcirculation pattern is correlated with sex and age [103].

Adherens Junctions and Epithelial Mesenchymal Transition

Intercellular adhesiveness is mediated by a family of glycoproteins named cadherins. This kind of intercellular junctions is very important for the epithelial cells and essential to maintain tissue homeostasis and to prevent the organism from bacterial infections or other type of injuries [108].

The intracellular domain of cadherins binds to proteins called catenins [109]. Among the proteins called cathenins, α-catenin, β-catenin and γ-catenin may be found [110]. These catenins play a crucial role in cadherin function and were recently shown to be involved in other functions such as signaling and activation of transcription factors [111] and cell motility [112].

Of the most studied cadherins involved with cell adhesion is the E-cadherin. There are two distinct E-cadherin complexes that can be found in the same cell, one composed of E-cadherin, alpha- and β-catenin, and the other of E-cadherin, alpha- and γ-catenin [113].

Concerning to β-catenin, it is a multifunctional protein important for adherens junction formation [114], cell migration [115], and signal transduction [116]. In normal cells, the APC, GSK3beta (serine threonine glycogen kinase) and Axin complex promotes the degradation of β-catenin [117], which leaves a small pool of free cytoplasmic β-catenin protein. Elevation of the free pool of cytoplasmic β-catenin is the result of the inactivation of the APC-system or activation of the WNT-1 pathway [118, 119].

Normally, the WNT signal stabilizes free β-catenins, while mutation in APC or β-catenin [119,120,121,122,123,124,145,126] mimic WNT signaling, stimulating cell proliferation or antagonizing apoptosis [127].

The uncomplexed β-catenins that were not degraded by the above mentioned complex, accumulate in the cytoplasm and then bind to other newly synthesized proteins, especially transcription factors, such as LEF-1 and TCF-4, that are transported to the nucleus [128,129,130,131], and activate gene expression [132,133,134]. This gene expression drives cell proliferation or inhibits apoptosis [135,136,137,138,139,141,142,143]. In summary, β-catenin protein is normally not present in the nucleus, but enters nuclei during the transduction of WNT signaling, while LEF-1/TCF is a nuclear protein that is imported constitutively. Binding to LEF-1/TCF, β-catenin may lead to the transcription of genes involved in human carcinogenesis, such as E-cadherin gene [129], cyclin-D1 [144,145], c-MYC [146], c-jun.

Loss of adherens junctions is far related to the so-called epithelial-mesenchymal transition (EMT), an orchestrated series of events in which cell-cell and cell-extracellular matrix (ECM) interactions are altered. By these means, epithelia passing through EMT tend to release epithelial cells from the surrounding tissue [147].

As noticed, when the cells participate in the EMT process, their cytoskeleton is reorganized to confer the ability to move through a three-dimensional ECM, and a new transcriptional program is induced to maintain the mesenchymal phenotype [147]. EMT also changes cellular morphology, leads to loss of cellular polarization, and to the decrease in cell-extracellular matrix adhesion and production of metalloproteinases [148]. Thus, EMT mechanisms are integral components of the progression of many tumors of epithelial tissues. Indeed, when invasive clones of head and neck squamous cell carcinoma (HNSCC) cell lines were obtained through invasion chamber assays to simulate tumor progression, the cytoskeleton organization of these clones was found to be distinct from the non-invasive cancer cells. This cytoskeleton different organization could be observed with the new network-like pattern of a specific protein related to carcinogenesis and involved with EMT [149]. Several intra-cellular signaling pathways participate in cancer establishment and progression through tumor growth and metastasis. These signaling pathways are constantly cross-talking, and due to this intensive participation of different proteins and pathways, cancer tends to be a difficult disease to treat and study.

The PI3K/Akt signaling pathway is among the most studied pathways and is directly linked to EMT in human oral squamous cell carcinoma. Akt activation induces the decrease of specific proteins expressed in epithelial cells such as cadherin-E and β-catenin. As said, this is essential to tumor organization and progression, once the cells need to lose these adherent junctions in order to begin their process of migration to the underneath connective tissue [148].

Akt activation also triggers the expression of mesenchymal markers such as vimentin. The vimentin protein is among several and exhaustively studied proteins related to invasion, cell proliferation and metastasis. Vimentin is a type III intermediate filament protein found in most mesenchymal cells and is also expressed when tumor cells undergo EMT [150,151,152]. Interestingly, this protein contributes to the cancer cells aggressive pattern in breast, cervix and bladder carcinomas [153,154]. Vimentin has been suggested as a predictive of the recurrence and invasive potential of prostate cancer cells as well [155].

Briefly, the PI3K/Akt signaling pathway plays essential role in maintaining physiological cell division and behavior. In normal cells, this pathway is down-regulated by the far known tumor suppressor PTEN. By these means, the aberrant PI3K activation through the loss of the PTEN expression activates mutations in PI3K. This kind of event results in phosphatidylinositol (3,4,5)P3 accumulation at cell membranes, and serves as a docking site for Akt localization and activation. Following membrane recruitment, PDK1 phosphorylates Akt at one site (T308 in Akt1) in the kinase domain while mTOR phosphorylates Akt at another site (S473 in Akt1) [156,157,158]. Therefore, after membrane activation of Akt the signal transduction is moved downstream where other important proteins take part. The mTOR kinase, for example, is the mammalian target of the drug rapamycin and assembles into at least two distinct complexes, mTOR complex 1 (mTORC1-raptor), and mTOR complex 2 (mTORC2-rictor) [156]. Extensive work with rapamycin indicates that the raptor–mTOR complex positively regulates cell growth and that its inhibition causes a large decrease in cell size. Once mTOR participates in the PI3k/Akt signaling pathway it also plays a role in promoting cell proliferation and survival [159], which is so important when studying cancer. In fact, the widespread activation of the PI3K/Akt/mTOR pathway in HNSCC progression has raised the possibility of using specific mTOR inhibitors and their derivatives for the treatment of HNSCC patients [160,161]. In this regard, blockade of mTOR with rapamycin exerts a potent antitumoral effect and even prevents minimal residual disease in several human HNSCC xenograft models [162,163,164].

The NF-kB family of transcription factors can also be activated by the PI3k/Akt signaling pathway [165] and plays important roles in breast and oral cancers [166,167,168,169]. The target genes of this family are related to key cellular processes such as innate and adaptive immunity [170], lymphoid organ development [171] and cell proliferation and survival [172]. Among these genes, cyclin D1 can be mentioned and relates to cancer cells permanent cycling [173].

Epithelial Dysplasia and Squamous Cell Carcinoma

The transition from a normal epithelium to an oral carcinoma, passing through a pre-malignant stage is a result of accumulated genetic and epigenetic alterations in a multi-step process and the clonal expansion of selected cell populations [174]. In the case of HNSCC, tumor progression involves genetic alterations leading to dysplasia (9p21, 3p21, 17p13), carcinoma *in situ* (11q13, 13q21, 14q31) and invasive tumors (4q26–28, 6p, 8p, 8q) [175,176]. Multiple known tumor suppressor genes in HNSCC, such as p16 and p14ARF (9p21), responsible for the G1 cell cycle regulation and MDM2 mediated degradation of p53, are involved in tumor progression [177].

In fact, P53 is a tumor suppressor gene that is regulated (degraded) by the MDM2. And accordingly, the overexpression of the MDM2 protein may result in uncontrolled cell growth and subsequent tumorigenesis [178]. Oral squamous cell carcinoma cells exhibit overexpression of MDM2, while normal oral keratinocytes should present normal levels of this protein [179].

Among the genes involved in oral dysplasia development, loss of chromosomal region 9p21 is found in 70–80% of these lesions. This along with the inactivation of the remaining alleles of p16 and 14ARF by promoting hypermethylation also represents one of the earliest and most frequent events in HNSCC progression [175,180].

Thus, oral epithelial dysplasia is often taken as a precursor of OSCC and typically presents itself as a predominantly white, or red, or mixed white and red mucosal lesion. The leucoplakia is a white lesion that cannot be wiped away or characterized as any other specific disease. On the other hand, red surface lesions are designated as erythroplakias; and not otherwise specified nonwipeable lesions that contain both red and white features are named erythroleukoplakias (speckled leukoplakias) [181]. All above mentioned lesions should be well distinguished from oral carcinomas or oral epithelial dysplasia; however, they are the oral lesions in which dysplastic or truly malignant changes occur more frequently when compared to normal oral mucosa [182]. When these lesions are examined with the help of a microscope, several aspects can be found, from nondysplastic epithelial hyperplasia and/or hyperkeratosis to maturation disturbance features that characterize some level of dysplasia [181]. According to the World Health Organization (WHO, 2005), the epithelial dysplasia should be divided as follows: Mild dysplasia: architectural disturbance generally limited to the lower third of the epithelium accompanied by cytological atypia define the minimum criteria of dysplasia (Figure 4); Moderate dysplasia: architectural disturbance extending into the middle third of the epithelium as the initial criterion for recognizing this category. However, consideration of the degree of cytologic atypia may require upgrading (Figure 5); Severe dysplasia: recognition of severe dysplasia starts with greater than two thirds of the epithelium showing architectural disturbance with associated cytologic atypia. Nevertheless, as noted in the previous paragraph, architectural disturbance extending into the middle third of the epithelium with sufficient cytologic atypia may be upgraded from moderate to severe dysplasia (Figure 6). As said, the degree of cytologic atypia needs to be taken in consideration when examining the hystopathological features of oral epithelial dysplasia. These atypia are represented by increased nuclear/cytoplasmic ratios, nuclear pleomorphism, nuclear hyperchromatism, increased numbers of mitoses above the basal cell stratum, as well as abnormal mitotic figures; and epithelial architectural disturbances including stratification disarray and droplet-shaped rete pegs [181]. After correct histopathological diagnosis, the evidence of dysplasia in a clinically termed leukoplakia is considered an important predictor of the potential of this lesion to undergo transformation to OSCC; thus, the treatment of oral epithelial dysplastic lesions should be well conducted and the patient should be followed-up accordingly. Nevertheless, spontaneous regressions of these lesions have been reported in the literature, mainly when the patients decided to abandon the habits that relate to oral cancer (smoking, drinking and diet) [184,185,186].

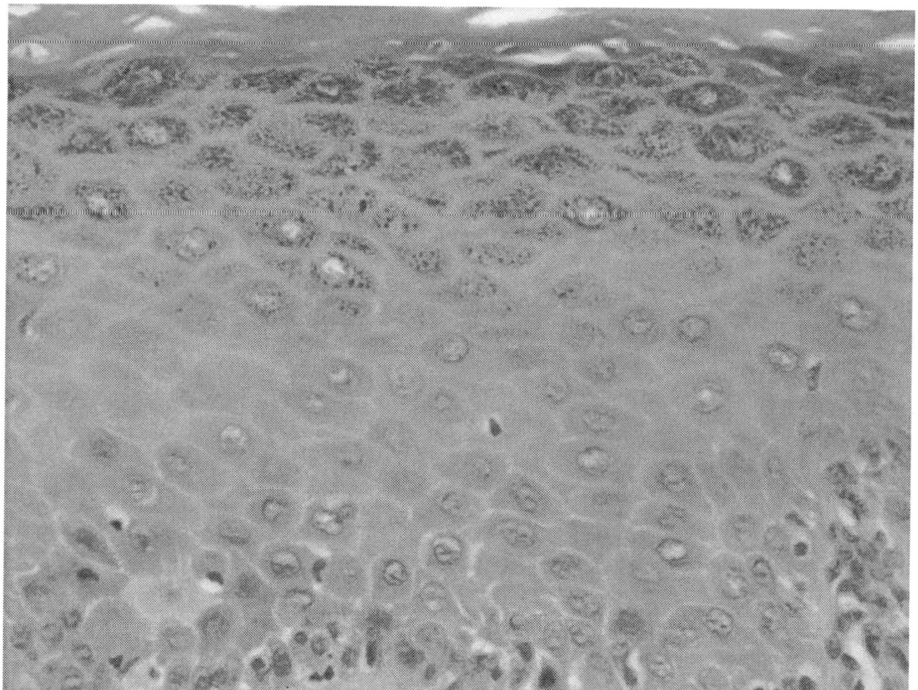

Figure 4. Oral mild dysplasia: lower third of the epithelium presents hypercromatic basal cells and several mitoses.

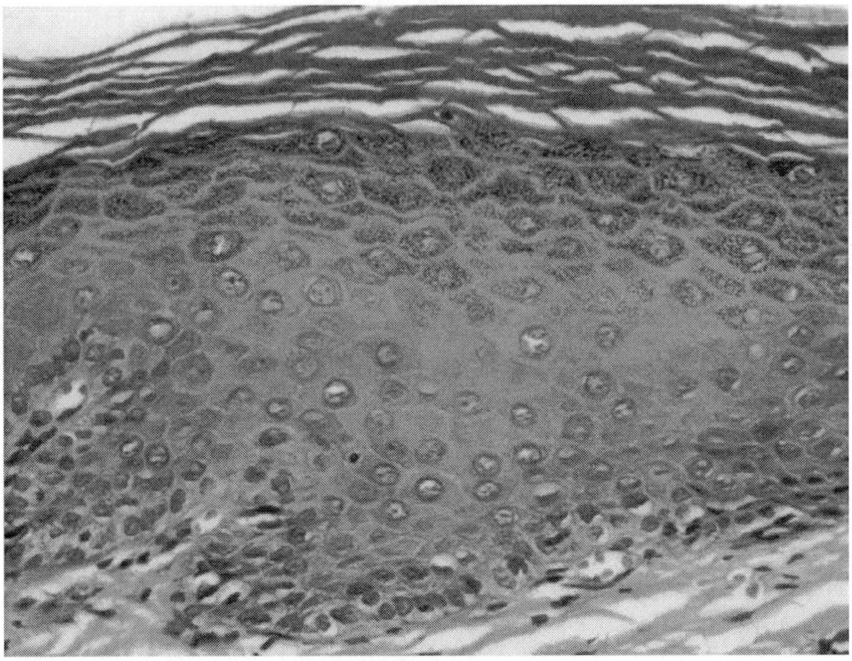

Figure 5. Oral moderate dysplasia: about two thirds of the epithelium presents cytological atypia. Hypercromatic cells, some mitoses above the basal layer, hyperkeratinization.

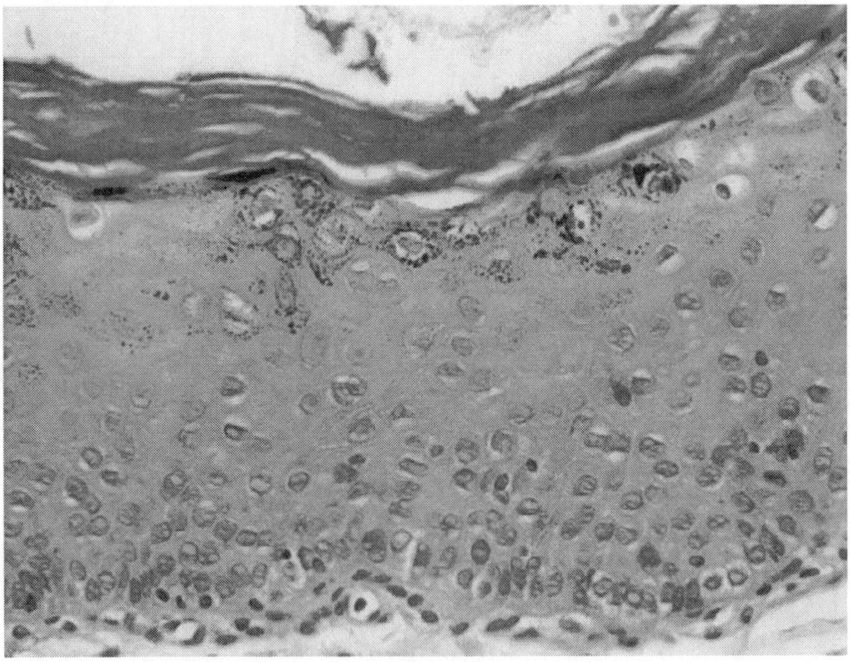

Figure 6. Oral severe atypia: more than two thirds of the epithelium shows cytological atypia. Hypercromatic cells are present in all layers, as well as several and atypical mitoses and hyperkeratinization.

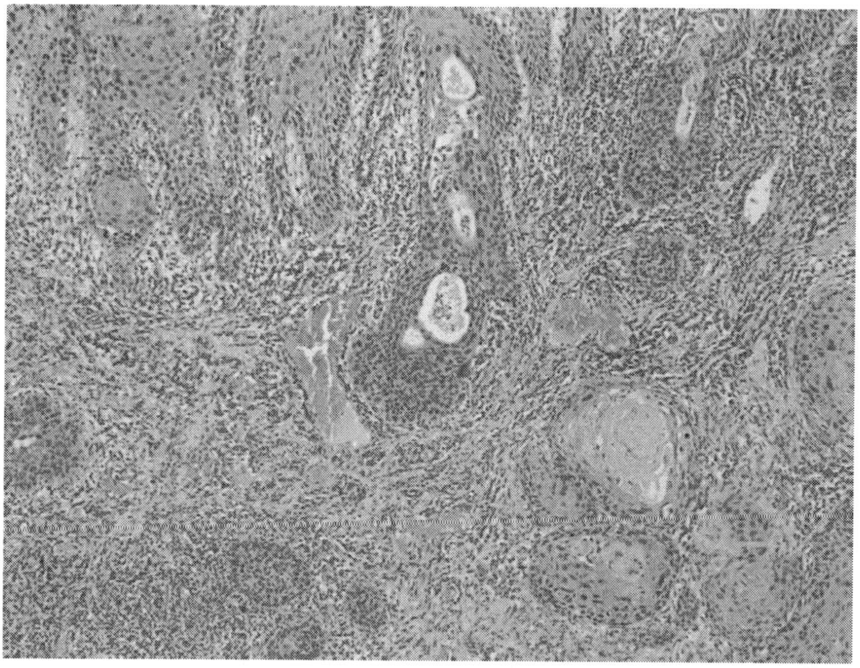

Figure 7. Oral squamous cell carcinoma: increased numbers of mitoses in all layers, abnornal mitotic figures, nuclear/cytoplasmic ratios, epithelial stratification disarray, droplet-shaped rete pegs, epithelial isle and cords invaded the parenchyma of the tumor. Several blood vessels are present and the cells are pleomorfic and hypercromatic.

Still according to the World Health Organization (WHO, 2005), a diagnosis of carcinoma *in situ* is delivered when that malignant transformation has occurred but invasion is not present. It is not always possible to recognize this morphologically and full thickness or almost full thickness architectural abnormalities in the viable cellular layers accompanied by pronounced cytologic atypia that are recommended to perform the diagnosis. Atypical mitotic figures and abnormal superficial mitoses are commonly seen in carcinoma *in situ* as well.

A diagnosis of OSCC, on the other hand, is made when there is evidence of nests of epithelial cells isolated from the epithelial-stromal interface and invading the underlying lamina propria and deeper submucosa [181] (Figure 7). With approximately 500,000 new cases annually, HNSCC, represent one of the six most common cancers in the world [187]. The malignant lesions arising in the oral cavity, larynx and pharynx, results in nearly ~11,000 deaths each year in the United States alone [188]. When it comes to the oral cavity, cancer often arises from the retromolar region, boarder of the tongue, floor of mouth, lips, palate and less frequently other sites [189].

The high mortality rate in cancer patients results from the dissemination of tumor cells resistant to conventional therapy throughout the orgasnism [190] and although there are many advances in the prevention and treatment of other types of cancers, the five-year survival rate after diagnosis for HNSCC remains approximately 50%, which is considerably lower than for other cancers, such as colorectal, cervix and breast [188].

Preventive care seems to be a very good choice, once the most common and well known risk factors involved in HNSCC development are represented by alcohol and tobacco use and betel nut chewing [191,192,193]. And along with the accumulation of genetic and epigenetic changes in a limited set of genes that may lead to cancer, dietary habits and infection with the human papillomavirus (HPV) are also widely studied and may share some relationship to these lesions. Several papers have shown that HPV may be involved at least in the pathogenesis of a subgroup of oral and cervical SCC [194].

Normal cells proliferate only when needed and, as mentioned above, there are stem cells within the basal cell layer of the oral epithelium. In cancer, however, the controlling mechanisms follow their own internal program for timing their reproduction. Hence, as it is known now the majority of cancer cells are derived from the clonal expansion and aberrant growth of a single stem cell present in the tissue or from a few tumor-initiating cells that have re-acquired self-renewal capacity [195]. Besides growing in an unrestricted manner, these cancer cells can escape cell senescence and death programs becoming immortal. Cancer progression is than characterized by promotion of new blood vessels formation, acquirement of the ability to migrate from their original site, invade nearby tissues, and metastasize to distant anatomical sites [174,196].

Concerning to the epithelial markers highly used for the normal epithelial cells, their expression may vary when studied in dysplastic or carcinoma cells. As previously explained, the adherent junctions and their respective proteins (cadherins and catenins) should have attenuated appearance when cells lose their connectivity and tend to migrate and spread, what happens to cancer cells. In accordance to that, immunocytochemistry based results have demonstrated that most normal cell cultures express more E-cadherin, integrin b-1, keratin (K) 14, K19, and involucrin than SCC cultures [21].

The keratins K13 and K4 are expressed in the suprabasal compartment of the oral epithelium [10]. They can be also focally detected in head and neck squamous cell carcinoma, being their expression varied, especially in the poorly differentiated tumors [197]. Besides, other keratins such as K6 and K16 are expressed in non-keratinized epithelium [38] and may be abundantly found in the squamous cell carcinoma [34].

The expression of E-cadherin decreases in oral cancers, oral cancer cell lines, and oral immortalized keratinocytes [198,199,200]. Actually, normal human oral keratinocytes express high E-cadherin levels and also increased involucrin protein levels. The increased levels of E-cadherin in these cultures may thus be associated with increased differentiation, which means normal cells entering senescence and apoptosis, differently from neoplastic cells [21]. Moreover, the non-degraded β-catenin accumulation in the cytoplasm of oral cancer cells appears to correlate with tumor stage, and increased proliferative ability of these cells [201]. Thus, in undifferentiated cancers of the oral cavity, β-catenin membrane staining decreases in intensity and the staining appears cytoplasmatic [199,202,203].

Integrins represent a family of heterodimeric cell membrane adhesion molecules, which bind extracellular matrix components. Integrin β-1 has been implicated as an adhesion and signaling molecule in a variety of biological phenomena. In normal stratified epithelia, integrins containing subunit β-1 are localized at the basal layer [204] and decrease in the suprabasal layers, as the keratinocyte terminal differentiation begins [205]. During the progression of some types of cancer though, two integrins containing β-1 ($\alpha_v\beta_1$, $\alpha_5\beta_1$) and b-6 subunits have been reported to increase, favoring cell spreading and tissue invasion [206,207,208]. Other two integrins containing the β-1 subunit ($\alpha_2\beta_1$, $\alpha_3\beta_1$), however, decrease in oral cancers [209].

Normal cells with high E-cadherin content also have higher levels of saturated fatty acids when compared to SCC cells. SCC cells show, conversely, higher proportions of unsaturated fatty acids. Whether this difference is related to the decreased growth, as suggested by Calder et al [210] or differentiation is not clear.

References

[1] Walker, WB. The Oral Cavity and Associated Structures. In: Walker, HK; Hall, WD; Hurst, JW, editors. *Clinical Methods: The History, Physical, and Laboratory Examinations.* 3rd edition. Boston: Butterworths; 1990. Chapter 129, 616-621.

[2] Squier, CA; Johnson, NW; Hackemann, M. Structure and Function of Normal Human Mucosa. In: Meyer, J; Squier, CA; Gerson, SJ, editors. *Oral Mucosa in Health and Disease,* A.E. Dolby. Ed. Oxford: Blackwell; 1975. Chapter 1, 1-101.

[3] Squier, CA; Kremer, M.J. Biology of Oral Mucosa and Esophagus. *Journal of the National Cancer Institute Monographs.* 2001, 29, 7-15.

[4] Koss, LG. *Diagnostic Cytology and its Histopathologic Bases* 4th edition. Philadelphia, Lippincott Co JB, 1992.

[5] Squier, CA. Introduction. In: Squier, CA; Meyer, J. editors. *Current Concepts of the Histology of Oral Mucosa.* Illinois, Thomas Springfield Ed.: C C Book, 1971. Chapter 1, 5-33.

[6] Bath-Balogh, M; Fehrenbach, MJ. *Ilustrated Dental Embriology, Histology, and Anatomy.* Philadelphia, W. B. Saunders Co, 1997.

[7] Davis, WL. The Oral Mucosa. In Davis, WL. *Oral Histology, Cell Structure and Function.* Philadelphia, W. B. Saunders Co, 1986. Chapter 13, 179-198.

[8] Schroeder, HE. *Oral Structural Biology.* New York, Thieme Medical Publishers, 1991. Chapter 6, 350-387.

[9] Ten Cate, AR. *Histologia Bucal, Desenvolvimento, Estrutura e Função.* Rio de Janeiro, Guanabara Koogan, 2001.

[10] Squier, CA; Johnson, NW; Hopps, RM. *Human Oral Mucosa. Development, Structure and Function,* Oxford, Blackwell Scientific Publications, 1976.

[11] Pierce, GBJr; Midgely, ARJr; Sri Ram, JJ. The histogenesis of the Basement Membrane. *Exp. Med.* 1963,1:339-348.

[12] Hay, Ed; Revel, JP. Autoradiographic studies of the origin of the basement lamella in Ambystoma. *Dev. Bio.* 1963, 7:152-168.

[13] Clausen, H; Vedtofte, P; Moe, D; Dabelsteen, E; Sun, TT; Dale, B. Differentiation-dependent expression of keratins in human oral epithelia. *J. Invest. Dermatol.* 1986, 86: 249–254

[14] Smith, SA; Dale, BA. Immunologic localization of filaggrin in human oral epithelia and correlation with keratinization. *J. Invest. Dermatol.* 1986, 86:168–172

[15] Steinert, PM; Marekov, LN. The proteins elafin, filaggrin, keratin intermediate filaments, loricrin, and small proline-rich proteins 1 and 2 are isodipeptide cross-linked components of the human epidermal cornified cell envelope. *J. Biol. Chem.* 1995, 270: 17702-17711.

[16] Smack, DP; Korge, BP; James, WD. Keratin and keratinization. *J. Am. Acad. Dermatol.* 1994, 30:85-102.

[17] Sun, TT; Tseng, SC; Huang, AJ; Cooper, D; Schermer, A; Lynch, MH; Weiss, R; Eichner, R. Monoclonal antibody studies of mammalian epithelial keratins: a review. *Ann. NY Acad. Sci.* 1985, 455:307–329.

[18] Fuchs, E; Weber, K. Intermediate filaments: structure, dynamics, function, and disease. *Annu. Rev. Biochem.* 1994, 63:345–382.

[19] Bloor, BK; Tidman, N; Leigh, IM; Odell, E; Dogan, B; Wollina, U; Ghali, L; Waseem, A. Expression of Keratin K2e in Cutaneous and Oral Lesions Association with Keratinocyte Activation, Proliferation, and Keratinization. *Am. J. Pathol.* 2003, 162:963-975.

[20] Freshney, RI. Introduction. In: Freshney, RI. editors. *Culture of epithelial cells*. New York, Wiley-Liss, 1992, 1-23.

[21] Gasparoni, A; Fonzib, L; Schneidera, GB; Wertza, PW; Johnsona, GK; Squier, CA. Comparison of differentiation markers between normal and two squamous cell carcinoma cell lines in culture. *Archives of Oral Biology* 2004, 49:653-664.

[22] Bowden, PE. Keratins and other epidermal proteins. In: Priestley, GC. editors. *Molecular aspects of dermatology*. Chichester, Juhn Wiley and Sons Ltd, 1993, 19-54.

[23] Silverman, S. Nonkeratinization and Keratinization: the Extremes of the Human Range. In: Squier, CA; Meyer, J. editors *Current Concepts of the Histology of Oral Mucosa.* Illinois, Thomas Springfield Ed.: C C Book, 1971. Chapter 4, 80-96.

[24] Wilke, MS; Hsu, BM; Wille Jr, JJ; Pittelkow, MR; Scott, RE. Biologic Mechanisms for the Regulation of Normal Human Keratinocyte Proliferation and Differentiation. *American Journal of Pathology* 1988, 131:171-181.

[25] Lagtha, LG. Stem cell concepts. *Differentiation* 1979, 14:23-34.

[26] Lavker, RM; Sun, TT. Epidermal Stem Cells. *The Journal of Investigative Dermatology* 1983, 81:121s-127s.

[27] Winning, TA; Townsend, GC. Oral mucosal embryology and histology. *Clin Dermatol* 2000, 18:499–511.

[28] van Oijen, MG; Gilsing, MM; Rijksen, G; Hordijk, GJ, Slootweg, PJ. Increased number of proliferating cells in oral epithelium from smokers and ex-smokers. *Oral Oncol.* 1998, 34:297–303.

[29] Karring, T; Lang, NP; Loe, H. The role of gingival connective tissue in determining epithelial differentiation. *J. Periodontal. Res.* 1975, 10:1–11.

[30] Mackenzie, IC; Hill, MW. Connective tissue influences on patterns of epithelial architecture and keratinization in skin and oral mucosa of the adult mouse. *Cell Tissue Res.* 1984 , 235:551–559.

[31] Chung, JH, Cho, KH, Lee, DY; Kwon, OS, Sung, MW, Kim, KH, Eun, HC. Human oral buccal mucosa reconstructed on dermal substrates: a model for oral epithelial differentiation. *Arch. Dermatol. Res.* 1997, 289:677–685.

[32] Presland, RB; Jurevic, RJ. Making sense of the epithelial barrier: what molecular biology and genetics tell us about the functions of oral mucosal and epidermal tissues. *Journal of Dental Education* 2002, 66:564-574.

[33] Hansson, A; Bloor, BK; Sarang, Z; Haig, Y; Morgan, PR; Stark, H-J; Fusenig, NE; Ekstrand, J; Graström, RC. Analysis of proliferation, apoptosis and keratin expression in cultured normal and immortalized human buccal keratinocytes. *European Journal of Oral Science* 2003, 111:43-41.

[34] Moll, R; Divo, M; Langbein, L. The human keratins: biology and pathology. Review. *Histochem. Cell Biol.* 2008, 129:705–733.

[35] Waschke, J. The desmosome and pemphigus. *Histochem. Cell Biol.* 2008, 130:21–54.

[36] Uitto, J; Richard, G; McGrath, JA. Diseases of epidermal keratins and their linker proteins. *Exp. Cell Res.* 2007, 313:1995–2009.

[37] Schweizer, J; Bowden, PE; Coulombe, PA; Langbein, L; Lane, EB; Magin, TM; Maltais, L; Omary, MB; Parry, DA; Rogers, MA; Wright, MW. New Consensus Nomenclature for mammalian keratins. *J. Cell Biol.* 2006 17, 174:169-174.

[38] Moll, R; Franke, WW; Schiller, DL. The Catalog of Human Cytokeratins: Patterns of Expression in Normal Epithelia, Tumors and Cultured Cells. Review. *Cell* 1982, 31:11-24.

[39] Kirfel, J; Magin, TM; Reichelt, J. Keratins: a structural scaffold with emerging functions. *Cell Mol. Life Sci.* 2003, 60:56–71.

[40] Porter, RM, Lane, EB. Phenotypes, genotypes and their contribution to understanding keratin function. *Trends Genet.* 2003, 19:278–285.

[41] Bosch, FX; Leube, RE; Achtstätter, T; Moll, R; Franke, WW. Expression of simple epithelial type cytokeratins in stratifed epithelia as detected by immunolocalization and hybridization in situ. *J. Cell Biol.* 1988, 106:1635–1648.

[42] Moll, R. Cytokeratins as markers of differentiation: expression profiles in epithelia and epithelial tumors. *Progress in pathology* (series), vol 142. Stuttgart, Gustav Fischer: 1–197, 1993.

[43] Morley, SM; Lane, EB. The keratinocyte cytoskeleton. In: Leigh, I; Lane, B; Watt, F, editors. *The keratinocyte handbook*. Cambridge, Cambridge University Press, 1994, 293-321.

[44] Hall, DE; Reichardt, LF; Crowley, E; Holley, B; Moezzi, H; Sonnenberg A et al. The a1/b1 and a6/b1 integrin heterodimers mediate cell attachment to distinct sites on laminin. *J. Cell Biol.* 1990,110:2175-2184.

[45] Lewis, JE; Jensen, PJ; Wheelock, MJ. Cadherin function is required for human keratinocytes to assemble desmosomes and stratify in response to calcium. *J. Invest. Dermatol.* 1994,102:870-877.

[46] Nicholson, J; Watt, FM. Decreased expression of fibronectin and the a5b1 integrin during terminal differentiation of human keratinocytes. *J. Cell Sci.* 1991, 98:225-232.

[47] Rice, RH; Green, H. The cornified envelope of terminally differentiated human epidermal keratinocytes consists of cross-linked protein. *Cell* 1977, 11:417-422.

[48] Madison, KC; Wertz, W; Strauss, JS; Downing, DT. Lipid composition of cultured murine keratinocytes. *J. Invest. Dermatol.* 1986, 87:253-259.

[49] Bader BL, Magin TM, Hatzfeld M, Franke WW (1986) Amino acid sequence and gene organization of cytokeratin no. 19, an exceptional tail-less intermediate Wlament protein. *EMBO J.* 5(8):1865–1875.

[50] Hesse M, Franz T, Tamai Y, Taketo MM, Magin TM (2000) Targeted deletion of keratins 18 and 19 leads to trophoblast fragility and early embryonic lethality. *EMBO J.* 19(19):5060–5070.

[51] Lindberg, K; Rheinwald, JG. Suprabasal 40 kd keratin (K19) expression as an immunohistologic marker of premalignancy in oral epithelium. *Am. J. Pathol.* 1989, 34:89-98.

[52] Fuchs, E. Epidermal differentiation: the bare essentials. *J. Cell Biol.* 1990, 111:2807—2814.
[53] Dale, BA; Salonen, l; Jones, AH. New Approaches and concepts in the study of differentiation of oral epithelia. *Crit. Rev. Oral Biol. Med.* 1990,1:167-90.
[54] Mackenzie, IC; Gao, Z. Patterns of cytokeratin expression in the epithelia of inflamed human gingiva and periodontal pockets. *J. Periodont. Res.* 1993, 28:49-59.
[55] Tomakidi, P; Fusenig, NE; Kohl, A; Komposch, G. Histomorphological and biochemical differentiation capacity in organotypic co-cultures of primary gingival cells. *J. Periodont. Res.* 1997, 32:388-400.
[56] Fuchs, E; Green, H. Changes in keratin gene expression during terminal differentiation of the keratinocyte. *Cell,* 1980, 19:1033–1042.
[57] Oda, D; Watson, E. Human oral epithelial cell culture. II. Improved conditions for reproducible culture in serum-free medium. *In Vitro Cell Dev. Biol.* 1990, 26:596-603.
[58] Sawaf, MH; Ouhayoun, JP; Forest, N. Cytokeratin profiles in oral epithelial: a review and a new classification. *J. Biol. Buccale,* 1991, 19:187–198.
[59] Bloor, BK; Su, L; Shirlaw, PJ; Morgan, PR. Gene expression of differentiation- specific keratins (4/13 and 1/10) in normal human buccal mucosa. *Lab. Invest.* 1998, 78:787–795.
[60] Collin, C; Moll, R; Kubicka, S; Ouhayoun, JP; Franke, WW. Characterization of human cytokeratin 2, an epidermal cytoskeletal protein synthesized late during differentiation. *Exp. Cell Res.* 1992a, 1:132–141.
[61] Collin, C; Ouhayoun, JP; Grund, C; Franke, WW. Suprabasal marker proteins distinguishing keratinizing squamous epithelia: cytokeratin 2 polypeptides of oral masticatory epithelium and epidermis are different. *Differentiation,* 1992b, 51:137–148.
[62] Rogers MA, Edler L, Winter H, Langbein L, Beckmann I, Schweizer J (2005) Characterization of new members of the human type II keratin gene family and a general evaluation of the keratin gene domain on chromosome 12q13.13. *J. Invest. Dermatol.* 124(3):536–544.
[63] Magin, TM; Vijayaraj, P; Leube, RE. Structural and regulatory functions of keratins. *Exp. Cell Res.* 2007, 313:2021–2032.
[64] Oriolo, AS; Wald, FA; Ramsauer, VP; Salas, PJ. Intermediate filaments: a role in epithelial polarity. *Exp. Cell Res.* 2007 313(10):2255–2264.
[65] Moll, R; Lowe, A; Laufer, J; Franke, WW. Cytokeratin 20 in human carcinomas. A new histodiagnostic marker detected by monoclonal antibodies. *American Journal of Pathology* 1992, 140:427±447.
[66] Barrett, AW; Cort, EM; Patel, P; Berkovitz, BKB. An immunohistological study of cytokeratin 20 in human and mammalian oral epithelium. *Archives of Oral Biology* 2000, 45:879±887.
[67] Alvares, OF; Meyer, J. Introduction. In: Meyer, J; Squier, CA; Gerson, SJ, editors. *The Structure and Function of Oral Mucosa.* New York, Pergamon Press. 1984. Chapter 1, 1-6.
[68] Chen, SY; Meyer, J. Regional Differences in Tonofilaments and Keratohyaline Granules. In: Squier, CA; Meyer, J. editors *Current Concepts of the Histology of Oral Mucosa.* Illinois, Thomas Springfield Ed.: C C Book, 1971. Chapter 6, 114-128.
[69] Candi, E; Schmidt, R; Melino, G. The cornified envelope: a model of cell death in the skin. *Nat. Rev. Mol. Cell Biol.* 2005, 6:328-340.

[70] Chen, SY; Squier, CA. The Ultrastructure of the Oral Epithelium. In: Meyer, J; Squier, CA; Gerson, SJ, editors. *The Structure and Function of Oral Mucosa.* New York, Pergamon Press. 1984. Chapter 2, 7-31.

[71] Silverman, S. Nonkeratinization and Keratinization: the Extremes of the Human Range. In: Squier, CA; Meyer, J. editors *Current Concepts of the Histology of Oral Mucosa.* Illinois, Thomas Springfield Ed.: C C Book, 1971. Chapter 4, 80-96.

[72] Kuechle, MK; Presland, RB; Lewis, SP; Fleckman, P; Dale, BA. Inducible expression of filaggrin increases keratinocyte susceptibility to apoptotic cell death. *Cell Death Differ.* 2000, 7:566-573.

[73] Resing, KA; Walsh, KA; Haugen-Scofield, J; Dale, BA. Identification of proteolytic cleavage sites in the conversion of profilaggrin to filaggrin in mammalian epidermis. *J. Biol. Chem.* 1989, 264:1837±1845.

[74] Dale, BA; Holbrook, KA; Steinert, PM. Assembly of stratum corneum basic protein and keratin filaments in macrofibrils. *Nature,* 1978, 276:729 ± 731

[75] Dale, BA; Presland RB; Fleckman, P; Kam, E; Resing, KA. Phenotypic expression and processing of filaggrin in epidermal differentiation. In: *Molecular biology of the skin.* Darmon, M; Blumenberg, M. editors. San Diego, Academic Press 1993, 79-106.

[76] Kuechle, MK; Thulin, CD; Presland, RB; Dale, BA. Profilaggrin requires both linker and filaggrin peptide sequences to form granules: Implications for profilaggrin processing in vivo. *J. Invest. Dermatol.* 1999, 112:843 ± 852.

[77] Steinert, PM; Marekov, LN. The proteins elafin, filaggrin, keratin intermediate filaments, loricrin, and small proline-rich proteins 1 and 2 are isodipeptide cross-linked components of the human epidermal cornified cell envelope. *J. Biol. Chem.* 1995, 270:17702-17711.

[78] Dale, BA; Presland, RB; Lewis, SP; Underwood, RA; Fleckman, P. Transient expression of epidermal filaggrin in cultured cells causes collapse of intermediate filament networks with alterations of cell shape and nuclear integrity. *J. Invest. Dermatol.* 1997, 108:179 ± 187.

[79] Ruoslahti, E; Reed, J. Anchorage dependence, integrins, and apoptosis. *Cell,* 1994, 77:477 ± 478.

[80] Presland, RB; Dale, BA. Epithelial Structural Proteins of the Skin and Oral Cavity: Function in Health and Disease. *Crit. Rev. Oral Biol.* 2000, 11:383-408.

[81] Hohl, DA. New star in the heavens of epidermal proteins: loricrin--what is it? *Hautarzt* 1990, 41:299-301.

[82] Folk, JE; Finlayson, JS. The epsilon-(gamma-glutamyl) lysine crosslink and the catalytic role of transglutaminases, *Adv. Protein Chem.* 1977, 31:1–133.

[83] Hohl, DA; de Viragh, PA; Amiguet-Barras, F; Gibbs, S; Backendorf, C; Huber, M. The small proline-rich proteins constitute a multigene family of differentially regulated cornified cell envelope precursor proteins. *J. Invest. Dermatol.* 1995, 104:902-9.

[84] Gosden, R; Spears, N. Programmed cell death in the reproductive system. *Br. Med. Bull.* 1997, 53: 644–661.

[85] Watt, FM. Involucrin and other markers of keratinocyte terminal differentiation. *J. Invest. Dermatol.* 1983, 81:100s-3s.

[86] Ishida-Yamamoto, A; Kartasova, T; Matsuo, S; Kuroki, T; Iizuka, H. Involucrin and SPRR are synthesized sequentially in differentiating culture epidermal cells. *J. Invest. Dermatol.* 1997, 108:12-6.

[87] Scully, C; Bagan, J-V; Black, M; Carrozzo, M; Eisen, D; Escudier, M; Farthing, P; Kuffer, R; Lo Muzio, L; Mignogna, M; Porter, SR. Epithelial Biology. *Oral Diseases,* 2005, 11:58–71.
[88] Sasaki, H. Freeze-fracture analysis of renal-epithelial tight junctions, *Methods in Molecular Medicine,* 2003, 86: 155–166.
[89] Yeager, M; Harris, AL. Gap junction channel structure in the early 21st century: facts and fantasies. *Current Opinion in Cell Biology,* 2007, 19:521–528.
[90] Nagafuchi A, "Molecular architecture of adherens junctions," *Current Opinion in Cell Biology,* vol. 13, no. 5, pp. 600–603, 2001.
[91] Greenand, KJ; Jones, JCR. Desmosomes and hemidesmosomes: structure and function of molecular components. *The FASEB Journal,* 1996, 10:871–881.
[92] Lourenco, SV; Kapas, S. Integrin expression in developing human salivary glands. *Histochemistry and Cell Biology* 2005, 124:391–399.
[93] Lourenco, SV; Kapas, S; Williams, DM; Leite, K; Araújo VC. Expression patterns of integrins on pleomorphic adenoma and adenoid cystic carcinoma: study on specimens and in vitro investigation of the effects of extracellularmatrix on the expression of these adhesion molecules. *J. Oral. Pathol. Med.* 2004, 33:574–580.
[94] Barnett, ML; Szabo, G. Gap junctions in human gingival keratinized epithelium. *J. Periodontal. Res.* 1973, 8:111-126.
[95] Schroeder, HE. *Differentiation of human oral stratified epithelium.* Basel: Karger; 1981, 33–156.
[96] Farquhar, MG; Palade, GE. Junctional complexes in various epithelia. *J. Cell Biol.* 1963, 17:375-412.
[97] Bizzozero, G. Delle cellule cigliate, del reticulo Malpighiani d'ell epiderme. *Ann. Univ. Med.* 1864, 190.
[98] Fawcett, DW; Selby, CC. Observations on the fine structure of the turtle atrium. *J. Biophys. Biochem. Cytol.* 1958, 4:63-72.
[99] North, AJ; Bardsley, WG; Hyam, J; Bornslaeger, EA; Cordingley, HC; Trinnaman, B; Hatzfeld, M; Green, KJ; Magee, AI; Garrod DR. Molecular map of the desmosomal plaque. *J. Cell Sci.* 1999, 112:4325-4336.
[100] Borysenko, JZ; Revel, JP. Experimental Manipulation of Demosome Structure. *Am. Journal Anat.* 1973, 137:403-421.
[101] Papini, S; Cecchetti, D; Campani, D; Fitzgerald, W; Grivel, JC; CHEN, S; Margolis, L; Revoltella, R. Isolation and clonal analysis of human epidermal keratinocyte stem cells in long-term culture. *Stem Cells* 2003, 21:481–494.
[102] Kerdvongbundit, V; Vongsavan, N; Soo-Ampon, S; Hasegawa, A. Microcirculation and micromorphology of healthy and inflamed gingivae. *Odontology,* 2003, 91:19–25.
[103] Scardina, GA; Cacioppo, A; Messina, P. Anatomical evaluation of oral microcirculation: Capillary characteristics associated with sex or age group. *Ann. Anat.* 2009, 191:371-378.
[104] Scardina, GA; Fuca, G; Messina, P. .Microvascular characteristics of the human interdental papilla. *Anat. Histol. Embryol.* 2007, 36:266–268.
[105] Kano, M; Shimizu, Y; Okayama, K; Kikuchi, M. Quantitative study of ageing epiglottal taste buds in humans. *Gerodontology,* 2007, 24:169–172.

[106] Ogrin, R; Darzins, P; Khalil, Z. Age-related changes in microvascular blood flow and transcutaneous oxygentension under Basal and stimulated conditions. *J. Gerontol. A Biol. Sci. Med. Sci.* 2005, 60:200–206.

[107] Kelly, RI; Pearse, R; Bull, RH; Leveque, JL; de Rigal, J; Mortimer, PS. The effects of aging on the cutaneous microvasculature. *J. Am. Acad. Dermatol.* 1995, 33:749–756.

[108] Takeichi, M; Hatta, K; Nose, A; Nagafuchi, A. Identification of a gene family of cadherin cell adhesion molecules. *Cell Differ. Dev.* 1988, 25:91–94.

[109] Gumbiner, BM; McCrea, PD. Catenins as mediators of the cytoplasmic functions of cadherins. *J. Cell Sci. Suppl.* 1993, 17:155–158.

[110] Herrenknecht K, Ozawa M, Eckerskorn C, Lottspeich F, Lenter M, Kemler R (1991). The uvomorulin-anchorage protein alpha catenin is a vinculin homologue. *Proc. Natl. Acad. Sci. USA* 88: 9156–9160.

[111] Karnovsky, A; Klymkowsky, MW. Anterior axis duplication in Xenopus induced by the over-expression of the cadherin-binding protein plakoglobin. *Proc. Natl. Acad. Sci. USA,* 1995, 92:4522–4526.

[112] Tao, YS; Edwards, RA; Tubb, B; Wang, S; Bryan, J; McCrea, PD. Beta-catenin associates with the actin-bundling protein fascin in a noncadherin complex. *J. Cell Biol.* 1996, 134:1271–1281.

[113] Butz, S; Kemler, R. Distinct cadherin–catenin complexes in Ca(2+)-dependent cell–cell adhesion. *FEBS Lett,* 1994, 355:195–200.

[114] Bullions, LC; Levine, AJ. The role of beta-catenin in cell adhesion, signal transduction, and cancer. *Curr. Opin. Oncol.* 1998, 10:81–87.

[115] Nathke, IS; Adams, CL; Polakis, P; Sellin, JH; Nelson, WJ. The adenomatous polyposis coli tumor suppressor protein localizes to plasma membrane sites involved in active cell migration. *J. Cell Biol.* 1996, 134:165–179.

[116] Vleminckx, K; Wong, E; Guger, K; Rubinfeld, B; Polakis, P; Gumbiner, BM. Adenomatous polyposis coli tumor suppressor protein has signaling activity in Xenopus laevis embryos resulting in the induction of an ectopic dorsoanterior axis. *J. Cell Biol.* 1997, 136:411–420.

[117] Rubinfeld, B; Robbins, P; El Gamil, M; Albert, I; Porfiri, E; Polakis, P. Stabilization of beta-catenin by genetic defects in melanoma cell lines. *Science,* 1997, 275:1790–1792.

[118] Hinck, L; Nelson, WJ; Papkoff, J. Wnt-1 modulates cell–cell adhesion in mammalian cells by stabilizing beta-catenin binding to the cell adhesion protein cadherin. *J. Cell Biol.* 1994, 124:729–741.

[119] Kawanishi, J; Kato, J; Sasaki, K; Fujii, S; Watanabe, N; Niitsu, Y. Loss of E-cadherin-dependent cell–cell adhesion due to mutation of the beta-catenin gene in a human cancer cell line, HSC-39. *Mol. Cell Biol.* 1995, 15:1175–1181.

[120] Munemitsu, S; Albert, I; Rubinfeld, B; Polakis, P. Deletion of an amino-terminal sequence beta-catenin in vivo and promotes hyperphosporylation of the adenomatous polyposis coli tumor suppressor protein. *Mol. Cell Biol.*(1996 16: 4088–4094.

[121] Robbins, PF; El Gamil, M; Li, YF. et al A mutated betacatenin gene encodes a melanoma-specific antigen recognized by tumor infiltrating lymphocytes. *J. Exp. Med.* 1996, 183:1185–1192.

[122] Yost, C; Torres, M; Miller, JR; Huang, E; Kimelman, D; Moon, RT. The axis-inducing activity, stability, and subcellular distribution of beta-catenin is regulated in Xenopus embryos by glycogen synthase kinase 3. *Genes Dev.* 1996, 10:1443–1454.

[123] Barth, AI; Pollack, AL; Altschuler, Y; Mostov, KE; Nelson, WJ. NH2-terminal deletion of beta-catenin results in stable colocalization of mutant beta-catenin with adenomatous polyposis coli protein and altered MDCK cell adhesion. *J. Cell Biol.* 1997, 136:693–706.
[124] Morin, PJ; Sparks, AB; Korinek, V et al.. Activation of beta-catenin-Tcf signaling in colon cancer by mutations in beta-catenin or APC. *Science,* 1997, 275:1787–1790.
[125] Pai, LM; Orsulic, S; Bejsovec, A; Peifer, M. Negative regulation of armadillo, a Wingless effector in Drosophila. *Development* 1997, 124:2255–2266.
[126] Rubenstein, A; Merriam, J; Klymkowsky, MW. Localizing the adhesive and signaling functions of plakoglobin. *Dev. Genet.* 1997, 20:91–102.
[127] Peifer, M. Beta-catenin as oncogene: the smoking gun. *Science,* 1997, 275:1752–1753.
[128] Behrens, J; von Kries, JP; Kuhl, M et al. Functional interaction of beta-catenin with the transcription factor LEF-1. *Nature,* 1996, 382:638–642.
[129] Huber, O; Korn, R; McLaughlin, J; Ohsugi, M; Herrmann, BG; Kemler, R. Nuclear localization of beta-catenin by interaction with transcription factor LEF-1. *Mech. Dev.* 1996, 59:3–10.
[130] Molenaar, M; van de Wetering, M; Oosterwegel, M et al. XTcf-3 transcription factor mediates beta-catenin-induced axis formation in Xenopus embryos. *Cell,* 1996, 86:391–399.
[131] Aoki, M; Hecht, A; Kruse, U; Kemler, R; Vogt, PK. Nuclear endpoint of Wnt signaling: neoplastic transformation induced by transactivating lymphoid-enhancing factor 1. *Proc. Natl. Acad. Sci. USA,* 1999, 96:139–144.
[132] Korinek, V; Barker, N; Morin, PJ et al. Constitutive transcriptional activation by a beta-catenin-Tcf complex in APC-/- colon carcinoma. *Science,* 1997, 275:1784–1787.
[133] Pories, SE; Hess, DT; Swenson, K et al. Overexpression of pp60c-src elicits invasive behavior in rat colon epithelial cells. *Gastroenterology,* 1998, 114: 1287–1295.
[134] Young, CS; Kitamura, M; Hardy, S; Kitajewski, J. Wnt-1 induces growth, cytosolic beta-catenin, and Tcf/Lef transcriptional activation in Rat-1 fibroblasts. *Mol. Cell Biol.* 1998, 18:2474–2485.
[135] Funayama, N; Fagotto, F; McCrea, P; Gumbiner, BM. Embryonic axis induction by the armadillo repeat domain of beta-catenin: evidence for intracellular signaling. *J. Cell Biol.* 1995,128:959–968.
[136] Schneider, S; Steinbeisser, H; Warga, RM; Hausen, P. Beta-catenin translocation into nuclei demarcates the dorsalizing centers in frog and fish embryos. *Mech. Dev.* 1996, 57:191–198.
[137] Hao, X; Tomlinson, I; Ilyas, M; Palazzo, JP; Talbot, IC. Reciprocity between membranous and nuclear expression of beta-catenin in colorectal tumours. *Virchows Arch.* 1997, 431:167–172.
[138] Brabletz, T; Jung, A; Hermann, K; Gunther, K; Hohenberger, W; Kirchner, T. Nuclear overexpression of the oncoprotein beta-catenin in colorectal cancer is localized predominantly at the invasion front. *Pathol. Res. Pract.* 1998, 194: 701–704.
[139] Fagotto, F; Gluck, U; Gumbiner, BM. Nuclear localization signal-independent and importin/karyopherin-independent nuclear import of beta-catenin. *Curr. Biol.* 1998, 8:181–190.

[140] Sheng, H; Shao, J; Williams, CS et al. Nuclear translocation of beta-catenin in hereditary and carcinogen-induced intestinal adenomas. *Carcinogenesis,* 1998, 19:543–549.

[141] Simcha, I; Shtutman, M; Salomon, D et al. Differential nuclear translocation and transactivation potential of beta-catenin and plakoglobin. *J. Cell Biol.* 1998, 141:1433–1448.

[142] Rimm, DL; Caca, K; Hu, G; Harrison, FB; Fearon, ER. Frequent nuclear/cytoplasmic localization of beta-catenin without exon 3 mutations in malignant melanoma. *Am. J. Pathol.* 1999, 154:325–329.

[143] Yokoya, F; Imamoto, N; Tachibana, T; Yoneda, Y. Betacatenin can be transported into the nucleus in a Ranunassisted manner. *Mol. Biol. Cell,* 1999, 10: 1119–1131.

[144] Shtutman, M; Zhurinsky, J; Simcha, I et al. The cyclin D1gene is a target of the beta-catenin/LEF-1 pathway. *Proc. Natl. Acad. Sci. USA,* 1999, 96:5522–5527.

[145] Tetsu, O; McCormick, F. Beta-catenin regulates expression of cyclin D1 in colon carcinoma cells. *Nature,* 1999, 398:422–426.

[146] He, TC; Sparks, AB; Rago, C et al. Identification of c-MYC as a target of the APC pathway. *Science,* 1998, 281:1509–1512.

[147] Radisky, DC. Epithelial-mesenchymal transition. *J. Cell Sci.* 2005, 1;118:4325-4326. No abstract available.

[148] Grille, SJ; Bellacosa, A; Upson, J et al. The protein kinase Akt induces epithelial mesenchymal transition and promotes enhanced motility and invasiveness of squamous cell carcinoma lines. *Cancer Res.* 2003, 63:2172–2178.

[149] Giudice, FS; Vechio, AM; Abrahão, AC; Sperandio, FF; Pinto-Junior DD. Different expression patterns of pAkt, NF-κB and cyclin D1 proteins during the invasion process of head and neck squamous cell carcinoma: an in vitro approach. *J. Oral Pathol. Med.* 2011, 40:405-411.

[150] Araujo, V; Junior, DP; Sousa, S; Nunes, F; Araujo, N. Vimentin in oral squamous cell carcinoma. *Eur. Arch. Otorhinolaryngol.* 1993, 250:105-109.

[151] Ivaskaa, J; Pallarib, HM; Nevoa, J; Eriksson, J. Novel functions of vimentin in cell adhesion, migration, and signaling. *Exp. Cell Res.* 2007, 313:2050-2062.

[152] Mendez, MG; Kojima, SI; Goldman, RD. Vimentin induces changes in cell shape, motility, and adhesion during the epithelial to mesenchymal transition, *FASEB J.* 2010, 22. [Epub ahead of print].

[153] Taki, M; Kamata, N; Yokoyama, K; Fujimoto, R; Tsutsumi, S; Nagayama, M. Down-regulation of Wnt-4 and up-regulation of Wnt-5a expression by epithelialmesenchymal transition in human squamous carcinoma cells. *Cancer Sci.* 2003, 94:593-597.

[154] Gilles, C; Polette, M; Zahm, JM; Tournier, JM; Volders, L; Foidart, JM; Birembaut, P. Vimentin contributes to human mammary epithelial cell migration. *J. Cell Sci.* 1999, 112:4615-4625.

[155] Liu, LK; Jiang, XY; Zhou, XX; Wang, DM; Song, XL; Jiang, HB. Upregulation of vimentin and aberrant expression of E-cadherin/beta-catenin complex in oral squamous cell carcinomas: correlation with the clinicopathological features and patient outcome. *Mod. Pathol.* 2010, 23:213-224.

[156] Guertin, DA; Sabatini, DM. Defining the Role of mTOR in Cancer. *Cancer Cell,* 2007, 12:9–22.

[157] Hanada, M; Feng, J; Hemmings, BA. Structure, regulation and function of PKB/AKT–a major therapeutic target. *Biochim. Biophys. Act.* 2004, 1697:3–16.
[158] Manning, BD; Cantley, LC. AKT/PKB signaling: navigating downstream. *Cell,* 2007, 129:1261–1274.
[159] Shaw,RJ; Cantley, LC. Ras, PI(3)KandmTOR signalling controls tumour cell growth. *Nature,* 2006, 441:424-430.
[160] Amornphimoltham, P; Sriuranpong, V; Patel, V et al. Persistent activation of the Akt pathway in head and neck squamous cell carcinoma: a potential target for UCN-01. *Clin. Cancer Res.* 2004, 10:4029-4037.
[161] Molinolo, AA; Hewitt, SM; Amornphimoltham, P et al. Dissecting the Akt/mammalian target of rapamycin signaling network: emerging results from the head and neck cancer tissue array initiative. *Clin. Cancer Res.* 2007, 13:4964-4973.
[162] Amornphimoltham, P; Patel, V; Sodhi, A et al. Mammalian target of rapamycin, a molecular target in squamous cell carcinomas of the head and neck. *Cancer Res.* 2005, 65:9953-9961.
[163] Nathan, CO; Amirghahari, N; Rong, X et al. Mammalian target of rapamycin inhibitors as possible adjuvant therapy for microscopic residual disease in head and neck squamous cell cancer. *Cancer Res.* 2007, 67:2160-2168.
[164] Wislez, M; Spencer, ML; Izzo, JG et al. Inhibition of mammalian target of rapamycin reverses alveolar epithelial neoplasia induced by oncogenic K-ras. *Cancer Res.* 2005, 65:3226-3235.
[165] Kane, LP; Shapiro, VS; Stokoe, D; Weiss, A. Induction of NF-kappaB by Akt/PKB kinase. *Curr. Biol.* 1999, 9:601–604.
[166] He, D; Xu, Q; Yan, M et al. The NF-kappa B inhibitor, celastrol, could enhance the anti-cancer effect of gambogic acid on oral squamous cell carcinoma. *BMC Cancer,* 2009, 9:343.
[167] Cogswell, PC; Guttridge, DC; Funkhouser, WK; Baldwin, AS Jr. Selective activation of NF-kappaB subunits in human breast cancer: potential roles for NF-kappaB2/p52 and for Bcl-3. *Oncogene,* 2000, 19:1123–1131.
[168] Basseres, DS; Baldwin, AS. Nuclear factor-kappaB and inhibitor of kappaB kinase pathways in oncogenic initiation and progression. *Oncogene,* 2006, 25:6817–6830.
[169] Belguise, K; Sonenshein, GE. PKCtheta promotes c-Rel driven mammary tumorigenesis in mice and humans by repressing estrogen receptor alpha synthesis. *J. Clin. Invest.* 2007, 117:4009–4021.
[170] Bonizzi, G; Karin, M. The two NF-kappaB activation pathways and their role in innate and adaptive immunity. *Trends Immunol.* 2004, 25:280–288.
[171] Weih, F; Caamano, J. Regulation of secondary lymphoid organ development by the nuclear factor-kappaB signal transduction pathway. *Immunol. Rev.* 2003, 195:91–105.
[172] Papa, S; Bubici, C; Zazzeroni, F et al. The NF-kappaB mediated control of the JNK cascade in the antagonism of programmed cell death in health and disease. *Cell Death Differ.* 2006, 13:712–729.
[173] Witzel, II; Koh, LF; Perkins, ND. Regulation of cyclin D1 gene expression. *Biochem. Soc. Trans.* 2010, 38:217–222.
[174] Hanahan, D; Weinberg, RA. The hallmarks of cancer. *Cell,* 2000, 100:57–70.

[175] Ha, PK; Chang, SS; Glazer, CA; Califano, JA; Sidransky, D. Molecular techniques and genetic alterations in head and neck cancer. *Oral Oncol.* 2009, 45:335-339. Epub 2008 Jul 31. Review.

[176] Califano, J; van der Riet, P; Westra, W; Nawroz, H; Clayman, G; Piantadosi, S et al. Genetic progression model for head and neck cancer: implications for fieldcancerization. *Cancer Res.* 1996, 56:2488–2492.

[177] Hunter, KD; Parkinson, EK; Harrison, PR. Profiling early head and neck cancer. *Nat. Rev. Cancer,* 2005, 5:127–135.

[178] Marques, YM; de Lima, MD; Melo Alves, SM Jr; Soares, FA, Araújo, VC; Pinto, SJr, Mantesso, A. Mdm2, p53, p21 and pAKT protein pathways in benign neoplasms of the salivary gland. *Oral Oncol.* 2008, 44:903-908. Epub 2008 May 15.

[179] Klingbeil, MFG; Mathor, MB; Giudice, FS; Yoshito, D; Pinto Jr, DS. Is it safe to utilize in vitro reconstituted human oral epithelium? An oncogenetic pathway study. *Cell Tissue Bank.* 2010, 21. [Epub ahead of print]

[180] Forastiere, A; Koch, W; Trotti, A; Sidransky, D. Head and neck cancer. *N. Engl. J. Med.* 2001, 345:1890–1900.

[181] Pindborg, JJ; Reichart, PA; Smith, CJ; van der Waal, I. Histological typing of cancer and precancer of the oral mucosa. Berlin: Springer-Verlag; 1997.

[182] Brennan, M; Migliorati, CA; Lockhart, PB; Wray, D; Al-Hashimi, I; Axéll, T; Bruce, AJ et al. Management of oral epithelial dysplasia: a review. *Oral Surg. Oral Med. Oral Pathol. Oral Radiol. Endod.* 2007, 103 Suppl:S19.e1-12. Epub 2007 Jan 25.

[183] Barnes, L; Eveson, JW; Reichart, P; Sidranskym, D editors. World Health Organization classification of tumours. *Pathology and genetics. Head and neck tumours.* International Agency for Research on Cancer (IARC). Lyon, France: IARC Press; 2005.

[184] Reibel J. Prognosis of oral pre-malignant lesions: significance of clinical, histopathological, and molecular biological characteristics. *Crit. Rev. Oral Biol. Med.* 2003, 14:47-62.

[185] Gupta, PC; Murti, PR; Bhonsle, RB; Mehta, FS; Pindborg, JJ. Effect of cessation of tobacco use on the incidence of oral mucosal lesions in a 10-yr follow-up study of 12,212 users. *Oral Dis.* 1995, 1:54-58.

[186] Sankaranarayanan, R; Mathew, B; Varghese, C; Sudhakaran, PR; Menon V; Jayadeep, A et al. Chemoprevention of oral leukoplakia with vitamin A and beta carotene: an assessment. *Oral Oncol.* 1997, 33:231-236.

[187] Parkin, DM; Bray, F; Ferlay, J; Pisani, P. Global cancer statistics. *CA Cancer J. Clin.* 2005, 55:74–108.

[188] Jemal, A; Siegel, R; Ward, E; Hao, Y; Xu, J; Murray, T et al. Cancer statistics. *CA Cancer J. Clin.* 2008, 58:71–96.

[189] Marocchio, LS; Lima, J; Sperandio, FF; Corrêa, L; de Sousa, SO. Oral squamous cell carcinoma: an analysis of 1,564 cases showing advances in early detection. *J. Oral Sci.* 2010, 52:267-273.

[190] Monteiro, J; Fodde, R. Cancer stemness and metastasis: therapeutic consequences and perspectives. *Eur. J. Cancer,* 2010, 46:1198–1203.

[191] Mao, L; Hong, WK; Papadimitrakopoulou, VA. Focus on head and neck cancer. *Cancer Cell,* 2004, 5:311–316.

[192] Forastiere, A; Koch, W; Trotti, A; Sidransky, D. Head and neck cancer. *N. Engl. J. Med.* 2001, 345:1890–1900.
[193] Bagan, JV; Scully, C. Recent advances in Oral Oncology 2007: epidemiology, aetiopathogenesis, diagnosis and prognostication. *Oral Oncol.* 2008, 44:103–108.
[194] Pannone, G; Santoro, A; Papagerakis, S; Lo Muzio, L; De Rosa, G; Bufo, P. The role of human papillomavirus in the pathogenesis of head and neck squamous cell carcinoma: an overview. *Infect. Agent Cancer,* 2011, 29;6-4.
[195] Lobo, NA; Shimono, Y; Qian, D; Clarke, MF. The biology of cancer stem cells. *Annu. Rev. Cell Dev. Biol.* 2007, 23:675–699.
[196] Molinolo, AA; Amornphimoltham, P; Squarize, CH; Castilho, RM; Patel, V; Gutkind, JS. Dysregulated molecular networks in head and neck carcinogenesis. *Oral Oncol.* 2009, 45:324-334. Epub 2008 Sep 19. Review.
[197] Moll, R. Cytokeratins as markers of differentiation in the diagnosis of epithelial tumors. *Subcell Biochem.* 1998,31:205–262.
[198] Hoteiya, T; Hayashi, E; Satomura, K; Kamata, N; Nagayama, M. Expression of E-cadherin in oral cancer cell lines and its relationship to invasiveness in SCID mice in vivo. *J. Oral Pathol. Med.* 1991, 28:107—111.
[199] Bagutti, C; Speight, PM; Watt, FM. Comparison of integrin, cadherin, and catenin expression in squamous cell carcinomas of the oral cavity. *J. Pathol.* 1998, 186:8—16.
[200] Kandikonda, S; Oda, D; Niederman, R; Sorkin, BC. Cadherin-mediated adhesion is required for normal growth regulation of human gingival epithelial cells. *Cell Adhes. Commun.* 1996, 4:13—24.
[201] Gasparoni, A; Chaves, A; Fonzi, L; Johnson, GK; Schneider, GB; Squier CA. Subcellular localization of beta-catenin in malignant cell lines and squamous cell carcinomas of the oral cavity. *J. Oral Pathol. Med.* 2002, 31:385—394.
[202] Williams, HK; Sanders, DS; Jankowski, JA; Landini, G; Brown, AM. Expression of cadherins and catenins in oral epithelial dysplasia and squamous cell carcinoma. *J. Oral. Pathol. Med.* 1998, 27:308—317.
[203] Lo Muzio, M; Staibano, S; Pannone, G; Grieco, M; Mignona, MD; Cerrato A et al. Beta- and gamma-catenin expression in oral squamous cell carcinomas. *Anticancer Res.* 1999, 19:3817—3826.
[204] Hall, DE; Reichardt, LF; Crowley, E; Holley, B; Moezzi, H; Sonnenberg, A et al. The $\alpha1/\beta1$ and $\alpha6/\beta1$ integrin heterodimers mediate cell attachment to distinct sites on laminin. *J. Cell Biol.* 1990,110:2175—2184.
[205] Nicholson, J; Watt, FM. Decreased expression of fibronectin and the $\alpha5\beta1$ integrin during terminal differentiation of human keratinocytes. *J. Cell Sci.* 1991, 98:225—232.
[206] Koivisto, L; Grenman, R; Heino, J; Larjava, H. Integrins $\alpha5\beta1$, $\alpha_v\beta1$, and $\alpha5\beta6$ collaborate in squamous carcinoma cell spreading and migration on fibronectin. *Exp. Cell Res.* 2000, 255:10—17.
[207] Thomas, GJ; Lewis, MP; Hart, IR; Marshall, JF; Speight, PM. $\alpha_v\beta6$ integrin promotes invasion of squamous carcinoma cells through up-regulation of matrix metalloproteinase-9. *Int. J. Cancer,* 2001,92:641—650.
[208] Downer, CS; Watt, FM; Speight, PM. Loss of $\alpha6$ and $\beta4$ integrin subunits coincides with loss of basement membrane components in oral squamous cell carcinomas. *J. Pathol.* 1993,171:183—190.

[209] Thomas, GJ; Jones, P; Speight, PM. Integrins and oral cancer. *Oral Oncol.* 1997, 33:381—388.

[210] Calder, PC; Yaqoob, P; Harvey, DJ; Watts, A; Newsholme, EA. Incorporation of fatty acids by concanavalin A-stimulated lymphocytes and effects on fatty acid composition and membrane fluidity. *Biochem. J.* 1994, 300:509—518.

In: Endothelium and Epithelium
Editors: J. Carrasco and M. Mota, pp. 35-60
ISBN 978-1-61470-874-2

Chapter II

Pathophysiology of Endothelium: A Multidisciplinary Approach

Luciana Le Sueur Maluf, Regina Célia Spadari, Carolina Prado de Franca Carvalho, Camila Aparecida Machado de Oliveira and Daniel Araki Ribeiro
Department of Biosciences, Federal University of São Paulo, UNIFESP, Santos, SP, Brazil

Abstract

The endothelial cells (ECs) are squamous cells that form a simple epithelium and line the interior surface of the circulatory system. Different from the first idea of endothelium as an inert tissue, it has emerged as a dynamic tissue with metabolic and synthetic functions, exerting significant autocrine, paracrine and endocrine functions. Vasoconstrictor (endothelin-1) and vasodilator factors (nitric oxide, prostacyclin and endothelium-derived hyperpolarizing factor) are synthesized and released by ECs. Vasodilators acts primarily by lowering cytosolic calcium concentration in the adjacent vascular smooth musccle cells, whereas endothelin-1 has the opposite effect. Besides regulating vasomotor tone, these vasoactive substances as well as others produced by ECs have functions such as regulation of the recruitment and activity of inflammatory cells and proliferation of smooth muscle cells. Importantly, ECs also produce antithrombogenic agents such as plasminogen activator factor, thrombomodulin and prostacyclin, preventing thromboembolic diseases by keeping blood fluidity. However, in pathologic conditions ECs also produce thrombogenic factors such as platelets activator factor (PAF) and von Willebrand factor. The capillary endothelial cells act as a semi-permeable barrier, regulating the transfer of small and large molecules between blood and tissue. Ultrastructurally, these capillaries can be classified in continuous, fenestrated and sinusoidal, according to the continuity of the endothelial layer and its basal lamina. The distribution of these capillary types among different tissues is related to its functions. The transport through the endothelium occurs by specific mechanisms (i.e. facilitative glucose transporters and Na + dependent and independent amino acid transporters), and it is also mediated by transcellular and paracellular pathways that regulate the passage of ions, water and macromolecules, as well as cell polarity. In the heart, the endothelial cells play an obligatory role in regulating and maintaining cardiac function, in particular, at the endocardium and in the myocardium capillaries where endothelial cells directly interact with adjacent cardiomyocytes. Cardiac endothelial cells, like vascular endothelium, also express and release a variety of auto- and paracrine agents which directly influence heart vasculature density, as well as heart metabolism, contractile performance, rhythmicity, and hypertrophy. A blood-heart barrier has been proposed. At the early stages of cardiac failure, all major cardiovascular risk factors may cause cardiac endothelial activation as an adaptive response often followed by cardiac endothelial dysfunction. In addition, it will be presented the specific features of the brain capillary endothelial cells that form the blood-brain barrier. The blood-brain barrier constitutes the physical and metabolic interface between blood and brain. It is responsible for providing a suitable environment for neuronal function, regardless fluctuations in blood composition. The transcellular transport through the endothelial cells is very selective while the paracellular passage is restricted. The specific phenotype of the brain endothelium is induced and maintained by factors released mainly by astrocytic end-feet that surround the abluminal surface of the capillaries besides the influence of pericytes, neurons and microglias that together form the neurovascular unit.

Introduction

Different from the first idea of endothelium as an inert tissue lining the blood vessels, it is now considered as an essential structural and functional component of the cardiovascular system, contributing to cardiovascular homeostasis by regulating vascular permeability, adjusting the caliber of blood vessels to hemodynamic and hormonal demands, and by maintaining blood fluidity. The substances released by the endothelial cells include vasoconstrictor and vasodilator factors, pro- and anticoagulant factors, pro- and antitrombotic factors, growth and antigrowth substances, factors that contribute to angiogenesis and tissue remodeling, as well as to immune reactions and tissue inflammation. Some of these properties of the endothelial cells are described in detail in this chapter.

Embryological Origin and Morphological Features of Endothelial Cells

The internal surface of heart and blood vessels are lined by a continuous layer of a simple squamous epithelium called endothelium. The endothelial cells can show some changes in cell shape that can be related to the differences in blood pressure and its distending forces exerted in these cells along venous and arterial blood systems. The endothelial cell dimensions can range among 5-30 µm in length, 10-15 µm in width and 0.2-0.4 µm in thickness. In arteries, where the pressure is higher the endothelial cells are much flattened as compared to veins. In contrast, the veins are low pressure vessels and its endothelial cells may be more elliptical. The long axes of endothelial cells also seem to be aligned with the direction of blood flow (Bloom and Fawcett, 1986; Abramson, 1962). Besides the relative influence of blood pressure in endothelial cell shape, the cytoskeleton also plays a crucial role to maintain the characteristic shape of these cells. The maintenance of the flattened shape of endothelial cells is achieved through the centrifugal tension generated by interconnected microtubule struts, focal adhesions, and interendothelial junctions that resists the centripetal tension (or counteradhesive forces) generated by action-myosin cross-bridging of stress fibers (Mehta and Malik, 2006).

Concerning on the embryological origin of endothelial cells it has well established that hematopoietic and endothelial lineages arise within "blood islands" in the extraembyronic yolk sac mesoderm during gastrulation. These "blood islands" derive from aggregates of mesodermal cells that colonize the yolk sac at approximately 7.5 days postcoitum (dpc) in mouse embryos and between the second and third week of human gestation. The central cells within these "blood islands" give rise to the embryonic hematopoietic cells while the peripheral population differentiates in endothelial cells which form the first vascular structures that surround the inner blood cells. Because the hematopoietic and endothelial lineages share the same spatio-temporal place within the "blood islands" it has been proposed that they arise from a common precursor, the hemangioblast. Several studies from 90's decade have reported that both cell lineages share the expression of a number of different genes and these findings give support for the hemangioblast hypothesis (Baron, 2003, Choi et al., 1998).

In comparison with other blood vessels, capillaries have the simplest structure with a very small caliber and a thin wall. These thin vessels have an average diameter of around 7 to 12 μm. In a cross section of a small capillary, just a single endothelial cell is found lining its lumen, whereas in larger capillaries up to three endothelial cells can be observed. The capillary wall is formed by a layer of endothelial cells and a basal lamina (Bloom and Fawcett, 1986). The main extracellular matrix (ECM) components found in the basal lamina and secreted by endothelial cells are collagen IV, fibronectin, entactin, laminin, chondroitin sulfate, and heparan sulfates, perlecan and syndecan. The interaction between ECM components with endothelial cell surface integrins generates signals that inhibit endothelial cell proliferation and migration; in contrast it stimulates cell-cell and cell-matrix adhesion. The cell-matrix adhesion is crucial for the formation of the intact endothelial barrier. In vivo and in vitro studies have shown the increase in endothelial permeability after disruption or degradation of different components of ECM (Mehta and Malik, 2006). The wall of larger blood vessels has accessory coats of smooth muscle and connective tissues.

Endothelial cells have a flattened nucleus which is protruding to the vessel lumen. Around the nucleus, the cytoplasm is thicker than in peripheral portions where it is very thin (0.2 to 0.4 μm). Small golgi complexes and a few mitochondria are found in juxtanuclear region. Endoplasmic reticulum has a more scattered spatial distribution, including the peripheral cytoplasm. Lysosomes are rare but the pinocytosis vesicles are very common and are directly implicated with transcellular transport pathways in endothelial cells. The edges of adjacent cells are frequently overlapped and a thin marginal ridge may be projected into the lumen. Pericytes are mesenchymal-originated cells which are found associated with the capillary wall. These cells have been implicated to play a role in control the endothelial cell proliferation and differentiation during angiogenesis, by a paracrine and cell-cell contact manner. Moreover, the pericytes may contribute to control the capillary diameter through its contractile ability and also integrate signals along the vessel length due to the numerous endothelial cells that can be contacted by them (Gerhardt and Betsholtz, 2003; Bloom and Fawcett, 1986; Bailey et al., 1984)

The endothelial cytoskeleton has crucial role to maintain a flattened endothelial cell phenotype and cell-cell contacts required for the endothelium barrier function. It has been shown that the permeability-increasing agonists (such as thrombin) increase the formation of actin stress fibers and destabilize microtubules. Stress fibers are composed by bundles of polymerized actin and myosin filaments and they are the primary elements of the contractile machinery of endothelial cells being associated with rounding of endothelial cells. Microtubules are presumed to resist compression of cells by opposing actin-myosin contractility. The intermediate filaments found in endothelial cells are formed by polymers of vimentin and cytokeratin proteins. The intermediate filaments also provide the maintenance of endothelial cell shape, but the role of these filaments in the regulation of endothelial permeability is less understood. Studies with endothelial cells have also suggested the role of actin cytoskeleton and microtubule tracks for vesicle trafficking during transcytosis (Mehta and Malik, 2006).

Classification of Capillaries (Types of Capillaries)

Based on ultrastructural and functional characteristics, the capillaries can be basically divided into three categories. The *continuous* capillaries are found in muscle, nervous tissue and connective tissues. The endothelium in this capillary type forms a thin uninterrupted layer around the entire circumference of the vessel. The basal lamina of continuous capillaries is also continuous (uninterrupted) (Bloom and Fawcett, 1986; Bailey et al., 1984). Among the examples of continuous capillaries, those found in brain have specific features which will be better discussed in a separated section of this chapter.

In contrast, in *fenestrated* capillaries the endothelium varies in thickness, and the thinnest regions are interrupted by circular fenestrations or pores with 80 to 100 nm in diameter. These small pores are closed by a pore diaphragm which is thinner than a cell membrane. Thus, these fenestraes are just permeable to molecules smaller than 10 nm. Scanning micrographs or freeze-fracture preparations show a very uniform distribution of these pores (fenestraes) which have a regular spacing of around 50 nm among them. However the areas exhibiting pores make up only a fraction of the vessel wall. In this way, the fenestrated capillaries show both fenestrated and unfenestrated areas along its walls, and the relative proportions of these areas vary in different organs. The basal lamina of fenestrated capillaries forms a continuous layer. The fenestrated capillaries are found pancreas, intestinal tract an in endocrine glands. There is a special type of fenestrated capillary that is found in renal glomerulus which presents pores exceptionally not closed by pore diaphragms. These renal glomerulus fenestrated capillaries also present a thicker basal lamina than other capillaries (Bloom and Fawcett, 1986; Bailey et al., 1984).

The *sinusoids* are wider capillaries with 30 to 40 μm in diameter that have an irregular outline. As compared to fenestrated capillaries, the sinusoid fenestraes are larger and without a pore diaphragm. Moreover, the sinusoids may have a discontinuous endothelium layer (or with intercellular gaps) and a scanty and discontinuous basal lamina. The endothelial cells of sinusoids are more active in endocytosis and have a better-developed lysosomal system than endothelial cells found in other capillary types. The most typical example of sinusoids is the hepatic sinusoid, found in liver, but sinusoids are also found in spleen and bone marrow. The capillaries from lymphoid organs and some endocrine glands, such as adenohypophysis and adrenal cortex for some authors may be considered sinusoidal (or sinusoid-like) capillaries, but they are not as typical as liver sinusoids (Bloom and Fawcett, 1986; Bailey et al., 1984)

The Structure of Interendothelial Junctions and Focal Adhesion

In this section it will be presented a concise overview of the structure of endothelial cell junctions (adherens, tight and gap junctions) and its functional role for the permeability barrier function. The adherens and tight junctions are those more commonly attributed to the interendothelial junction-denomination, but gap junctions and other junctional adhesion molecules (JAMs) also participate to the endothelial cell-cell contact. This section also briefly presents the structure of endothelial focal adhesion, a cell-extracellular matrix interaction.

The vascular endothelial (VE) cadherin is the main structural protein of endothelial adherens junctions (AJs). VE-cadherin mediates a homophilic adhesion of adjacent cells in a Ca^{2+}-dependent manner. The juxtamembrane cytoplasmic domain of VE-cadherin binds p120-catenin (p120), whereas the C-terminal domain binds with β-catenins, α-catenins and γ-catenin (plakoglobin). This catenin-complex is implicated to link AJs to the actin cytoskeleton. The protein p120 have been reported as the most important determinant of AJ stability and can also regulates the cadherin expression through its interaction with the molecular motor kinesin and via transcription factor kaiso. Dynamic interactions between the AJ proteins and the actin cytoskeleton are crucial to the regulation of junctional permeability. The re-organization of actin into contractile stress fibers disrupts AJs. Actin-mediated endothelial cell contraction is the result of myosin light chain (MLC) phosphorylation which drives myosin-actin cross-bridge contraction cycling (Vandenbroucke et al., 2008; Mehta and Malik, 2006).

Disassembly of AJs can be modulated by changes in phosphorylation of AJ proteins, such as VE-cadherin and β-catenin. The microtubule disassembly has also been reported as a mechanism of AJ disassembly whereas the stabilization of microtubules is associated to the inhibition of contraction and prevention of AJs-redistribution in response to permeability agonists (Vandenbroucke et al., 2008).

The endothelial tight junctions (TJs) consist in sealing strands formed by the homotypic interaction of occludin or claudins. The COOH-terminal domain of occludin is associated to the zonula occludens protein (ZO-1) in cytoplasm. ZO-1 links occludin and claudins to the actin filaments through cingulin and can also interacts with other cytoplasmic proteins as α-catenin, spectrin, and Cx43. These interactions allow actin rearrangement to regulate TJ stability. Other zonula occludens members (ZO-2 and ZO-3) are also found in endothelial cells. The claudin-5 is the only claudin subtype expressed in endothelial cells. The phosphorylation of TJ proteins has a different effect on endothelial permeability depending on whether TJs are in the assembled or disassembled state. The PKC activation has been implicated with the assembly of TJs, but in established TJs can impairs its integrity (Vandenbroucke et al., 2008; Mehta and Malik, 2006).

Junctional adhesion molecules (JAMs) are members of the IgG superfamily of proteins that interact with TJs proteins, mainly with ZOs proteins. JAMs form homotypic interactions with neighboring endothelial cells and contribute to regulate the paracellular permeability. The members JAM-A, JAM-B, and JAM-C are expressed in endothelial cells. The JAM-A shows colocalization with occludin, claudins, ZOs, and cingulin at the level of TJs. It has been established that JAM-A plays a regulatory role recruiting and binding these proteins for the junctional assembly. JAM-C regulates paracellular permeability by inducing inhibition of small GTPase Rap1 to allow disassembly of AJs. Moreover, JAM upregulation can be related to chronic inflammation and atherosclerotic vessels (Vandenbroucke et al., 2008; Mehta and Malik, 2006).

The gap junction (GJ) is made up of two connexons from adjacent cells. The connexon is a hexamer of connexin subunits. The members of connexins (Cxs) expressed by endothelial cells are Cx37, Cx40, and Cx43. GJs provide electrotonic and metabolic pathways for direct cell-to-cell transfer of signaling molecules and ions. Phosphorylation regulates the rate of channel assembly and turnover of connexins. Some studies have demonstrated the role of gap junction in regulating nitric oxide (NO) production with possible consequence on endothelial permeability (Mehta and Malik, 2006).

Focal adhesions (FAK) are a highly specialized complex array of integrins that are connected to actin filaments by several associated proteins (such as talin, filamin, tensin, α-actinin, tensin, and SHP2). These junctions provide the primary means of anchoring of the endothelial cell monolayer to the extracellular matrix components. Upon thrombin stimulation, the role of focal adhesions is to strengthen cell-substrate adhesion. FAK can also strengthen the endothelial barrier function by promoting the assembly of AJs. For review the arrangement of focal adhesion components and the signaling pathways involved in modulation processes for the endothelial barrier function see Mehta and Malik, 2006.

The Endothelial Barrier Function and the Transport Pathways

The endothelium consists in a semipermeable barrier which is crucial for the establishment of the transendothelial protein gradient (the colloid osmotic gradient) required for tissue fluid homeostasis. The endothelial barrier function depends on several mechanisms that include the selective permeability based on size and charge molecular properties besides the appropriate cell-cell and cell-extracellular matrix contacts. The role of cytoskeleton network and the cell to cell or cell to extracellular matrix interactions in endothelial permeability have already been discussed in the previous sections.

In a continuous endothelium, there are basically two types of transport routes: the transcellular and the paracellular pathways, the latter corresponds to the interendothelial junction pathway. The vascular endothelium transports solutes with a range of molecular radii from 0.1 nm (sodium ion) to 11.5 nm (immunoglobulin IgM). Under basal conditions, the transcellular pathway can mediate the transport of plasma proteins and molecules with molecular radii over than 3 nm, such as albumin. These transcellular transports can be mediated by caveolae, via an absorptive receptor-mediated pathway, by a fluid-phase pathway or by a transendothelial channel formation. The transendothelial channels are transiently formed by a fusion of a multiple caveolae and can allow the albumin passage. The water can be transported by aquaporin channels or via paracellular. Small molecules with up to 3 nm molecular radii, including urea, glucose and amino acids are transported by paracellular pathway (Mehta and Malik, 2006).

Albumin can act as circulating chaperone, facilitating the co-transport of hydrophobic substances, fatty acids, and some amino acids, hormones, and drugs. The transport of albumin can be mediated by caveolae in either the fluid phase or bound to gp60 (a cell surface albumin-docking protein). Caveolae are specialized invaginations of the plasma membrane also known as "lipid rafts" that forms microdomain resulted from localized accumulation of cholesterol, glycosphinogolipids, and structural proteins. The caveolae seem to flask-shaped vesicles attached by a neck to the plasma membrane and has the caveolae-specific protein caveolin. The vesicle formation in a caveolae-mediated transport is clathrin-independent. Signaling pathways activated by the interaction of gp60 and caveolin-1 may regulate albumin permeability in endothelial cells via transcytosis. In this pathway a heterotrimeric GTP binding protein Gi, and the Src tyrosine kinase are involved. In case of fluid-phase transcytosis, most of the albumin transported to the basal side is free within vesicles (Mehta and Malik, 2006; Minshall et al., 2002).

Charge-selective permeability properties in endothelial cells are attributed to glycocalix. The glycocalyx is a negatively charged, surface coat of membrane-bound glycolipids, glycoproteins, proteoglycans, glycosaminoglycans, and adsorbed plasma proteins lining the luminal surface of the endothelium. The gycocalyx plays a crucial role regulating the permeability of endothelial barrier function by a charge-selective manner. The endothelial glycocalyx may also functions as a fluid shear-stress sensor regulating the production of NO (Mehta and Malik, 2006; Pries et al., 2000).

Endothelial Cell Function

Anticoagulant and Procoagulant Function

A central physiologic function of endothelial cells is to facilitate blood flow, providing an antithrombotic surface through secretion of factors that prevent coagulation, adhesive processes and inhibit platelet function (Cines et al., 1998, Pate et al., 2010). One of these factors is tissue factor pathway inhibitor (TFPI), which prevents the formation of thrombin. Thrombin is a serine protease with diverse functions in coagulation, including the activation of platelets, coagulation enzymes, and cofactors, and stimulation of procoagulant pathways on the endothelial cells. The endothelium also helps to restrain thrombin activity through the expression of thrombomodulin. Binding of thrombomodulin to thrombin increases activation of the anticoagulant protein C, and reduces thrombin activation of platelets, factor V, factor XIII, and fibrinogen. Other factors produced by endothelial cells with antithrombogenic properties are tissue plasminogen activator (t-PA), nitric oxide and prostacyclin (Cines et al., 1998; Galley and Webster, 2004; Zanesco and Antunes, 2005).

However, when the endothelium is damaged (i.e. by trauma, infection and inflammation) a procoagulation response is initiated. In this context, endothelial cell membrane is transformed into a procoagulant surface. A key event in this process is the induction of tissue factor (TF) expression which occurs rapidly after vascular injury. TF synthesis in vitro is stimulated by thrombin, endotoxin, several cytokines, hypoxia, oxidized lipoproteins, and many other irritants. Some of TF actions include accelerated activation of coagulation factors X and IX. Other procoagulant factors produced by endothelial cells are von Willebrand factor and platelet activator factor (PAF) (Cines et al., 1998; Galley and Webster, 2004; Zanesco and Antunes, 2005).

A dynamic equilibrium normally exists between anticoagulant and procoagulant states, allowing the endothelium to return to its physiologic, anthitrombotic condition once the damaging stimulus has been removed. Such balance is fundamental, since a disproportion between anticoagulation/procoagulation factors may cause either excessive bleeding or thromboembolic disease, a major global cause of death.

Endothelial Cells and Inflammation

The endothelium of precapillary arterioles controls local blood supply, whereas the endothelium of the postcapillary venules controls vascular permeability and leukocytes adhesion and diapedesis. The increase in vascular permeability is initiated when the inflammatory mediators generated locally (as histamine, PAF, leukotrienes and others) activate specific receptors at endothelial cells, leading to widening of endothelial cell-cell junctions and contraction of the endothelial cells cytoskeleton, with consequent formation of intercellular pores. Plasmatic proteins, especially albumin, escape to interstitial space through these pores, followed by liquid extravasation from capillaries caused by the increase in osmotic pressure exerted by the proteins accumulated in the interstitial space. These processes result in an inflammatory edema. The extravasation of proteins seems to occur through a synergistic action between the vasodilators and permeability mediators; the former potentiates the venular plasma exudation due to its potent vasorelaxing properties at the precapillary arterioles level (Zanesco and Antunes, 2005).

In addition, endothelial cells can be activated by infections, pathogen recognition pathways (such as toll-like receptors and lipopolysacharide), molecules generated in the inflammatory cascade such as kinins, complements, cytokines (IL-1, TNF-α) and by-products of the coagulation pathway such as thrombin and fibrinogen. Endothelial-platelet interactions, endothelial interactions with neutrophils, lymphocytes, mast cells, eosinophils and mononuclear cells may also contribute to endothelial cells activation. Once activated, endothelial cells synthesize and release inflammatory mediators, including cytokines (TNF-α, IL-1, IL-6, IL-8, IL-11), growth factors (VEGF, bFGF, PDGF) chemokines (CXC, CC, CX3C), hematopoietins (granulocyte-macrophage colony stimulating factor, granulocyte colony stimulating factor, macrophage colony stimulating factor, thrombopoietin), lipid mediators (leukotrienes, prostanoids) and molecules important for the regulation of coagulation. IL-1, along with TNF-α, stimulates the expression of cell adhesion molecules on endothelial cells, which is pivotal for leukocyte migration to the inflamed area, an event of the inflammatory process where the role of endothelial cells is fundamental (Galley and Webster, 2004; Zanesco and Antunes, 2005; Pate et al., 2010).

Regulation of Vascular Tone

The vasoconstrictors endothelin-1, prostacyclin A_2 and thromboxane A_2 and the vasodilators nitric oxide, prostacyclin and endothelium-derived hyperpolarizing factor are the most relevant vasoactive substances produced by endothelial cells. Vasodilators acts primarily by lowering intracellular calcium concentration ($[Ca^{2+}]_i$) in the adjacent vascular smooth muscle cells (VSMC), whereas vasoconstrictors have the opposite effect, as shown in

Figure 1. Besides regulating vasomotor tone, these substances have other functions, such as regulation of the recruitment and activity of inflammatory cells and proliferation of VSMC.

Endothelium-Derived Vasoconstrictors

Endothelin-1 (ET-1)

Endothelins (ETs) are a potent long-acting vasoconstrictors peptide family formed by ET-1, ET-2, and ET-3. Vascular endothelial cells produce only ET- 1, but ET-2 and ET-3 are expressed in other tissues such as brain, lungs, heart and kidney. The ETs are synthesized as prepro-ETs, which are cleaved by endopeptidases generating pro-ETs, also called big ET. Big ET must be cleaved by the endothelin-converting enzyme to produce ET (Masaki, 1998).

ET is secreted by a constitutive pathway; thus ET synthesis and release from endothelial cells is largely controlled at a transcriptional level. Expression of ETs is regulated in response to various chemical and mechanical stimuli such as thrombin, angiotensin II, vasopressin and hemodynamical shear stress, suggesting the existence of a cardiovascular control system mediated by this peptide (Inagami et al., 1995; Galley and Webster, 2004).

Two isoforms of ET receptors, ET_A and ET_B, are known. ET_A has a high affinity to ET-1 and ET-2, but low affinity to ET-3, and is predominantly found in vascular smooth muscle cells. ET_B binds the three ETs equally and is located in endothelial cells and VSMC. In vascular tissues, ET_A and ET_B mediate paracrine vasoconstrictor effects of ET -1, whereas ET_B on the endothelial cells mediates autocrine functions of ET-1 that result in the release of nitric oxide (Masaki, 1998).

The primary action of ETs in VSMC is the activation of phospholipase C, inducing the formation of inositol trisphosphate (IP_3) with a consequent transient increase in $[Ca^{2+}]i$. ETs also activate calcium channels, leading to calcium influx in the VSMC. These steps are followed by a long sustained vascular contraction (Inagami et al., 1995; Zanesco and Antunes, 2005).

In addition to the vasoconstrictor functions, ET-1 promotes proliferation of vascular endothelial cells, VSMC, and fibroblasts, which may result in structural changes of the vascular wall. It also promotes contraction of nonvascular smooth muscles, inhibition of renin release from the glomeruli, stimulation of catecholamine and aldosterone secretion from the chromaffin cells and zona glomerulosa cells, respectively, by the adrenal glands, and modulation of neuronal transmission in the central nervous system (Inagami et al., 1995).

Prostaglandin H_2 (PGH_2) and Thromboxane A_2 (TXA_2)

Vascular endothelium as well as platelets synthesize and release vasoconstrictor prostanoids such as PGH_2 and TXA_2. PGH_2 biosynthesis starts with arachidonic acid, which is oxidized to PGH_2 by the enzyme cyclooxygenase (COX). Two isoforms of COX have been identified to date. COX-1, a constitutive enzyme expressed in the vascular endothelium, is thought to contribute to the maintenance of vascular homeostasis (a splice variant of COX-1, also referred to as COX-3, has also been reported). COX-2 is an inducible isozyme expressed in the cardiovascular system (and in immune cells) only during pathogenic situations, although it is suggested that this isoform is constitutively present in human endothelial cells.

TXA_2 is generated from PGH_2 by the enzyme thromboxane-A synthase. TXA_2 is very unstable and is rapidly hydrolyzed to the biologically inactive TXB_2. Due to its short half life, its action is limited to a small area. Besides its vasoconstrictor effect, TXA_2 strongly stimulates platelet aggregation. Both vasoconstriction and platelet aggregation effects of these prostanoids require their binding to TP subclass receptors, activating phospholipase C with the production of diacylglycerol (DAG) and IP_3, the latter increasing intracellular calcium concentration in vascular smooth muscle cells or platelets (Zanesco and Antunes, 2005; Kawabe, 2010).

Endothelium-Derived Vasodilators

Nitric Oxide (NO)

NO, also known as endothelium-derived relaxing factor (EDRF), is a lipid-soluble gas synthesized from L-arginine through the action of the enzyme NO synthase (NOS). It has a half-life of seconds, being rapidly inactivated by oxygen, hemoglobin, and reductant agents (Inagami et al., 1995; Vanhoutte and Mombouli, 1996).

There are three isoforms of NOS, all produced by different genes: brain NOS (bNOS), endothelial NOS (eNOS), and inducible NOS (iNOS). bNOS and eNOS are classified as constitutive NOS (cNOS) since they are constitutively expressed. Differently, iNOS is expressed in the presence of pathological stimulus, such as cytokines and lipopolysaccharides. IFN-y is a potent inducer of iNOS expression, whereas TNF-a, IL-6, IL-1, or EGF produce weaker responses individually but pronounced responses when more than one of these substances is present. bNOS is expressed in neuronal tissues as well as other structures such adrenal glands, lung epithelial cells, pancreatic islets cells and skeletal muscle. In the central nervous system, NO seems to be involved in memory and learning. bNOS and iNOS exist in the soluble fraction, whereas eNOS exists in the membrane-bound particulate fraction of cells (Inagami et al., 1995; Villar et al., 2006).

Despite being constitutively active, eNOS activity increases within seconds in response to the activation of specific cell surface receptors and shear stress. After binding of the agonists (i.e. acetylcholine, bradykinin, angiotensin II, vasopressin, histamine, norepinephrine, serotonin, thrombin) to their specific G protein-coupled receptors, phospholipase C is activated, producing IP_3 and DAG; the former increases the cytosolic Ca^{2+} concentration and then activates calmodulin, which activates eNOS. Phosphorylation by protein kinase A, on the other hand, attenuates cNOS activity. Phosphorylation regulates, in a reversible way, the intracellular localization and consequently the biological activity of eNOS. During exercise, the increase in eNOS activity evoked by shear stress contributes to the phenomenon of flow-mediated vasodilatation, an important autoregulatory mechanism by which exercises increase blood flow (Inagami et al. 1995).

NO shows diverse biological actions. The sustained secretion of NO from resistance vessels regulates blood flow and blood pressure under steady-state conditions. Endothelium-derived NO induces vascular smooth muscle relaxation through activation of soluble guanylate cyclase, increasing the production of the second messenger cyclic GMP (cGMP). NO/cGMP reduces intracellular calcium concentration through inhibition of IP_3 formation, increased ER calcium uptake, inhibition of extracellular calcium influx, stimulation of membrane Ca^{2+}-ATPase and opening of potassium channels (Inagami et al., 1995; Zanesco and Antunes, 2005).

In addition, endothelial cell-derived NO suppresses platelet functions including adhesion, aggregation and secretion of vasoconstrictive substances into the vascular lumen, in part through a cGMP-dependent mechanism, and promotes platelet disaggregation. Leukocyte and monocyte adhesion to the endothelium, ET-1 production, VSMC proliferation and cell adhesion molecules expression are also inhibited by NO. However, NO action is limited to a small area because of its rapid inactivation (Inagami et al., 1995; Pate et al., 2010).

Prostacyclin (PGI_2)

PGI_2 is thought to be one of the most important prostanoid involved in the regulation of the homeostasis of the cardiovascular system. Platelet precursors, megakaryocytes, as well VSMC are the major targets of PGI_2. PGI_2 is a potent vasodilator and an inhibitor of platelet aggregation, leukocyte adhesion and VSMC proliferation. It is synthesized from the instable cyclic endoperoxides PGG_2 and PGH_2, produced from arachidonic acid by COX enzymes. The endoperoxides are then converted in PGI_2 through the action of the enzyme prostacyclin synthetase (Kawabe et al., 2010).

PGI_2 has a short half-life, and after its release from ECs, it elicits its effects by activating specific cell-surface receptors that are G-protein-coupled to adenylyl cyclase and thereby elevate cyclic adenosine monphosphate (cAMP) levels. In VSMC, cAMP reduces intracellular Ca^{2+} availability, causing a potent vasorelaxion (Villar et al., 2006; Kawabe et al., 2010).

Endothelium-Derived Hyperpolarizing Factor (EDHF)

Endothelium-dependent vasodilation is generally attributed to the release of NO and/or PGI2. However, this response cannot be totally explained by the release of these two mediators, since it can still occur in the presence of NOS and COX inhibitors. The alternative mechanism leading to vascular relaxion is often associated with hyperpolarization of the underlying VSMCs and was initially attributed to a non-characterized endothelial factor termed EDHF. But in fact, EDHF has become a generic term referring to the many signaling pathways involved in communications between the endothelial cells and myocytes which result in VSMC hyperpolarization. Some of these do involve the release of factors from endothelial cells but it also seems to occurs through electrical and even metabolic coupling between endothelial cells and myocytes via myoendothelial gap junctions (Félétou and Vanhoutte, 2009; Edwards et al., 2010).

EDHF responses are classified in two broad categories. The classical pathway is associated with endothelial cells hyperpolarization. This follows an increase in intracellular $[Ca^{2+}]$ and the opening of endothelial calcium-dependent potassium channels (K_{Ca}). In some vessels, endothelial hyperpolarization is transmitted to myocytes through myoendothelial gap junctions without involving the synthesis and/or release of any factor. In others, the K^+ that effluxes from endothelial cells through K_{Ca} activates another type of potassium channels (Ba^{2+}-sensitive K_{IR}) and the Na^+/K^+ ATPases in VSMC and endothelial cells, leading to VSMC hyperpolarization. For the classical pathway, the hyperpolarizing "factor" involved is the K^+ that efflux through endothelial K_{Ca} channels. The second category of EDHF pathway does not require endothelial hyperpolarization. It involves the endothelial release of factors that include NO, HNO, H_2O_2 and vasoactive peptides as well as prostacyclin and epoxyeicosatrienoic acids. These hyperpolarize myocytes by opening various populations of VSMC potassium channels (Félétou and Vanhoutte, 2009; Edwards et al., 2010). Independent of the mechanism involved, the ensuing hyperpolarization decreases Ca^{2+} influx in the VSMC, thereby causing vasodilation. Besides, these substances acts not only as vasodilators, but they seem to play a multi-faceted role in vascular homeostasis, including inhibition of mitogenesis, platelet aggregation, and the extravasation of leukocytes. The importance of EDHF pathways increases as the diameter of a blood vessel diminishes, whereas NO effects are more pronounced in larger caliber vessels (Villar et al., 2006).

The Endothelial Endocardium

In the heart, endothelial cells play an essential role in regulating and maintaining cardiac function by controlling cardiac metabolism, growth, contractile performance, and rhythm. Two different types of endothelial cells can be found in the heart, those belonging to endocardium and the endothelial cells of the intramyocardial capillaries. The close proximity of endothelial cells to adjacent cardiomyocytes allows a cross-talking among them even through molecules with short half-lives, such as nitric oxide (Brutsaert, 2003).

Cardiac Endothelial Morphology

The luminal surface of the vertebrate's heart is structurally complex, comprising furrows, cylinder- and sheetlike trabeculae, and papillary muscles. It is completely lined by the endocardial endothelium - a sheet of endothelial cells somewhat larger than endothelial cells in the other portions of the circulatory system.

The luminal surface of most of these cells possesses a scattered variety of micro appendages, or microvilli, projecting into the cardiac cavity, greatly amplifying the surface area. An increased internal heart surface area offers an astonishingly high ratio between the cavity surface area and the ventricular volume, particularly in the right ventricle. These special features suggest an important sensor role for the endocardial endothelium (Brutsaert, 1989). The intercellular clefts between the endocardium endothelial cells are highly tortuous with two or three cells imbricated and tightly adhered to each other by cell adhesion proteins.

Most cells contain one or two tight junction complexes. Thus, the permeability of the endocardial endothelium is mediated by: 1) the extent and complex structure of clefts which are, moreover, lined by a dense electrically charged glicocalix; 2) the presence of one or two tight junctions (zonula occludens); and 3) the presence of a well-organized zonula adherens with complexes interactions between a circumferential actin filament band and several connecting proteins, e.g. vinculin. Endocardial endothelium cells also present abundance of gap junctions, suggesting that these cells display an intimate electrochemical linkage with the neighboring cells (Beny, 1999; Brutsaert, 2003). The normal cardiac growth, contractile performance and rhythmicity are dependent on interactions between the endothelial cells and cardiomyocytes. A reciprocal signaling occurs by peptide growth factors through auto- and paracrine mechanisms.

Cardiac Growth

The impact of the cardiac endothelial cells on myocardial growth, e.g., whether and to what extent cardiac endothelial cells control cardiogenesis in the embrio or play a role in cardiac remodeling in the adult is still poorly understood. Many factors released by the endocardial endothelium cells have been implicated in the steps of cardiogenesis, such as VEGF, neuroregulin, serotonin, angiopoietin, transformin growth factor-β, insulin-like growth factor I, myocardial-derived hepatocytes growth factor, nuclear factor of activated T cells, nerofibromin, PDGF, pre-B-cell growth-stimulating factor, ET, NO, angiotensin (for a detailed review see Brutsaert, 2003).

Cardiac Contractile Performance

The endocardial endothelium has proved to influence myocardial contraction. Selective damage of it in isolated papillary muscle (Brutsaert et al., 1988) resulted in an immediate and irreversible abbreviation of the isometric twitch, with earlier onset of tension decline and concomitant decrease in peak twitch tension, with little change in the early contraction phase of the twitch or in maximal shortening velocity . This change in twitch pattern was due to a reduced responsiveness of the contractile proteins to the $[Ca^{2+}]_i$ (De Keulenaer and Sys, 1995, Wang and Morgan,1992). This response has been ascribed predominantly to known auto- and paracrine signaling agents, released or activated by the cardiac endothelial cells, such as NO, ET, PGI_2, and angiotensin II. In relation specifically to the endocardial endothelium, an active transendothelial physicochemical gradient for various ions, or blood-heart barrier, has been postulated (Fransen et al., 2001). The potential participation of other endothelium-mediated signaling pathways, such as bFGF, VEGF, neuregulin, or angiopoietin, in modulating myocardial inotropic performance remains to be determined.

Autocrine and Paracrine Signaling from Cardiac Endothelial Cells

Cardiac endothelial cells like all other endothelial cells express and release a variety of auto- and paracrine agents, which directly influence cardiac metabolism, growth, contractile performance, and rhythmicity. The effects of NO produced by the endocardial endothelium cells on myocardial contraction, relaxation, and heart rate have been much studied. Many of the actions of NO on cardiac performance are mediated by the activation of myocardial soluble guanylate cyclase to produce cGMP, but some are cGMP independent. The underlying mechanisms are however still not fully understood (Fleming and Busse, 1999; Mayer and Hemmes, 1997; Shah and MacCarthy, 2000). NO seems to exert a dual action on myocardial contractile performance. In the lower concentration range, corresponding to endogenously generated NO through eNOS, NO exerts a slight positive inotropic action, contributing to maintenance of basal contractile performance of the heart under physiological conditions. It is unclear, however, whether the observed positive or negative inotropic and lusitropic actions of NO are beneficial or detrimental to overall pump performance of the heart.

NO also decrease myocardial tissue oxygen consumption (Locke et al., 1999; Trochu et al., 2000), what suggests a potentially cardioprotective mechanism, perhaps in part due to the decrease in cardiac loading and in cardiac systolic work output leading to increased myocardial metabolic efficiency (Shen et al., 2001); presumably also by influencing substrate (free fatty acids vs. glucose) utilization (Tada et al., 2000) or by regulating mitochondrial metabolism (Stumpe et al.,2001).

Another endothelium derived factor that plays a role in the heart physiology is ET. This peptide has been shown to be involved in cardiac development, growth, remodeling, and also in the control of cardiac contractile performance and rhythmicity. ET_A and ET_B receptors are expressed on cardiomyocytes, conduction tissue, and coronary vascular and endocardial endothelium cells. ET is one of the most potent positive inotropic agents known. Its inotropic action has been partly explained by its binds to ET_B receptor on the endothelial surface, thereby stimulating the release of NO and PGI_2 in an autocrine manner. The ET-1-induced release of NO could then affect the binding of ET-1 to its receptors and alter ET-1-stimulated Ca^{2+} mobilization and rearrangement of the cytoskeleton in the endothelial cells and enhanced affinity of the contractile proteins to calcium (Wang et al., 1991), secondary to changes resulting from activation of the sarcolemmal Na^+/H^+ exchanger (Kramer et al., 1991). Similar autocrine interactions of ET may also exist with endothelium-derived PGI_2 (de Nucci et al., 1988) and angiotensin II (De Hurtado et al., 2000). McClella et al. (1994) proposed ET storage and release from cardiac endothelial cells as a supportive cardioregulatory mechanism. Low ET concentrations may have important protective activities in the normal heart.

Endothelium has also an important role in suppressing arrhythmias. An antiarrhythmic action was attributed to the endothelium-derived NO and PGI_2 and a pronounced cardiac endothelial dysfunction preconditioned for severe arrhythmias (Hassanabad et al., 1998).

Beyond the reciprocal interplay among the various endothelium-mediated auto-/paracrine signalings, a still higher scale of complexity in the in vivo intact heart may ensue from their interaction with other important cardiomodulatory pathways, such as the β-adrenergic or cholinergic pathways in the heart, atrial and brain natriuretic peptide activity, and circulating thyroid and aldosterone hormones. The synthesis, secretion, and activities of these endothelium-derived substances are closely linked, interrelated, and interactive. Many of these agents modulate the actions of the other endothelium-derived agents on the same target cell, actions which may be mutually additive, synergistic, or inhibitory. They may even result in novel effects, not seen with either agent alone, and as determined by the milieu in which the interactions occur. It may therefore be simplistic to try and define their properties independently from one another. Too little attention is given to such interactions and interdependence (Brutsaert, 2003).

The Brain Endothelium

The Blood-Brain Barrier and Neurovascular Unit

In this section, the specific features of the brain endothelial cells that form the blood-brain barrier (BBB) and distinguish the cerebral endothelium from those found in peripheral tissues will be presented. The BBB constitutes the physical and functional interface between blood and brain. It is originally located at the level of brain capillaries, and it is responsible for providing a suitable environment for neuronal function, regardless of fluctuations in blood composition.

The BBB capillaries are composed by one to three continuous-type endothelial cells that display well-developed intercellular junctions. The cytoplasm of the cerebral endothelium is characterized by very low rate of transcytotic vesicles and large number of mitochondria, which reflect high energy demands of these cells (Reese and Karnovsky, 1967; Brightman and Reese, 1969; Rowland et al., 2001). Together, the intercellular junctions and the small number of transcytotic vesicles are responsible for the restrictive paracellular passage and selective transcellular transport through the brain endothelial cells, respectively.

Besides the efficient physical barrier the BBB also constitutes a metabolic barrier. The brain endothelium contains a number of enzyme systems that support the protective and detoxifying roles of the BBB in addition to the glycocalyx that covers the luminal surface of these cells. The former is represented by cytoplasmic enzymes such as lysosomal and extra-lysosomal ones (monoamine oxidase and cytochrome P450) and by enzymes associated with the outer surface of the plasma membrane such as peptidases, responsible for the metabolism of potentially neuroactive substances (bradykinin, enkephalins and substance P). The glycocalyx is rich in sialic acid residue, which confers the anion property to the endothelium. It acts as a sieve, selecting substances based on their molecular mass and ionic charge (Allt and Lawrenson, 2000).

The cerebral endothelium is underlined by a thick basal lamina. It is common to the ones of pericytes and the basal lamina of perivascular astrocytes. Many of the BBB properties, including formation of tight junctions and the polarized expression of transporters, depend on a close association of brain capillaries with astrocytic endfeet (Rubin et al., 1991; Abbott, 2002; Wolburg et al., 2009). These cells have stellate shape and contain multiple processes that expand toward neurons and vessels. The ends of the astrocytes are so-called endfeet. Astrocytic endfeet surround about 95% of the abluminal surface of the brain capillaries (Rowland et al., 1991) and they are responsible for releasing several factors that induce and maintain the BBB phenotype. Among these factors, TGF-b (Tran et al., 1999), GDNF (Igarashi et al., 1999), bFGF (Sobue et al., 1999), IL-6, hydrocortisone (Hoheisel et al., 1998), angiopoetin 1 (Lee et al., 2003) and angiotensin II (Wosik et al., 2007) have been identified. Reciprocally, endothelial cells release leukemia inhibitory factor (LIF) that together with the agrin (a heparan sulfate proteoglycan of the endothelial basal lamina) induce the differentiation and functional maintenance of the astrocytes (Mi et al. 2001; Wolburg and Lippoldt, 2002). In this way, the exchange of signaling molecules and induction of some features via the extracellular matrix or close cell-cell interaction add further complexity to the induction process (Abbott et al., 2010).

Recent studies have demonstrated that besides astrocytes, multiple agents and cell types including pericytes, neurons and perivascular microglias are involved in the modulation of the BBB permeability. Together, these cells interact with microvessels regulating the local blood flow and the vascular tone. This functional concept has been widely accepted among neuroscientists and since the beginning of this century it has been termed as "neurovascular unit" (Abbott et al., 2006; Zlokovic, 2008).

Transport System and Therapeutics Approach

The molecular organization of tight junction associated-proteins in the brain endothelium is more complex than the ones found in non-neural endothelial cells. Together with adherens junctions, they form junctional complexes that occupy not only the apical location but also entire length of the interendothelial contact, forming networks of multiple barriers (Schulze and Firth, 1993; Abbott et al., 2006). The complexity of tight junction strains has been related to an increased transepithelial electrical resistance or low conductance of ions through the cell-cell contact (Claude, 1978). At the BBB, the transendothelial electrical resistance is ~1800 Ω cm2 against 2–20 Ω cm^2 in peripheral capillaries (Butt et al., 1990). It restricts the passage of ions (such as Na^+, Cl^- and K^+), peptides and immune cells from the bloodstream into the central nervous system (CNS). In this way, almost 100% of large molecules (>400 Da) and more than 98% of small molecules do not cross the BBB by the paracellular route (Pardridge, 2005). Reports have demonstrated that tight junction-associated protein occludin is not required for the formation of tight junction strands. However, posttranslational modifications of this protein and the ZO-1, such as phosphorylation in its cytoplasmic domains, seem to be associated with TJ regulation and modulation of the paracellular permeability (Rubin and Staddon, 1999; Staddon et al., 1995).

While the paracellular passage is restrictive, the transcellular transport through the endothelial cell is very selective. Only small lipophilic substances, such as O_2 and CO_2, diffuse freely from blood to brain across endothelial membranes, according to the concentration gradient. On the other hand, large lipophilic and hydrophilic components should pass through the BBB exclusively by transcellular transport. This system is bidirectional and it can be classified into five main categories: carrier-mediated transport, caveolae-mediated transport, ion transport, and active efflux transport (for review see Zlokovic, 2008).

Carrier-mediated transport systems facilitate transport of nutrients into the neural tissue such as hexoses (glucose, galactose); essential amino acids (leucine, isoleucine, valine, tryptophan, tyrosine, phenylalanine, threonine, methionine and lysine) and non-essential amino acids (arginine and ornithine); monocarboxylic acids (lactate, pyruvate, ketone bodies); nucleosides (adenosine, guanosine, uridine); purines (adenine, guanine); amines (choline); and vitamins. The concentration gradients for nutrients are generally in the direction from blood to brain. These are regulated by brain metabolic needs, and by the concentrations of substrates in plasma (Zlokovic, 2008).

Caveolae-mediated transport involves the internalization of ligands bounded to specific receptors. At the BBB, by caveolar-transport system substances as iron ion (by transferrin receptor), insulin, albumin, LDL, HDL, leptin and interleukin-1 can be transported into the brain. In this way, caveolae system constitutes control sites of transendothelial transport by regulating endocytosis and transcytosis, besides coordinating intracellular signaling events. The latter has been described as being mediated by G proteins, members of the MAPK pathway, src tyrosine kinase, protein kinase C, and eNOS. In addition, there are other rafts in the BBB, which are associated with interendothelial junctions, focal adhesion and adhesion/migration of immune cells across the brain endothelium (Zlokovic, 2008; Dodelet-Devillers et al., 2009).

Among the ion transporters localized on the endothelial membrane can be cited the Na^+/K^+ -ATPase pump. The location of this pump is mainly abluminal, where it is related to brain homeostasis by regulating sodium influx and removing excess potassium from the interstitial fluid, since it is responsible for increasing neuronal excitability. In addition, the sodium-hydrogen and the chloride-bicarbonate exchangers play critical roles in regulating intracellular pH, which is necessary for proper enzymatic action (Zlokovic, 2008).

Conversely, BBB endothelium has specific carriers that meditate the efflux from the CNS of potentially harmful endogenous and exogenous compounds. An important example of neurotoxic metabolite is the acidic amino acid glutamate. Glutamate concentration in blood is 50-100 μM, while in the brain interstitial fluid it is normally kept below 2 μM. This is because it can act on metabotropic NMDA receptors leading to intracellular Ca^{2+} increase and neuronal injury or death. In addition, the glutamate released during neurotransmission is normally rapidly taken up by neurons and neighboring astrocytes (Redzic, 2011). The mainly family of efflux transporters at the BBB is the ATP Binding Cassette (ABC) that transports a large panel of lipophilic molecules against a concentration gradient. The most active ABC-transporter at the BBB is P-glycoprotein (P-gp), which is encoded by the multidrug resistance gene (MDR1 or ABCB1). P-glycoprotein has been localized at the luminal and abluminal endothelial membrane, and plays a major role in protecting the brain from xenobiotics. In addition, multidrug resistance proteins (MRPs, or ABCC proteins) and the breast cancer resistance protein (BCRP or ABCG2) have also been described (Löscher and Potschka, 2005; Weiss et al., 2008).

Considering the BBB ensures a privileged environment for neural functions, on the other hand, it constitutes a major obstacle for therapeutic agent delivery to the brain. Many promising drugs for treatment of neurological diseases have failed in clinical trials due to poor BBB permeability. In this context, intra-cerebrospinal fluid or intra-cerebroventricular injections, local drug administration via intracerebral implants and temporary BBB disruption by intracarotid infusions of hyperosmotic solutions (such as mannitol) have been used as traditional approaches designed to circumvent the BBB. However, they are aggressive techniques that have side effects and toxicity risks (Weiss et al., 2008; Redzic, 2011).

More recently, progresses in the understanding of the specific structural and functional BBB features have contributed to the design of new therapeutic approaches for drug delivery into the brain. Among them can be cited: (i) transient BBB opening by ultrasound, an alternative strategy that seems limit neuronal damage by targeting a restricted brain area for BBB opening and drug delivery; (ii) pharmacological approaches based on the known effect of vasoactive mediators (such as bradykinin) on the BBB permeability; (iii) modulation of efflux transporters activity; (iv) control of receptor-mediated transport through the BBB. This strategy has been known as Trojan horses and uses well-characterized receptors (such as transferrin and insulin receptors) for mediating transcytosis of ligands to the abluminal side of the BBB (Misra et al., 2003; Weiss et al., 2008; Redzic, 2011).

The Role of Nitric Oxide on some Pathological Conditions

Accumulating evidence suggests that the reduced NO biological activity, caused both by synthase reduction and by increased degradation due to oxidation stress, is the most relevant mechanism in the multifactorial process in endothelial dysfunction and in the participation in the main cardiovascular dysfunctions, such as atherosclerosis (Pepine et al., 2009). Atherosclerosis is the major risk factor for the development of cardiovascular disease based on arterial endothelial dysfunction caused by the impairment of endothelial dependent dilation (Tarik and Ramahi, 2010; Boni et al., 2010). Atherosclerosis is a chronic inflammatory syndrome with a predilection for sites within the walls of arteries. Common factors involved in the mechanism of atherosclerosis include endothelial cells, VSMC, macrophages, platelets, and protein substances such as low-density lipoproteins (LDL), growth factors, and cytokines. This is a dynamic process, exhibiting constant changes in size and composition of the blood vessels (Chang et al., 2010).

Thus, reductions in nitric oxide bioavailability and subsequent endothelial dysfunction trigger, in the vascular environment, events such as tone alterations, thrombotic dysfunctions, smooth muscle cell (SMC) proliferation and migration, as well as leukocyte adhesion (Madamanchi et al., 2005). When endothelial dysfunction occurs, there is also an increased production of reactive oxygen species (ROS) (Cai, 2005), and these can reduce endothelial NO availability via different pathways: direct nitric oxide inactivation by superoxide with peroxynitrite (ONOO-)formation (Gao and Mann, 2009); reduction in nitric oxide synthase expression and activity due to changes in their substrate or cofactors, and in the increase in asymmetric dimethylarginine (ADMA) levels (De Gennaro et al., 2009); and also eNOS uncoupling caused by increased tetrahydrobiopterin (BH4) oxidation (Sessa, 2004).

It has also been well established that chronic inflammation contributes to about one in four of all cancer cases worldwide (Coussens and Werb, 2002). The induction of mutations in cancer-related genes or post-translational modifications of proteins by nitration, nitrosation, phosphorylation, acetylation or polyADP-ribosylation are some key events that can increase the cancer risk. In particular, high levels of NO are genotoxic through formation of carcinogenic nitrosamines or by directly modifying DNA or DNA repair proteins (Muntane and la Mata, 2010). It was found that aerobic solutions of NO, NO_2 and N_2O_3 led to deamination of nucleic acids (Wink et al., 1991). NO and NO-derived reactive nitrogen species induce oxidative and nitrosative stress which results in DNA damage (such as nitrosative deamination of nucleic acid bases, transition and/or transversion of nucleic acids, alkylation and DNA strand breakage) and inhibition of DNA repair enzymes (such as alkyltransferase and DNA ligase) through direct or indirect mechanisms (Fukumura et al., 2006).

iNOS produces much larger amounts of NO and has been detected in many tumors, including in the oral cavity (Rao et al., 2006). In addition, a positive correlation between iNOS expression and tumor progression has been reported (Connelly et al., 2005). In neoplastic tissues, nitric oxide has been implicated as having a positive role in permitting tumor growth, in turn DNA damage, mutagenesis, angiogenesis and metastasis (Nakamura et al., 2006). We found that the percentage of detectable iNOS expression significantly increased from 0% of normal control, 10% of 'normal' oral mucosa exposed to 4NQO during 4 weeks, to 30% of pre-neoplastic lesions and 35% of oral squamous cell carcinomas (Ribeiro et al., 2007). Similar results have been reported (Connelly et al., 2005). Although not statistically significant, analysis of our data also revealed that there was a slight increase in the labeling index for iNOS after 20 weeks 4NQO treatment when compared to 12 weeks. These results suggest aberrant iNOS expression will correlate with disease progression (Ribeiro et al., 2007).

NO may also influence the carcinogenesis process by alteration of cell-cycle checkpoints (Pervin et al., 2001), apoptosis (Melino et al., 1997) and DNA repair (Jaiswal et al., 2001). NO donors sensitize tumor cells to chemotherapeutic compounds by nitrosilation of critical thiols in DNA repair enzymes in hepatoma cell line (Laval and Wink, 1994). Other studies have demonstrated increased susceptibility to chemotherapy to cisplatin (Wink et al., 1997) and melphalan (Cook et al., 1997) by NO donors in different cell lines. These results implied substantial modification of key biological target(s) including DNA repair proteins and transcription factor known to be inhibited by NO (Muntane and la Mata, 2010).

Specifically endothelial production of nitric oxide regulates blood flow and angiogenesis and reduces tumor cell adhesion to the endothelium (Tecder et al., 2010). A high concentration of nitric oxide and its metabolites causes DNA damage during nitration, nitrosation and deamination. Nevertheless, both positive and negative effects on carcinogenesis and tumor growth, apoposis, and cytotoxic mechanisms may be explained by differential susceptibility of tumor cells to nitric oxide-mediated reactions (Tecder et al., 2010). In fact, some authors have postulated that endothelial nitric oxide synthase 786T>C polymorphism may play a role in the development of gastric cancer (Tecder et al., 2010).

Experimental metastases in the brain of mice that are infiltrated by microglia, and parabiosis experiments of green fluorescent protein (GFP(+)) and GFP(-) mice revealed that these microglia are derived from circulating monocytes (GFP(+), F4/80(+), and CD68(+)). These findings raised the question as to whether microglia possesses tumoricidal activity. C8-B4 murine microglia cells were incubated in vitro in medium (control) or in medium containing both lipopolysaccharide and interferon-γ (Brantley et al., 2010). Control microglia was not tumoricidal against a number of murine and human tumor cells, whereas lipopolysaccharide/interferon-γ-activated microglia lysed murine and human tumor cells by release of nitric oxide. Neither activated microglia nor activated macrophages lysed nontumorigenic murine or human cells. Collectively, these data demonstrate that brain metastasis-associated microglia are derived from circulating mononuclear cells and exhibit selective and specific tumoricidal activity (Brandley et al., 2010). Furthermore, it is possible that macrophages may be effective for cancer treatment because once activated, lung macrophages have enhanced contact with lung tumor cells and have a cytotoxic effect (Sawachi et al., 20010). This is because nitric oxide (NO) produced by lung macrophages activated with lipopolysaccharide (LPS) suppressed cell growth of human lung adenocarcinoma cells (Sawachi et al., 2010). NO signaling in tumors and endothelial cells regulates vascular permeability and blood flow and therefore influences tumor uptake and response to therapeutic compounds. As delivery and efficacy of chemotherapy is impaired in central nervous system neoplasms due to a partially intact BBB, some authors have investigated the effects of NO released by the short-acting NO donor disodium 1-[2-(carboxylato)pyrrolidin-1-yl]diazen-1-ium-1,2-diolate methanolate (PROLI/NO) on BBB integrity and blood flow in C6 gliomas (Weyerbrook et al., 2011). Such findings revealed that PROLI/NO, and perhaps other NO-donating compounds, can be used to selectively increase BBB permeability in gliomas through the NO/cGMP pathway at doses that do not cause unwanted vasodilatory changes in blood flow and that do not affect the systemic circulation (Weyerbrook et al., 2011).

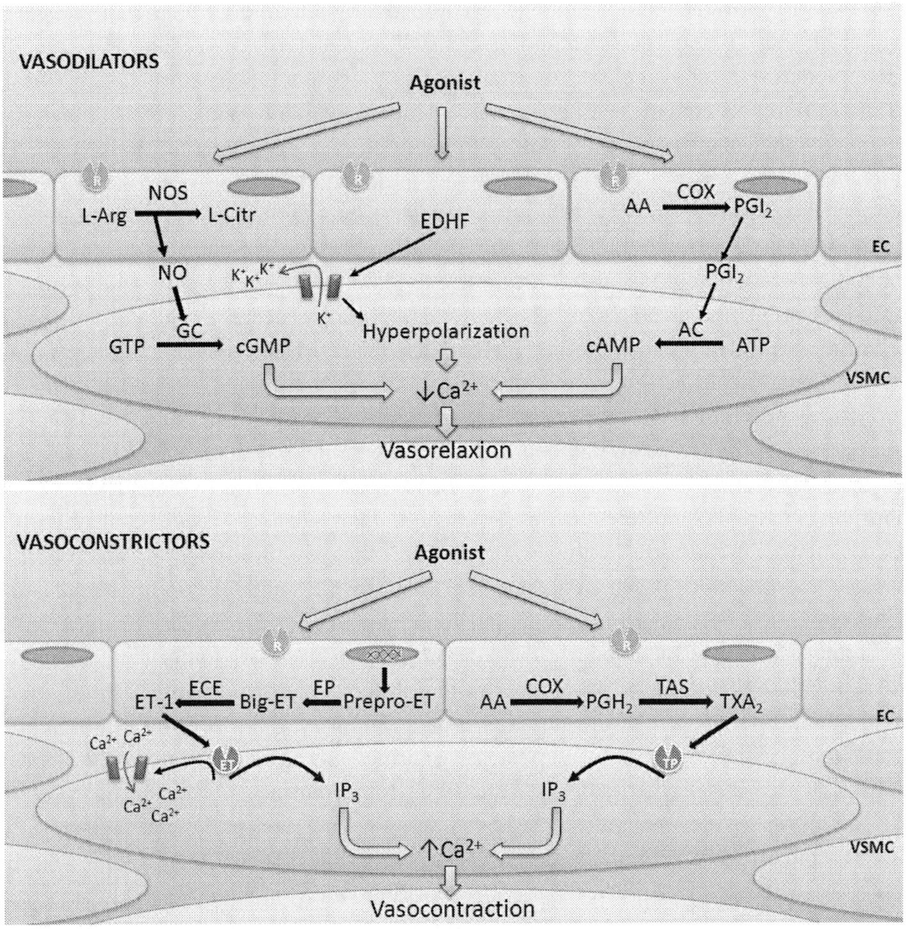

Figure 1. Simplified schematic view of the mechanism of action of the vasoactive substances to cause either vasodilation or vasocontraction. For details see the text. NO, nitric oxide, NOS, nitric oxide synthase, L-Arg, L-arginine; L-citr, L-citrulline; GC, guanylate cyclase; EDHF, endothelium-derived hyperpolarizing factor; AA, aracdonic acid; COX, cyclooxygenase; PGI2, prostacyclin; AC, adenylate cyclase, ECE, endothelin-converting enzyme; EP, endopeptidase; TAS, thromboxane-A-synthase; R, receptor; EC, endothelial cells; VSMC, vascular smooth muscle cells.

References

Abbott NJ. Astrocyte-endothelial interactions and blood-brain barrier permeability. *J. Anat*, 200: 629-638, 2002.

Abbott NJ, Patabendige AA, Dolman DE, Yusof SR, Begley DJ. Structure and function of the blood-brain barrier. *Neurobiol. Dis*, 37: 13-25, 2010.

Abbott NJ, Rönnbäck L, Hansson E. Astrocyte–endothelial interactions at the blood–brain barrier. *Nat. Rev. Neurosci,* 7: 41–53, 2006.

Abramson DI. Arterial and arteriolar systems. In: Abramson DI. Blood vessels and lymphatics. New York: Academic Press, 1962.

Allt G, Lawrenson JG. The blood-nerve barrier: enzymes, transporters and receptors - a comparison with the blood-brain barrier. *Brain Res. Bull*, 52: 1-12, 2000.

Bailey FR, Kelly DE, Wood RL, Enders AC. The circulatory systems. In: Bailey FR. Bailey's textbook of microscopic anatomy. 18th ed. Baltimore: Williams and Wilkins, 1984.

Baron MH. Embryonic origins of mammalian hematopoiesis. *Exp. Hematol*, 31: 1160–1169, 2003.

Batista AC, Silva TA, Chun JH, Lara VS. Nitric oxide synthesis and severity of human periodontal disease. *Oral. Dis*, 8: 254–260, 2002.

Bény JL. Information networks in the arterial wall. *NIPS*, 14: 68–73, 1999.

Bloom W, Fawcett DW. Blood and lymph vascular sysytems. In: Bloom W. A textbook of histology. 11st ed. Philadelphia: Saunders International, 1986.

Boni A, Lorenzoni R, Lazzari M, Gemignani C, Bovenzi F. Cardiovascular risk management: An overview. *Curr. News Cardiol*, 8: 407–416, 2007.

Brantley EC, Guo L, Zhang C, Lin Q, Yokoi K, Langley RR, Kruzel E, Maya M, Kim SW, Kim SJ, Fan D, Fidler IJ. Nitric oxide-mediated tumoricidal activity of murine microglial cells. *Transl. Oncol*, 3: 380-388, 2010.

Brightman MW, Reese TS. Junctions between intimately-apposed cell membranes in the vertebrate brain. *J. Cell Biol*, 40: 648-677, 1969.

Brutsaert D. Cardiac Endothelial-Myocardial Signaling: its role in cardiac growth, contractile performance, and rhythmicity. *Physiol. Rev*, 83: 59-113, 2003.

Brutsaert DL, Meulemans AL, Sipido KR, Sys SU. Effects of damaging the endocardial surface on the mechanical performance of isolated cardiac muscle. *Circ. Res*, 62: 358–366, 1988.

Brutsaert DL. The endocardium. *Annu. Rev. Physiol*, 51: 263–273, 1989.

Butt AM, Jones HC, Abbott NJ. Electrical resistance across the blood–brain barrier in anaesthetized rats: a developmental study. *J. Physiol*, 429: 47–62, 1990.

Cai H. NAD(P)H oxidase-dependent self-propagation of hydrogen peroxide and vascular disease. *Circ. Res*, 96: 818-22, 2005.

Chang JC, Kou SJ, Lin WT, Liu CS. Regulatory role of mitochondria in oxidative stress and atherosclerosis. *World J. Cardiol*, 2 :150-9, 2010.

Choi K, Kennedy M, Kazarov A, Papadimitriou JC. Keller G. A common precursor for hematopoietic and endothelial cells. *Development*, 125: 725-732, 1998.

Cines DB, Pollak ES, Buck CA, Loscalzo J, Zimmerman GA, McEver RP, Pober JS, Wick TM, Konkle BA, Schwartz BS, Barnathan ES, McCrae KR, Hug BA, Schmidt A, Stern DM. Endothelial cells in physiology and in the pathophysiology of vascular disorders. *Blood*, 91: 3527-3561, 1998.

Claude P. Morphologic factors influencing transepithelial permeability. A model for the resistance of the zonula occludens. *J. Membr. Biol.* 39: 219–232, 1978.

Connelly ST, Macabeo-Ong M, Dekker N, Jordan RC, Schmidt BL. Increased nitric oxide levels and iNOS over-expression in oral squamous cell carcinoma. *Oral. Oncol*, 41: 261–267, 2005.

Cook JA, Krishna MC, Pacelli R, DeGraff W, Liebmann J, Mitchell JB, Russo A, Wink DA. Nitric oxide enhancement of melphalan-induced cytotoxicity. *Br. J. Cancer*, 76: 325–334, 1997.

Coussens LM, Werb Z. Inflammation and cancer. *Nature*, 420: 860–867, 2002.

De Gennaro Colonna V, Bianchi M, Pascale V, Ferrario P, Morelli F, Pascale W, Tomasoni L, Turiel M. Asymmetric dimethylarginine (ADMA): an endogenous inhibitor of nitric oxide synthase and a novel cardiovascular risk molecule. *Med. Sci. Monit*, 15: RA91-101, 2009.

De Hurtado MC, Alvarez BV, Ennis IL, Cingolani HE. Stimulation of myocardial Na+-independent Cl--HCO3 - exchanger by angiotensin II is mediated by endogenous endothelin. *Circ. Res*, 86: 622–627, 2000.

De Keulenaer GW, Sys SU. Endothelial control of myocardial performance. In: Endocardial Endothelium: Control of Cardiac Performance, edited by Sys SU and Brutsaert DL. Austin, TX: Landes, 1995.

De Nucci G, Thomas R, D'Orleans JP, Antunes E, Walder C, Warner TD, Vane JR. Pressor effects of circulating endothelin are limited by its removal in the pulmonary circulation and by the release of prostacyclin and endothelium-derived relaxing factor. *Proc. Natl. Acad. Sci*, 85: 9797–9800, 1988.

Dodelet-Devillers A, Cayrol R, van Horssen J, Haqqani AS, de Vries HE, Engelhardt B, Greenwood J, Prat A. Functions of lipid raft membrane microdomains at the blood-brain barrier. *J. Mol. Med*, 87: 765-74, 2009.

Edwards G, Félétou M, Weston AH. Endothelium-derived hyperpolarising factors and associated pathways: a synopsis. *Pflugers Arch. – Eur. J. Physiol*, 459: 863–879, 2010.

Félétou M, Vanhoutte PM. EDHF: an update. *Clinical Science*, 117: 139–155, 2009.

Fleming I, Busse R. Signal transduction of eNOS activation. *Cardiovasc. Res*, 43: 532–541, 1999.

Fransen P, Demolder M. Physicochemical properties of the endocardial endothelium. In: Endocardial Endothelium: Control of Cardiac Performance, edited by Sys SU and Brutsaert DL. Austin,TX: Landes, 1995.

Fukumura D, Kashiwagi S, Jain RK. The role of nitric oxide in tumour progression. *Nat. Rev. Cancer*, 6: 521–534, 2006.

Galley HF, Webster NR. Physiology of the endothelium. *Br. J. Anaesth*, 93: 105-113, 2004.

Gao L, Mann GE. Vascular NAD(P)H oxidase activation in diabetes: a double-edged sword in redox signalling. *Cardiovasc. Res*, 82: 9-20, 2009.

Gerhardt H, Betsholtz C. Endothelial-pericyte interactions in angiogenesis. *Cell Tissue Res.* 314:15–23, 2003.

Hassanabad ZF, Furman BL, Parratt JR, Aughey E. Coronary endothelial dysfunction increases the severity of ischaemia-induced ventricular arrhythmias in rat isolated perfused hearts. *Basic Res. Cardiol*, 93: 241–249, 1998.

Hoheisel D, Nitz T, Franke H, Wagner J, Hakvoort A, Tilling T, Galla H. Hydrocortisone reinforces the blood–brain barrier properties in a serum free cell culture system. *Biochem. Biophys. Res. Comm*, 247: 312–315, 1998.

Igarashi Y, Utsumi H, Chiba H, Yamada-Sasamori Y, Tobioka H, Kamimura Y, Furuuchi K, Kokai Y, Nakagawa T, Mori M, Sawada N. Glial cell line-derived neurotrophic factor induces barrier function of endothelial cells forming the blood–brain barrier. *Biochem. Biophys. Res. Comm*, 261: 108–112, 1999.

Inagami T, Naruse M, Hoover R. Endothelium as an endocrine organ. *Annu. Rev. Physiol*, 57: 171-89, 1995.

Inoue A, Yanagisawa M, Takuwa Y, Mitsui Y, Kobayashi M, Masaki T. The human preproendothelin- 1 gene. *J. Biol. Chem*, 264: 14954-14959, 1989.

Jaiswal M, LaRusso NF, Nishioka N, Nakabeppu Y, Gores GJ. Human Ogg1, a protein involved in the repair of 8-oxoguanine, is inhibited by nitric oxide. *Cancer Res,* 61: 6388–6393, 2001.

Kawabe J, Ushikubi F, Hasebe N. Prostacyclin in vascular diseases – recent insights and future perspectives. *Circ. J,* 74: 836-843, 2010.

Kramer BK, Smith TW, Kelly RA. Endothelin and increased contractility in adult rat ventricular myocytes. Role of intracellular alkalosis induced by activation of the protein kinase C-dependent Na_-H_ exchanger. *Circ. Res,* 68: 269–279, 1991.

Laval F, Wink DA. Inhibition by nitric oxide of the repair protein, O6-methylguanine-DNA-methyltransferase. *Carcinogenesis*, 15: 443–447, 1994.

Lee SW, Kim WJ, Choi YK, Song HS, Son MJ, Gelman IH, Kim YJ, Kim KW. SSeCKS regulates angiogenesis and tight junction formation in blood-brain barrier. *Nat. Med*, 9: 900–906, 2003.

Loke KE, Laycock SK, Mital S, Wolin MS, Bernstein R, Oz M, Addonizio L, Kaley G, Hintze TH. Nitric oxide modulates mitochondrial respiration in failing human heart. *Circulation* 100: 1291–1297, 1999.

Löscher W, Potschka H. 2005. Blood-brain barrier active efflux transporters: ATP-binding cassette gene family. *Neuro Rx*. 2: 86-98, 2005.

Madamanchi NR, Vendrov A, Runge MS. Oxidative stress and vascular disease.*Arterioscler Thromb. Vasc. Biol,* 25: 29-38, 2005.

Mann GE, Yudilevich DL, Sobrevia L. Regulation of amino acid and glucose transporters in endothelial and smooth muscle cells. *Physiol. Rev*, 83: 183–252, 2003.

Masaki T. The discovery of endothelins. *Cardiovascular. Res*, 39: 530–533, 1998.

Mayer B, Hemmens B. Biosynthesis and action of nitric oxide in mammalian cells. *Trends Biochem. Sci.* 22: 477–481, 1997.

MccLellan G, Weisberg A, Rose D, and Winegrad S. Endothelia cell storage and release of endothelin as a cardioregulatory mechanism. *Circ. Res.* 75: 85–96, 1994.

Mehta D, Malik AB. Signaling mechanisms regulating endothelial permeability. *Physiol. Rev,* 86: 279 –367, 2006.

Melino G, Bernassola F, Knight RA, Corasaniti MT, Nistico G, Finazzi-Agro A. S-nitrosylation regulates apoptosis. *Nature*, 388: 432–433, 1997.

Mi H, Haeberle H, Barres BA. Induction of astrocyte differentiation by endothelial cells. *J. Neurosci*, 21: 1538-47, 2001.

Minshall RD, Tiruppathi C, Vogel SM, Malik AB. Vesicle formation and trafficking in endothelial cells and regulation of endothelial barrier function. *Histochem. Cell Biol*, 117:105–112, 2002.

Misra A, Ganesh S, Shahiwala A, Shah SP. Drug delivery to the central nervous system: a review. *J. Pharm. Pharm. Sci*, 6: 252-73, 2003.

Muntané J, la Mata MD. Nitric oxide and cancer. *World J. Hepatol*,2: 337-344, 2010.

Nakamura Y, Yasuoka H, Tsujimoto M, Yoshidome K, Nakahara M, Nakao K, Nakamura M, Kakudo K. Nitric oxide in breast cancer, induction of vascular endothelial growth factor-C and correlation with metastasis and poor prognosis. *Clin. Cancer Res,* 12: 1201–1207, 2006.

Pardridge WM. Molecular biology of the blood-brain barrier. *Mol. Biotechnol.* 30: 57-70, 2005.

Pate M, Damarla V, Chi DS, Negi S, Krishnaswamy G. Endothelial cell biology: role in the inflammatory response. *Adv. Clin. Chem*, 52: 109-130, 2010.

Pepine CJ The impact of nitric oxide in cardiovascular medicine: untapped potential utility. *Am. J. Med*, 122: S10-15, 2009.

Pervin S, Singh R, Chaudhuri G. Nitric oxide-induced cytostasis and cell cycle arrest of a human breast cancer cell line (MDA-MB-231): potential role of cyclin D1. *Proc. Natl. Acad. Sci*, 98: 3583–3588, 2009.

Pries AR, Secomb TW, Gaehtgens P. The endothelial surface layer. Pflügers *Arch.-Eur. J. Physiol,* 440: 653–666, 2000.

Ramahi TM. Cardiovascular disease in the Asia Middle East Region: Global trends and local implications. *Asia Pac. J. Public Health*, 22: 83-89, 2010.

Rao CV, Reddy BS, Steele VE, Wang CX, Liu X, Ouyang N, Patlolla JM, Simi B, Kopelovich LR, Rigas R. Nitric oxide-releasing aspirin and indomethacin are potent inhibitors against colon cancer in azoxymethane-treated rats, effects on molecular targets. *Mol. Cancer Ther,* 5:1530–1538, 2006.

Redzic Z. Molecular biology of the blood-brain and the blood-cerebrospinal fluid barriers: similarities and differences. *Fluids Barriers CNS*, 8: 3, 2011.

Reese TS, Karnovsky MJ. Fine structural localization of a blood-brain barrier to exogenous peroxidase. *J. Cell Biol*, 34: 207-217, 1967.

Ribeiro DA, Kitakawa D, Domingues MA, Cabral LA, Marques ME, Salvadori DM. Survivin and inducible nitric oxide synthase production during 4NQO-induced rat tongue carcinogenesis: a possible relationship. *Exp. Mol. Pathol*, 83: 131-137, 2007.

Rowland LP, Fink ME, Rubin L. Cerebrospinal fluid: blood-brain barrier, brain edema and hydrocephalus. In: Kandel ER, Schwartz JH, Jessel TM (eds) Principles of Neural Science. 3 rd ed. New York: Elsevier. 1991.

Rubbin LL, Hall DE, Porter S, Barbu K, Cannon C, Horner HC, Janatpour M, Liaw CW, Manning K, Morales J, Tanner LI, Tomaselli KJ, Bard F. A cell culture model of the blood-brain barrier. *J. Cell Biol*, 115: 1725-1735, 1991.

Rubin LL, Staddon JM. The cell biology of the blood-brain barrier. *Annu. Rev. Neurosci*, 22: 11-28, 1999.

Sawachi K, Shimada Y, Taniguchi H, Hirota K, Inagawa H, Kohchi C, Soma G, Makino K, Terada H. Cytotoxic effects of activated alveolar macrophages on lung carcinoma cells via cell-to-cell contact and nitric oxide. *Anticancer Res*, 30: 3135-3141, 2010.

Schulze C, Firth A. Immunohistochemical localization of adherens junction components in blood-brain barrier microvessels of the rat. *J. Cell Sci*, 104: 773-782, 1993.

Sessa WC. eNOS at a glance. *J. Cell Sci,* 117: 2427-2429, 2004.

Shah AM, Maccarthy PA. Paracrine and autocrine effects of nitric oxide on myocardial function. *Pharmacol. Ther,* 86: 49–86, 2000.

Shen W, Tian R, Saupe KW, Spindler M, Ingwall JS. Endogenous nitric oxide enhances coupling between O2 consumption and ATP synthesis in guinea pig hearts. *Am. J. Physiol. Heart Circ. Physiol.* 281: H838–H846, 2001.

Sobue K, Yamamoto N, Yoneda K, Hodgson ME, Yamashiro K, Tsuruoka N, Tsuda T, Katsuya H, Miura Y, Asai K, Kato T. Induction of blood–brain barrier properties in immortalized bovine brain endothelial cells by astrocytic factors. *Neurosci. Res*, 35: 155–164, 1999.

Staddon JM, Herrenknecht K, Smales C, Rubin LL. Evidence that tyrosine phosphorylation may increase tight junction permeability. *J. Cell Sci*, 108: 609-619, 1995.

Stumpe T, Decking UKM, Schrader J. Nitric oxide reduces energy supply by direct action on the respiratory chain in isolated cardiomyocytes. *Am. J. Physiol. Heart Circ. Physiol.* 280: H2350–H2356, 2001.

Suzuki R, Kohno H, Suzui M, Yoshimi N, Tsuda H, Wakabayashi K, Tanaka T. An animal model for the rapid induction of tongue neoplasms in human c-Ha-ras proto-oncogene transgenic rats by 4- nitroquinoline 1-oxide, its potential use for preclinical chemoprevention studies. *Carcinogenesis*, 27: 619–630, 2006.

Tada H, Thompson CI, Recchia FA, Loke KE, Ochoa M, Smith CJ, Shesely EG, Kaley G, Hintze TH. Myocardial glucose uptake is regulated by nitric oxide via endothelial nitric oxide synthase in Langendorff mouse heart. *Circ. Res,* 86: 270–274, 2000.

Tecder Ünal M, Karabulut HG, Gümüş-Akay G, Dölen Y, Elhan A, Tükün A, Unal AE. Endothelial nitric oxide synthase gene polymorphism in gastric cancer. *Turk. J. Gastroenterol*, 21: 338-44, 2010.

Tran ND, Correale J, Schreiber SS, Fisher M. Transforming growth factor-beta mediates astrocyte-specific regulation of brain endothelial anticoagulant factors. *Stroke,* 30: 1671–1678, 1999.

Trochu JN, Bouhour JB, Kaley G, Hintze TH. Role of endothelium- derived nitric oxide in the regulation of cardiac oxygen metabolism. Implications in health and disease. *Circ. Res.* 87: 1108–1117, 2000.

Vandenbroucke E, Mehta D, Minshall R, Malik AB. Regulation of Endothelial Junctional Permeability. *Ann. NY Acad. Sci,* 1123: 134–145, 2008.

Villar IC, Francis S, Webb A, Hobbs AJ, Ahluwalia A. Novel aspects of endothelium-dependent regulation of vascular tone. *Kidney International*, 70: 840–853, 2006.

Wang J, Morgan JP. Endocardial endothelium modulates myofilament Ca 2+ responsiveness in aequorin-loaded ferret myocardium. *Circ. Res,* 70: 754–760, 1992.

Wang JX, Paik G, and Morgan JP. Endothelin 1 enhances myofilament Ca 2+ responsiveness in aequorin-loaded ferret myocardium. *Circ. Res,* 69: 582–589, 1991.

Weiss N, Miller F, Cazaubon S, Couraud PO. The blood-brain barrier in brain homeostasis and neurological diseases. *Biochim. Biophys. Acta*, 1788: 842-857, 2009.

Weyerbrock A, Walbridge S, Saavedra JE, Keefer LK, Oldfield EH. Differential effects of nitric oxide on blood-brain barrier integrity and cerebral blood flow in intracerebral C6 gliomas. *Neuro Oncol*, 13: 203-211, 2011.

Wink DA, Cook JA, Christodoulou D, Krishna MC, Pacelli R, Kim S, DeGraff W, Gamson J, Vodovotz Y, Russo A, Mitchell JB. Nitric oxide and some nitric oxide donor compounds enhance the cytotoxicity of cisplatin. *Nitric Oxide*, 1: 88–94, 1997.

Wink DA, Kasprzak KS, Maragos CM, Elespuru RK, Misra M, Dunams TM, Cebula TA, Koch WH, Andrews AW, Allen JS. DNA deaminating ability and genotoxicity of nitric oxide and its progenitors. *Science,* 254: 1001–1003, 1991.

Wolburg H, Lippoldt A. Tight junctions of the blood–brain barrier: development, composition and regulation. *Vasc. Pharmacol*, 38: 323–337, 2002.

Wolburg H, Noell S, Mack A, Wolburg-Buchholz K, Fallier-Becker P. Brain endothelial cells and the glio-vascular complex. *Cell Tissue Res*, 335: 75–96, 2009.

Wosik K, Cayrol R, Dodelet-Devillers A, Berthelet F, Bernard M, Moumdjian R, Bouthillier A, Reudelhuber TL, Prat A. Angiotensin II controls occludin function and is required for blood brain barrier maintenance: relevance to multiple sclerosis. *J. Neurosci,* 27: 9032–9042, 2007.

Zanesco A, Antunes E. Células endoteliais. In: Carvalho HF and Collares-Buzato C (Org). Células. 1st ed. Barueri: Manole, 2005.

Zlokovic BV. The blood-brain barrier in health and chronic neurodegenerative disorders. *Neuron*, 57: 178–201, 2008.

In: Endothelium and Epithelium
Editors: J. Carrasco and M. Mota, pp. 61-86
ISBN 978-1-61470-874-2

Chapter III

The Novel Role of Epithelial Innate Immunity in Inflammatory Diseases

De-Quan Li,[1,*] Zhijie Li,[1,2] Zuguo Liu[3] and Guiqiu Zhao[4]
[1]Department of Ophthalmology, Baylor College of Medicine, Houston, TX, USA
[2] Key Laboratory for Regenerative Medicine of Ministry of Education and Department of Ophthalmology, Jinan University, Guangzhou, China
[3]Xiamen Eye Institute, Xiamen University, Xiamen, China
[4]Qingdao University Medical College, Qingdao, China

Abstract

Epithelium is the first line of defense. A long-recognized property of epithelial cells is their physical barrier function against potentially harmful substances and microbial pathogens. Increasing new evidence indicates an important role of epithelial cells as both initiators and regulators of innate and adaptive immune responses, as well as the transition from innate immunity to adaptive immunity. Epithelial cells have been found to functionally express various pattern-recognition receptors, including Toll-like receptors (TLRs), which recognize conserved microbial patterns that trigger inducible activation of innate immunity.

Recent studies have identified two novel epithelium-derived pro-allergic cytokines, thymic stromal lymphopoietin (TSLP) and interleukin (IL) 33, as key initiators of the Th2-dominant responses that cause allergic inflammatory disease such as atopic dermatitis, asthma and ocular allergy. TSLP could activate myeloid dendritic cells to produce OX40 ligand (OX40L) that triggers Th2 response. IL-33 has been recently identified as a functional ligand to ST2 receptor on Th2 cells that mediate allergic inflammation. In ocular surface, TSLP/OX40L/OX40 and IL-33/ST2 signaling has been also found to be significantly activated in atopic patients with allergic conjunctivitis, and in a murine model of experimental allergic conjunctivitis induced by short ragweed pollen, a ubiquitous allergen. Interestingly, TSLP and IL-33 have been observed to be largely induced by specific microbial pathogens, the TLR ligands, such as viral double-

[*]Tel: (713) 798-1123, Fax: (713) 798-1457, E-mail: dequanl@bcm.tmc.edu

stranded RNA (dsRNA) and bacterial component flagellin, via TLR/MyD88, TLR/TRIF and NF-κB pathways. These discoveries suggest that epithelium-derived TSLP and IL-33 serve as a bridge that links innate and adaptive immune responses. TSLP and IL-33 might be the potential biomarkers and novel therapeutic targets for allergic diseases.

Discovery of T helper (Th) 17 cells as a new T helper cell subset distinct from Th1 and Th2 cells is a recent breakthrough in T cell biology. The differentiation of Th17 cells from naïve CD4+ T cells is initiated by transforming growth factor-β (TGF- β) and IL-6; and the further expansion of Th17 cells is induced by cytokines IL-23, IL-1β and IL-21. Interestingly, Th17 cells promote inflammation, autoimmunity and allergy frequently at mucosal epithelial sites that are rich in TGF-β, IL-6, IL-23, and IL-1β, the major inducing cytokines for Th17 lineage. Th17 cells produce IL-17A and IL-17F, as well as IL-22 and chemokine attractant ligand CCL20, the pro-inflammatory cytokines presenting an intriguing new bridge between adaptive and innate immunity. In ocular surface, Th17 differentiation has been shown to be enhanced by cytokines produced by corneal epithelium that are exposed to hyperosmotic stress, inflammatory stimuli and microbial pathogens. Th17 has been identified as a major T cell disorder involved in pathogenesis of human dry eye syndrome, an inflammatory autoimmune ocular surface disease. The increased levels of IL-17-producing Th17 cells and associated cytokines have been observed in human dry eye and murine experimental dry eye model. The desiccating stress has been demonstrated to promote the production of Th17 inducing cytokines by corneal and conjunctival epithelia, and creates an environment priming Th17 differentiation through a dendritic cell-mediated pathway. Th17 signaling molecules become novel therapeutic targets for treating dry eye conditions.

Keywords: Epithelium, Toll-like receptor, innate immunity, thymic stromal lymphopoietin, interleukin 33, Th17

Introduction

Epithelium is the first line of defense. A long-recognized property of epithelial cells is their physical barrier function that defends against exposure to potentially harmful substances and microbial pathogens. Until recently, it has become increasingly clear that epithelial cells play important roles not only in host defense and inflammation, but also in regulation of immune responses. The mammalian immune system is comprised of two branches, the innate immune system and the adaptive immune system, that work in tandem to provide resistance to infection. The innate immune cells, represented primarily by monocytes, macrophages, dendritic cells and granulocytes, are the first line of host defense and are responsible for immediate recognition and control of microbial invasion. In contrast, the adaptive immune system, represented by B and T lymphocytes, has a delayed response, which is characterized by clonal expansion of cells that bind to a highly specific antigen and have immunological memory [1]. Substantial new evidence now indicates that epithelial cells are central participants, as initiators, mediators and regulators, in the innate and adaptive immune responses, as well as in the transition from innate immunity to adaptive immunity.

This new concept has been recently generated from studies mainly in skin and airway epithelial cells. In addition to its function as a physical barrier, human skin has been shown to be an important immune organ displaying various defense mechanisms, which can be divided into three major functional compartments: natural epithelial defense, innate immunity and antigen-elicited adaptive immunity [2-4]. Airway epithelial cells have been recognized to be at the interface of innate and adaptive immunity [5-7]. Airway epithelial cells produce antimicrobial host defense molecules and proinflammatory cytokines and chemokines in response to microbial pathogens. Recruitment of immune cells, including dendritic cells, T cells and B cells into the proximity of epithelium results in the enhancement of adaptive immunity through interactions with epithelial cells. Epithelial cells are also responsible for mucus production in both protective immune responses and allergic airway inflammatory diseases. The crucial roles of epithelial cells in the innate and adaptive immune responses for host defense have also been recognized in other epithelia, including gastrointestinal mucosa [8-10] and ocular surface [11-13]. This chapter is focused on recent advances of this new concept that epithelial cells play a central role in initiating and regulating innate and adaptive immune responses in inflammatory disease based on literature review and our new findings using ocular mucosal epithelium as a model. Advanced understanding on molecular regulation of epithelial immune responses would facilitate new therapeutic strategies for treating inflammatory diseases.

Epithelium and Innate Immune Recognition

The innate immune response relies on evolutionarily ancient germline-encoded receptors, the pattern-recognition receptors (PRRs) [14], which recognize highly conserved microbial structures. PRRs recognize microbial components, known as pathogen-associated molecular patterns. A breakthrough in the understanding of the ability of innate immune system to rapidly recognize pathogens occurred with the discovery of the Toll-like receptors (TLRs), which is the most important family among the PRRs. At least 10 human TLRs have been identified to date. Each TLR has unique ligand specificity. In general, TLRs 1, 2, 4, 5 and 6 present on the cell plasma membrane and respond to a variety of components of bacteria and fungi; and TLRs 3, 7, 8 and 9 mainly present on endosomal membranes inside cells and recognize viral nucleic acids [12]. Structurally, TLRs are type I transmembrane proteins with leucine rich repeats in the extracellular domain for ligand recognition, and a Toll/IL-1 receptor (TIR) domain in the cytoplasmic portion for intracellular signaling [12, 15-17]. TLRs are expressed on immune cells that are most likely to first encounter microbes, such as neutrophils, monocytes, macrophages, and dendritic cells [15]. Ligand recognition by TLRs facilitates the dimerisation of TLRs that triggers the activation of signaling pathways, which originates from the cytoplasmic TIR domain, and culminates in the activation of the nuclear transcription factor NF-κB (nuclear factor kappa-light-chain-enhancer of activated B cells), which leads to the expression of pro-inflammatory molecules [16, 18]. In addition to innate immune cells, an array of TLRs is expressed by epithelial cells at interfaces between host and environment including that of the skin [2, 19], respiratory tract [5, 6], gastrointestinal tract [9], and ocular surface [12, 13]. Strategic expression of TLRs at such host/environment interfaces appears to play an important role in the first line of defense against microbial invasion at these sites.

Ocular mucosal epithelial cells have been identified to express an array of functional TLRs [12, 13]. The production of pro-inflammatory cytokines, chemokines and antimicrobial peptides is stimulated via TLR2 by corneal epithelial cells exposed to yeast zymosan [20] and peptidoglycan of staphylococcus aureus [21]. This pathway may have a role in the pathogenesis of Gram positive bacterial keratitis. Signaling through TLR9 appears important in Pseudomonas aeruginosa keratitis, and silencing TLR9 signaling reduces inflammation, but likely contributes to decreased bacterial killing in the cornea [22]. Stimulation of TLR3 can induce the expression of proinflammatory cytokines, chemokines and antiviral genes that help to defend the cornea against viral infection [23, 24]. The role of ligand-stimulated TLR signaling in epithelium on regulation of innate and adaptive immunity remains to be elucidated.

A Novel Epithelial Cytokine TSLP Links Innate and Adaptive Immunity

Compelling evidence has been recently provided that thymic stromal lymphopoietin (TSLP) represents a key initiator of allergic inflammation at the interface of epithelial and dendritic cells, and TSLP may have a determinant role in the initiation and maintenance of the allergic immune response in atopic dermatitis and asthma [25-29].

Skin-derived TSLP was found to trigger progression from epidermal-barrier defects to asthma, the atopic march, in mice [29, 30]. TSLP is a 140-amino acid IL-7-like 4-helix bundle cytokine that was first isolated from a murine thymic stromal cell line and shown to support B-cell development in the absence of IL-7 [31]. Mouse and human TSLP share a poor homology of 43% amino acid identity.

The human TSLP gene is localized in chromosome 5q22, not far from the gene cluster encoding for all the Th2-related cytokines, IL-4, IL-5 and IL-13 [32]. The TSLP receptor (TSLPR) complex consists of a TSLP binding chain and the IL-7 receptor α chain (IL-7Rα). Like TSLP, human and mouse TSLPR share approximately 40% amino acid identity.

By interacting with the heterodimeric receptor TSLPR/IL-7Rα, TSLP appears to initiate phosphorylation of signal transducer and activator of transcription (STAT) 3 and STAT-5 [25, 33].

It was demonstrated that epithelial cell-derived TSLP could strongly activate human myeloid dendritic cells to be mature dendritic cells that produce OX40 ligand (OX40L) in the absence of IL-12 to induce an inflammatory Th2 response characterized by high TNF-α and low IL-10 production, distinct from the regulatory Th2 responses characterized by low TNF-α and high IL-10 production [25, 26]. This suggests that TSLP represents a key initiator of allergic inflammation at the interface of epithelial cells and dendritic cells. TSLP was also demonstrated to direct the innate phase of allergic immune responses through activating mast cells. Therefore, TSLP and OX40L may represent important targets for intervention in the initiation of allergic inflammatory responses. Epithelial cells appear to be the major potential producer of TSLP in both mice and humans, although fibroblasts, smooth muscle cells, and mast cells all have the potential to produce TSLP [33, 34]. The expression of TSLP by epithelial cells has been recently shown to be stimulated by microbial ligands, inflammatory and Th2 cytokines [35-37]. Using ocular mucosal epithelium as a model, Li and associates [38] have evaluated the expression and production of TSLP by primary human corneal epithelial cells in response to 11 extracted or synthetic microbial components that are ligands of TLRs 1-9 (Table 1).

Table 1. Human toll-like receptors (TLRs) and their ligands

TLR	Principal exogenous ligands	Ligands used for study
TLR1/TLR2	Triacylated lipopeptide (LP)	Pam3CSK4 (1-100 μg/ml)
TLR2*	Bacterial lipoprotein	
	Peptidoglycan,	PGN-BS (1-50 μg/ml)
	lipoteichoic acid	LTA-SA (0.1-10 μg/ml)
	Zymosan (fungi)	Zymosan (1-100 μg/ml)
TLR3	Double stranded RNA (viruses dsRNA)	Poly I:C (5-50 μg/ml)
TLR4	LPS (Gram negative bacteria)	LPS (1-50 μg/ml)
	Bacterial HSP60	
	Respiratory syncytial virus coat protein	
TLR5	Flagellin (flagellated bacteria)	Flagellin-ST (1-25 μg/ml)
TLR6/TLR2	Diacylated lipopeptides	FSL-1 (0.1-10 μg/ml)
TLR7	Imidazoquinolone antiviral drug	Imiquimod (R837, 1-50 μg/ml)
TLR8	Single stranded RNA (viruses ssRNA)	ssRNA40 (0.1-10 μg/ml)
	Imidazoquinolone antiviral drug	
TLR9	Unmethylated CpG motifs of bacterial DNA	C-CpG-ODN (1-50 μg/ml)
TLR10	Unknown	

*TLR2 forms heterodimers with TLR1 and TLR6: TLR1 associates with TLR2 to recognise tri-acyl lipopeptides; TLR6/TLR2 heterodimer recognises diacyl lipopeptides.

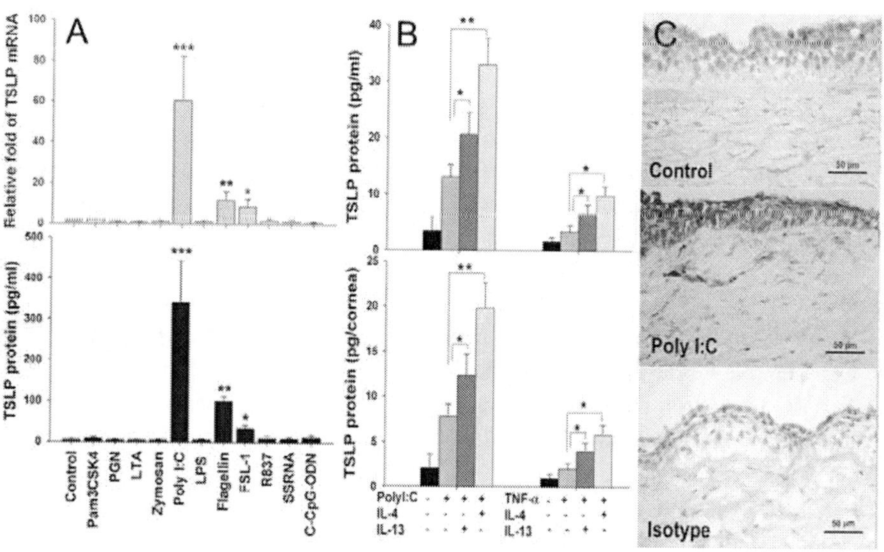

Figure 1. TLR-mediated induction of TSLP by microbial ligands, TNF-α and Th2 cytokines in HCECs. A. The confluent primary HCECs were incubated with 50μg/ml polyI:C or 10μg/ml of Pam_3CSK_4, PGN, LTA, Zymosan, LPS, flagellin, FSL-1, R837, ssRNA40 or C-CpG-ODN for 4 hours for TSLP mRNA expression by RT-qPCR, or for 48 hours for TSLP protein in the culture supernatants by ELISA; B. TSLP induction in an ex vivo model of human corneal tissues. A fresh corneoscleral tissue was cut into 4 equal size pieces. Each quarter of corneal tissue was placed into a well of 8-chamber slides in 150 μl of serum-free SHEM medium without or with polyI:C (50 μg/ml) or TNF-α (20 ng/ml) in the presence of IL-4 or IL-13 (100 ng/ml) for 24 hours. The culture supernatants were used for TSLP ELISA. Results shown are mean ± SD of 3 independent experiments. *, $p < 0.05$; **, $p < 0.01$; ***, $P <$

0.001. C. The corneal tissues were prepared for cryosections for TSLP immunohistochemical staining with isotype IgG as negative control.

As shown in Figure 1A, TSLP expression and production were found to be largely induced by polyI:C, Flagellin and FSL-1, the ligands for TLRs 3, 5 and 6, representing viral dsRNA and the bacterial components flagellin and lipopeptides, respectively. Polyinosine-polycytidylic acid (polyI:C) and flagellin were major TSLP inducers, stimulating TSLP production to 67- and 19-fold, respectively.

The TSLP mRNA reached the peak levels rapidly in 4 hours in response to these ligands. The specificity of this response was confirmed when respective antibody against TLR3 or TLR5 significantly blocked TSLP expression induced by polyI:C and flagellin, respectively. The pattern of TLR-dependant TSLP induction indicates that human corneal epithelial cells are able to rapidly initiate an innate immune response to virus or bacteria through TLR-mediated pathways.

TSLP was also moderately induced by pro-inflammatory cytokines at both mRNA and protein levels. TNF-α or IL-1β induced a concentration-dependent increase in the TSLP mRNA and protein, but their stimulatory effects were much weaker than that of polyI:C and flagellin, which stimulated TSLP protein production over 15- and 4-fold higher, respectively, than TNF-α.

These data suggest that TSLP induction is mainly through a TLR-dependant innate immune response to microbes in human ocular epithelium. Interestingly, IL-4 and IL-13, the major cytokines secreted by Th2 cells, not only moderately induced TSLP mRNA and protein, but also strongly synergized with microbial ligands, such as polyI:C, or pro-inflammatory cytokine TNF-α to promote TSLP expression and production (Figure 1B).

This synergized induction of TSLP was further confirmed in an ex vivo experiment model using fresh human corneal epithelial tissues (Figure 1C). These findings demonstrate that adaptive immunity-derived Th2 cytokines are capable of amplifying the TSLP expression and production by the corneal epithelium, which in turn has the capability of priming Th2 cell differentiation through dendritic cell activation [25, 26]. These findings suggest that blocking TSLP could be a novel strategy for treatment of allergic diseases or other TSLP-driven conditions.

Although the recognition of different ligands by specific TLRs leads to activation of an intracellular signaling cascade in a MyD88 dependent or independent fashion, all TLRs share NF-κB signal transduction pathways for activation of the transcription factors [18]. TNF-α is also well known to promote activation of the NF-κB signaling pathway [39]. TSLP induction has been observed through NF-κB activation in airway epithelial cells [28] and synovial fibroblasts [40].

As evaluated by Western blot analysis and immunofluorescent staining, NF-κB was found to be dramatically activated with p65 protein nuclear translocation in corneal epithelial cells exposed to polyI:C, flagellin or TNF-α for 4 hours. This p65 nuclear translocation and TSLP induction, stimulated by polyI:C, flagellin or TNF-α were markedly blocked by TLR3, TLR5 or TNF-α specific antibody, respectively, and also by quinazoline, a NF-κB activation inhibitor [38].

These findings confirm that TSLP is mainly induced by microbial components, proinflammatory cytokines and Th2 cytokines in human corneal epithelial cells via TLR and

NF-κB signaling pathways, suggesting that epithelium-derived TSLP links the innate and adaptive immune responses,

TSLP/OX40L/OX40 Signaling Pathways in Th2-Dominant Allergic Disease

TSLP has been identified as a key initiator in the development of human allergic disease [29, 41-43], including Asthma, atopic dermatitis and allergic conjunctivitis, a triad of common atopic IgE-dependent allergic diseases [44]. The direct link between TSLP expression and the pathogenesis of atopic dermatitis and asthma in vivo has been demonstrated [27]. TSLP was found to be highly expressed by keratinocytes in skin lesions of atopic dermatitis and was associated with dendritic cell activation in situ [45]. Evidence associating TSLP with human asthma has also been reported [27, 30]. Patients suffering from one member of the triad often show symptoms of one or both of the other members, suggesting a common genetic or initiating element in these diseases [29]. Using a well characterized murine model of experimental allergic conjunctivitis (EAC) induced by short ragweed (SRW) pollen [46, 47], Li and colleagues observed that the repeated topical challenges with ragweed pollen allergen generated typical signs of allergic conjunctivitis in the pollen-sensitized BALB/c mice, which developed lid edema, conjunctival redness, chemosis, and tearing, as well as frequent scratching of the eye lids [48]. They found that TSLP mRNA expression was significantly upregulated in the corneal and conjunctival epithelia from mice sensitized and challenged with pollen when compared with PBS alone and untreated normal controls.

Immunohistochemical staining confirmed an increase in TSLP production in the eyes challenged with SRW pollen. As shown in Figure 2A, the corneal and conjunctival epithelia in EAC mice displayed much stronger TSLP staining throughout the entire epithelium, especially the superficial epithelial layers of the conjunctiva, than the PBS-treated control. These data indicate the stimulated TSLP mRNA expression and protein production by ocular surface epithelia in the SRW-induced EAC murine model.

The accumulation of CD11c positive ($CD11c^+$) dendritic cells on the ocular surface was detected in this EAC model by RT-qPCR and immunostaining. The increased mRNA levels of CD11c, TSLPR and OX40L were observed in ocular surface, especially in conjunctival tissues, where their transcripts increased to 4-10 fold, respectively ($P< 0.05$ or <0.01), from SRW pollen-challenged mice, compared with PBS controls.

The large amount of CD11c+ dendritic cells was accumulated in the ocular surface of the pollen challenged eyes, primarily in the stroma subjacent to the conjunctival epithelia, in EAC mice, but not in the control mice treated with PBS (Figure 2A). These results suggest that ocular surface was infiltrated with TSLP activated dendritic cells that express TSLPR and produce OX40L in the EAC mice.

Th2-dominant inflammatory response was clearly observed on ocular surface in the mice challenged by pollen. The infiltration of T lymphocytes was evidenced by the increased CD4 mRNA expression and a markedly increased number of CD4 immunopositive ($CD4^+$) cells in the ocular surface, especially in the conjunctiva, of EAC mice (Figure 2A), when compared with PBS control mice. These $CD4^+$ T cells appear to be Th2 lineage because the transcripts of three key Th2 cytokines, IL-4, IL-5 and IL-13, were all found to be expressed at significantly higher levels in corneal and conjunctival tissues from the EAC mice than the PBS controls. Immunostaining data confirmed that the IL-4, IL-5 and IL-13-producing Th2 cells were largely infiltrated in conjunctival stroma (Figure 2A).

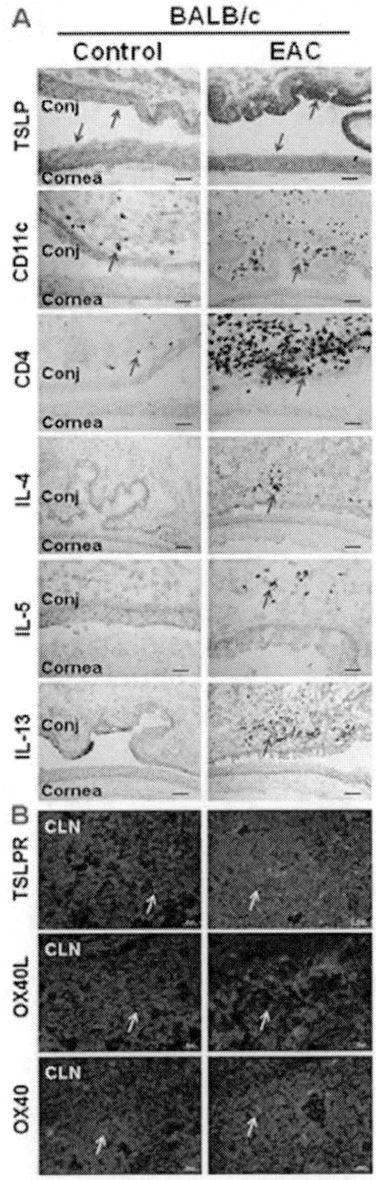

Figure 2. The stimulated production of TSLP signaling proteins and Th2-dominant inflammation in an experimental allergic conjunctivitis (EAC) model in BALB/c mice. A. The representative images showing immunohistochemical staining of epithelial TSLP, markers for dendritic (CD11c) and T cells (CD4), and Th2 cytokines (IL-4, IL-5 and IL-13) on cornea and conjunctiva (Conj) of EAC and PBS-treated control mice. B. The representative images showing immunofluorescent staining of TSLP activated signals, TSLPR, OX40L and OX40 on cervical lymph nodes (CLN) of EAC and control mice. Arrows: red or red brown positive staining signals; Bar: 20μm.

To confirm the TSLP signaling in SRW pollen-induced EAC mice, the ocular surface draining cervical lymph nodes (CLN) were collected for evaluation. Compared with PBS controls, the mRNA levels of TSLPR and OX40L were significantly stimulated to 7.5 and 4.1 fold respectively (both $P<0.01$) while CD11c expression only slightly increased in CLN from EAC mice, indicating $CD11c^+$ DCs were markedly activated by ocular surface epithelia-derived TSLP. Interestingly, the mRNA levels of OX40 (3.4 fold, $P<0.05$) and Th2 cytokines, IL-4, IL-5 and IL-13 (13.2, 12.8 and 5.5 fold, respectively, all $P<0.01$), significantly increased while CD4 expression was not changed in the CLN from EAC mice, indicating naive $CD4^+$ T cells were largely differentiated to Th2 cells that might be primed by OX40L produced by TSLP activated dendritic cells [26]. The increased $TSLPR^+$, $OX40L^+$ and $OX40^+$ cells in the draining CLN of the pollen challenged EAC mice were further confirmed by immunofluorescent staining that showed dramatically increased immunoreactivity of these 3 signaling proteins at cell membrane and cytoplasm in CLN (Figure 2B). All these results demonstrate that TSLP/OX40L/OX40 signaling plays a critical role in the development of Th2-dominant allergic inflammation in pollen induced EAC model in BALB/c mice, suggesting that TSLP signaling molecules could be novel therapeutic targets to treat allergic inflammatory disease.

Epithelium-Derived Interleukin 33 Initiates Allergic Inflammation

The IL-1 receptor family has several members, including the classical IL-1 receptor (IL-1R) and the IL-18 receptor (IL-18R). In 1989, one member of the family, ST2, a protein encoded by IL-1 receptor-like 1 (IL-1RL1) gene, was identified as an orphan receptor [49]. Investigation into the function of ST2 revealed its participation in inflammatory processes, particularly regarding mast cells, type 2 $CD4^+$ T helper cells and the production of Th2-associated cytokines. In fact, ST2 was characterized as a specific cellular marker that differentiated Th2 from Th1 T cells. Clinical and experimental observations led to the association of ST2 with disease entities such as asthma, pulmonary fibrosis, rheumatoid arthritis, collagen vascular diseases and septic shock [50]. In 2005, the discovery of IL-33 as a ST2 ligand provided new insights into ST2 signalling [51]. By binding to ST2 receptor, IL-33 can activate Th2 cells and mast cells to secrete the proinflammatory and Th2 cytokines and chemokines that lead to severe pathological changes in mucosal organs [52]. IL-33 is produced mainly by epithelial and endothelial cells, fibroblast, and others [52-54]. IL-33 expression has been found to be up-regulated by stimulation with inflammatory cytokines, TNF-α and IL-1β [55]. However, the expression and regulation of IL-33 by mucosal surface epithelia has not been well elucidated. Using fresh donor corneal tissues and primary human corneal epithelial cells (HCECs), we recently observed that IL-33 is mainly expressed by epithelium and largely induced by microbial products through TLR and NF-κB signaling pathways. The findings suggest that the mucosal epithelial cell-derived cytokine IL-33 may play an important role in allergic inflammatory diseases through innate immune responses.

As shown in Figure 3A, IL-33 expression and production were largely induced by polyI:C, lipopolysaccharides (LPS), flagellin, FSL-1 and R837, the ligands for TLR3, -4, -5, -6 and -7, representing viral dsRNA and the bacterial components flagellin and lipopeptides, respectively. PolyI:C and flagellin were major IL-33 inducers, stimulating IL-33 production by 7- and 4-fold, respectively, with the peak mRNA levels at 8 hours by HCECs. The IL-33 induction by these 2 ligands was further confirmed using an ex vivo donor corneal tissue model (Figure 3B). The specificity of TLR-dependent response by HCECs was also confirmed when an antibody against TLR3 or TLR5 significantly inhibited IL-33 expression by polyI:C or flagellin respectively (Figure 3C). The pattern of TLR-dependent IL-33 induction indicates that HCECs are able to rapidly initiate an innate immune response to virus or bacteria, and play an important role in allergic inflammatory disease.

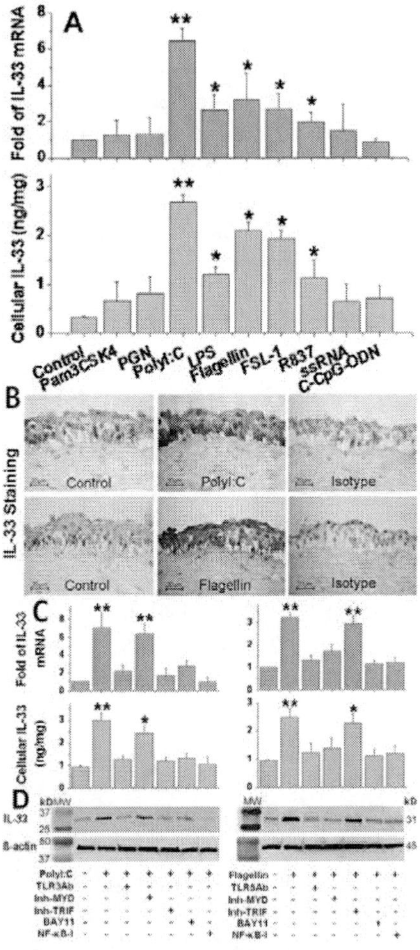

Figure 3. TLR-dependent induction of IL-33 by microbial ligands in human corneal epithelium. A. IL-33 mRNA and protein levels induced by primary human corneal epithelial cells (HCECs) exposed to 50μg/ml polyI:C or 10μg/ml of Pam3CSK4, PGN, polyI:C, LPS, flagellin, FSL-1, R-837, ssRNA40 or C-CpG-ODN for 8 or 48 hours, evaluated by quantitative real-time PCR or ELISA, respectively. Results shown are the mean ± SD of four independent experiments. $^*P< 0.05$; $^{**}P < 0.01$. B. The immunohistochemical staining showing IL-33 induction in an ex vivo human corneal tissues by polyI:C (50μg/ml) or flagellin (10μg/ml) for 24 hours with isotype IgG as a negative control. C. TLR and NF-κB signaling pathways involved in IL-33 induction by polyI:C or flagellin in HCECs exposed to polyI:C (50μg/mL) or flagellin (10μg/mL) in the absence or presence of preincubated rabbit TLR3Ab (10μg/mL), TLR5Ab (10μg/mL), BAY11-7082 (10μM) or quinazoline (10μM) for 1 hour, and Pepinh-MYD (40μM) or Pepinh-TRIF (40μM) for 6 hours. The cultures treated by ligands for 8 hours for IL-33 mRNA, or for 48 hours for IL-33 protein by ELISA and by Western blot with β-actin as control (D). Results shown are the mean ± SD of four independent experiments. $^*P< 0.05$; $^{**}P < 0.01$.

IL-33 is an extracellular inflammatory cytokine while it also acts as a nuclear transcription factor [56]. It has been shown that IL-33 is mainly localized to the nucleus of endothelial cells and bound to chromatin. IL-33 mRNA is primarily translated and synthesized in vivo as a 30-kDa precursor, a pro-IL-33 protein. As a member of IL-1 super family and like IL-1 and IL-18, pro-IL-33 protein sequence does not have a signal peptide that directs it for secretion via the ER–Golgi pathway [52, 57]. Like pro-IL-1β, human pro-IL-33 was reported to be cleaved by caspase-1 to generate active form, an 18-kDa fragment (mature IL-33), which is sufficient to activate signaling via the IL-33 receptor ST2. Recombinant mature IL-33 has been known to induce Th2-associated cytokines and inflammatory cytokines via its receptor, ST2 [57, 58]. However, processing of pro-IL-33 in vivo has not been clarified yet. It is not clear whether caspase-1 cleavage of pro-IL-33 occurs in vivo and whether, as for IL-1β, this cleavage is a prerequisite for IL-33 secretion and bioactivity. Our data have showed that no significantly changes of IL-33 protein levels can be detected by ELISA in the culture supernatants, other than in cell lysate, of the HCECs. IL-33 protein levels significantly increased in the cell lysate, but not in the culture supernatants, of HCECs exposed to polyI:C or flagellin for 24-48 hours, when compared with the untreated control. However, the stimulated cellular IL-33 protein was released outside the cells into culture supernatant after co-incubation with ATP for additional 30 minutes, as evaluated by ELISA and Western blot analysis. The finding supports a notion that caspase 1-dependant activation may involved in the release and secretion of IL-33 protein since ATP has been known to activate caspase-1 through triggering the P2X7 receptor [59]. Further studies are necessary to clarify the underlining mechanism. As shown in Figure 3C evaluated by RT-qPCR and ELISA, as well as in Figure 3D by Western blotting, synthetic dsRNA polyI:C-induced IL-33 mRNA expression and protein production were markedly blocked by TLR3 antibody and TRIF inhibitory peptide (Pepinh-TRIF), but not by MyD88 inhibitory peptide (Pepinh-MYD), while extracted bacterial component flagellin-induced IL-33 production was dramatically suppressed by TLR5 antibody and Pepinh-MYD, but not by Pepinh-TRIF, in corneal epithelial cells. Interestingly, the stimulated IL-33 induction by polyI:C or flagellin were also significantly blocked by IκB-α inhibitor BAY11-7082 or NF-κB activation inhibitor quinazoline. Further study has shown that NF-κB was dramatically activated with p65 protein nuclear translocation in corneal epithelial cells exposed to polyI:C or flagellin for 4 hours, as evaluated by Western blot analysis and immunofluorescent staining. BAY 11 selectively inhibits the phosphorylation and degradation of IκB-α, blocked the nuclear translocation of NF-κB p65 protein. NF-κB activation inhibitor quinazoline also blocked the p65 nuclear translocation. These findings demonstrate a novel phenomenon that a newly defined pro-allergic cytokine IL-33 is largely induced by microbial components through TLR and NF-κB signaling pathways in human corneal epithelium. This suggests that human ocular mucosal epithelium plays an important role in initiating Th2-dominant allergic inflammation via innate immune responses.

In a clinical study, we have tested conjunctival impression cytology specimens obtained from 8 patients with active atopic conjunctivitis and 8 normal subjects by RT-qPCR. The mRNA levels of TSLP, TSLPR, OX40L, OX40, IL-33 and ST2 were found to be significantly elevated in the atopic group compared to the normal control subjects, suggesting a potential role of TSLP/TSLPR/OX40L/OX40 and IL-33/ST2 signaling pathways in allergic conjunctivitis (Figure 4A).

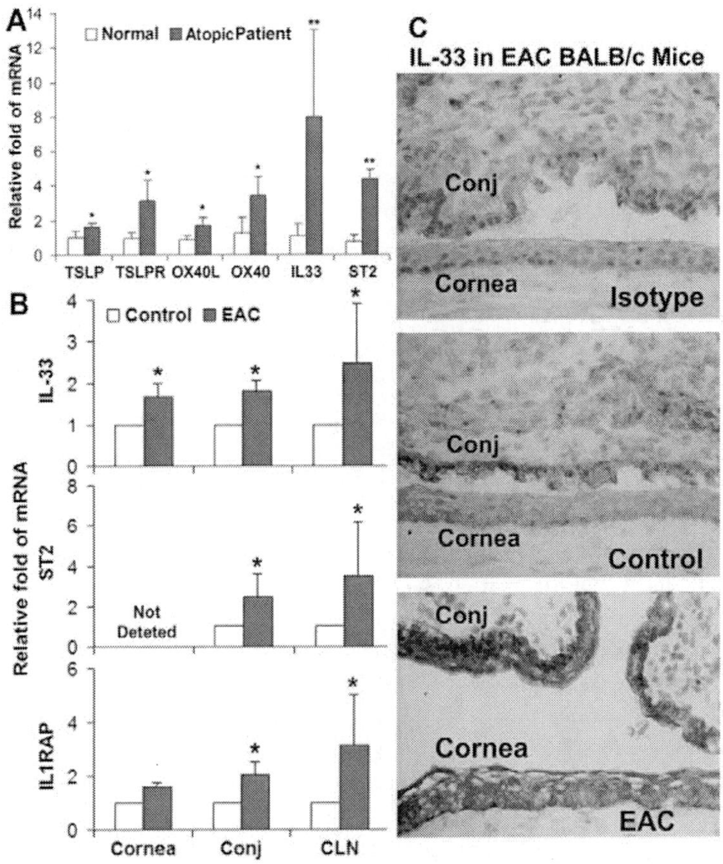

Figure 4. The role of IL-33 in allergic disease. A. Elevated mRNA levels of TSLP and IL-33 signaling molecules in conjunctiva of atopic patients compared with normal subjects by RT-qPCR. * $P<0.05$, ** $P<0.01$, n=8. B. The mRNA levels of IL-33, ST2 and IL1RAP in corneal epithelium (Cornea), conjunctiva (Conj) and cervical lymph nodes (CLN) of EAC in BALB/c mice with PBS-treated mice as controls. * $P<0.05$, n=3. C. Immunostaining images showing the stimulated IL-33 in ocular surface of EAC BALB/c mice with untreated mouse control and isotype IgG negative control.

In SRW pollen induced EAC mice, the transcripts and proteins of IL-33, ST2, and IL-1 receptor accessory protein (IL1RAP) were also found to be significantly increased in the corneal epithelium, conjunctiva and CLN, as evaluated by RT-qPCR (Figure 4B) and immunostaining (Figure 4C). IL-33, ST2 and TLRs could become novel biomarkers and molecular targets for the intervention to treat allergic inflammatory diseases.

Th17 Pathway Links Innate and Adaptive Immunity

Th17 has been recently identified as a new T helper cell subset. $CD4^+$ T helper (Th) cells now include three different types based on their cytokine signatures: interferon-γ (IFN-γ)-secreting Th1, IL-4, -5, and -13-secreting Th2 [60], and IL-17-producing Th17 cells [61].

Th17 cells have been recognized to be key effector T cells in a variety of human inflammatory and autoimmune diseases as well as experimental animal models [62, 63]. The IL-17 family includes 6 members (IL17A-F). IL-17A (also known as IL-17) and IL-17F are the founding members of the IL-17 cytokine family. The genes encoding IL-17A and IL-17F are localized in the same chromosomal region in mice and in humans. But IL-17F has significantly weaker biological activity than IL-17A [64, 65]. IL-17E, also known as IL-25, is produced by Th2 cells and mast cells. In contrast, IL-17B, IL-17C and IL-17D have not been well investigated [66, 67]. The receptor for IL-17A (IL-17R or IL-17RA) is a single-pass transmembrane protein of approximately 130 kDa. Four additional receptors (IL-17RB-RE) have been identified, but are not well characterized. IL-17RC was recently identified to be a receptor for IL-17F [68]. It has been reported that Th17 cells also produce IL-22 [69, 70] and CC-chemokine attractant ligand 20 (CCL20) [71] in mice and humans. Therefore, distinct from Th1 and Th2 cells, Th17 cells produce a unique and expanding array of pro-inflammatory cytokines.

Compelling evidence has demonstrated that the differentiation of Th17 cells from naïve $CD4^+$ T cells is initiated by cytokines IL-6 or TGF-β, and expanded by cytokines IL-23, IL-1β and IL-21. TGF-β and IL-6 are dominant in the initiation of Th17 cell differentiation although IL-6 or TGF-β as a single initiator was proposed to be necessary for antigen specific Th17 differentiation [72-74]. IL-23 was the first cytokine shown to selectively regulate IL-17 expression [75, 76], but it might not be required for the initial differentiation of Th17 cells in vivo [77]. Recently, IL-1β was found to promote Th17 cell development and proliferation in the presence of TGF-β and IL-6 [73]. IL-21, produced by activated T cells and natural killer (NK) cells [78], may be required for full commitment of Th17 cells [79, 80]. Hence IL-23, IL-1β and IL-21 may possibly maintain and expand the differentiated Th17 cells in the presence of IL-6 and TGF-β [73]. Furthermore, STAT3 has been found to mediate the initiation of Th17 cell differentiation by these inducing cytokines [72]. Activation of STAT3 induces the expression of retinoic-acid-receptor-related orphan receptor-α (RORα) and RORγt [81], two transcription factors that promote the Th17-cell-associated gene-expression program, leading to the production of IL-17, IL-17F, IL-22 and CCL20

Using peripheral $CD4^+$ T cell isolated from mouse spleen and cervical lymph nodes, our team evaluated the differential effects of these inducing cytokines in promoting Th17 differentiation. The results showed that IL-6 and TGF-β1, only minimally induced IL-17 production at both mRNA and protein levels. In the presence of IL-6 and TGF-β1, IL-23 was the strongest stimulator of the Th17 signature cytokines IL-17A and IL-17F, IL-22, and chemokine CCL20, as well as STAT3 among the 3 expanding cytokines IL-23, IL-1β and IL-21. Interestingly, in the 4 cytokine system, IL-1β stimulated much higher levels of IL-17 family cytokines, 1.5-2 fold greater than IL-21 in the presence of TGF-β1, IL-6 and IL-23. These findings suggest that TGF-β1 and IL-6 initiate low level differentiation of Th17 cells; and their maintenance and development need other expanding factors, among which IL-23 plays a potent role and IL-1β amplifies this expansion further in Th17 differentiation [82].

A variety of mucosal epithelia have been found to produce Th17 inducing cytokines, including TGF-β1, IL-6, IL-23, and IL-1β [83, 84]. IL-1β is a well recognized proinflammatory cytokine produced by mucosal epithelia in response to stress, infection or wounding. In an attempt to mimic known stressors of the ocular surface, we measured production of Th17 inducing cytokines in cultured HCECs in response to hyperosmotic stress, microbial components and inflammatory cytokines.

The ocular surface epithelium is subjected to hyperosmotic stress in dry eye conditions. Exposure of human epithelial cells to hyperosmotic stress has been noted to activate mitogen-activated protein kinase pathways and stimulate production of pro-inflammatory cytokines, such as IL-1β, TNF-α, and IL-8 [85]. We have observed that in addition to IL-1β, TGF-β1, IL-6 and IL-23 were highly induced in the corneal epithelium in response to hyperosmotic stress.

We evaluated the expression and production of Th17-inducing cytokines by HCECs in response to 9 extracted or synthetic microbial components that are ligands of TLRs 1-9, respectively. TGF-β1, IL-6, IL-23, and IL-1β expression and production were found to be largely induced by polyI: C, flagellin, and R837, the respective ligands for TLRs3, 5 and 7, representing viral or bacterial infections. Among these TLR agonists, polyI:C was the strongest stimulator of Th17 inducing cytokines by HCECs.

Hyperosmotic stress and microbial components also promoted production of pro-inflammatory cytokines including TNF-α, which plays an important role in ocular surface disease [85, 86]. Consequently, TNF-α stimulus was found to induce TGF-β1, IL-6, IL-23, and IL-1β markedly.

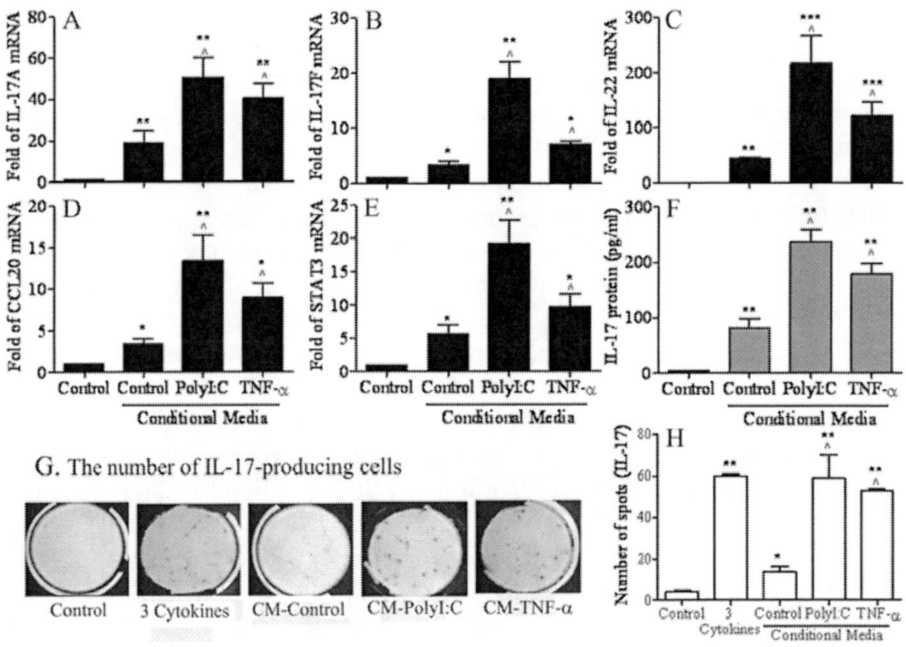

Figure 5. Induction of Th17 differentiation of murine CD4$^+$ T cells cultured in RPMI media containing 50% conditioned media (CM) of HCECs. A-E. Real-time PCR data showing the relative fold of mRNA of Th17 associated cytokines (IL-17A, IL-17F, IL-22, CCL-20) and regulator STAT3 in CD4$^+$ T cells incubated for 4 days with 50% of conditioned media of HCECs irritated by polyI:C (CM-PolyI:C) or TNF-α (CM-TNF-α) for 48 hrs. F. Luminex immunobead assay showing IL-17 concentration in the supernatant of CD4$^+$ T cells receiving the same treatment for 4 days. G and H. ELISPOT bioassay showing the spots/3x10^5 cells/well, representing the numbers of IL-17-producing T cells, in CD4$^+$ T cells treated with CM-PolyI:C, CM-TNF-α, or 3 cytokines (TGF-β1+IL-6+IL-23) for 7 days. Results shown are mean ± SD of 3-5 independent experiments. *, $P < 0.05$; **, $P < 0.01$; ***, $P < 0.001$ each treated groups vs. media control. ^, $P < 0.05$; ^^, $P < 0.01$; ^^^, $P < 0.001$ CM-PolyI:C or CM-TNF-α groups vs. CM-control group.

Based on our findings that among TLR ligands, polyI:C is the potent stimulator of Th17 inducing cytokines, and that TNF-α is a representative pro-inflammatory factor, we evaluated the Th17 inducing capacity of conditioned media (CM) of HCECs treated with polyI:C and TNF-α. It has been observed that Th17 cell differentiation was significantly stimulated in $CD4^+$ T cells exposed to the 50% conditioned media of HCECs challenged by polyI:C (CM-polyI:C) or TNF-α (CM-TNF-α), when compared with media from unstimulated cultures.

As shown in Figure 5A-E, the mRNA levels of IL-17A, IL-17F, IL-22, CCL-20 and STAT3 were significantly higher in $CD4^+$ T cells treated with CM-polyI:C or CM-TNF-α for 4 days compared with the control medium or conditioned media of HCEC culture without any stressors (CM-Control, all $P<0.05$, n=3). IL-17 protein levels in the supernatants of $CD4^+$ T cells exposed to CM-polyI:C or CM-TNF-α for 4 days (Figure 5F) were also significantly higher than the media (both $P<0.01$, n=3) and CM ($P<0.05$, n=3) controls. Furthermore, the number of IL-17-producing cells differentiated from $CD4^+$ T cells, determined by ELISPOT bioassay (Figure 5G, 5H), displayed the same pattern to the induction of IL-17 mRNA and protein. The numbers of IL-17-producing cells were stimulated by CM-polyI:C or CM-TNF-α, to the levels similar to that seen in the 3-cytokine system (TGF-β1+IL-6+IL-23), suggesting that cytokines in the conditioned media of HCECs exposed to polyI:C or TNF-α were capable to promote Th17 cell expansion to levels induced by IL-6+TGF-β1+IL-23. It suggests that the Th17 cells can be indeed promoted by factors produced by corneal epithelium in response to a variety of inflammatory stimuli [82]. Th17 differentiation was also found to be mediated through a dendritic cell-mediated pathway. Dendritic cells have an important function in Th17 cell differentiation. They are antigen-presenting cells specialized to activate $CD4^+$ T cells and through their interaction with $CD4^+$ T cells to initiate primary immune responses. Furthermore, when primed, certain DCs express a high-level of Th17 inducing cytokines, including IL-6, TGF-β, IL-23 and IL-1 [87].

Th17-mediated Inflammation in Dry Eye

Dry eye is the second most common problem of patients seeking eye care, and is characterized by eye irritation symptoms and blurred vision. The prevalence of dry eye increases with age, 6% at the age of 40, and 15-25% in the population over the age of 65. Among dry eye patients, 11% have been estimated to have the systemic autoimmune condition Sjögren's syndrome, a severe and potentially blinding condition. Dry eye is a potent stimulus of both innate and adaptive immune systems. At the nexus of the dry eye inflammatory/immune response is the dynamic interplay between the ocular surface epithelia and bone marrow derived immune cells. On one hand, the ocular surface epithelial cells play a key initiating role in this inflammatory reaction, while on the other hand they are targets of cytokines that are produced by activated T cells that are recruited to the ocular surface in response to dry eye. Dry eye has been demonstrated to cause inflammation on the ocular surface, evidenced by increased levels of inflammatory cytokines (IL-1, IL-6, and TNF-α) in the tear fluid, corneal and conjunctival epithelium, and an increased infiltration of dendritic cells and T lymphocytes in the conjunctiva [88-91]. Recently, increased levels of IL-17, IL-23 and IL-6 were found in saliva and salivary glands biopsies obtained from patients with the severe autoimmune dry eye condition, Sjögren's syndrome [92]. The increased matrix metalloproteinase (MMP) 9 and disrupted barrier function were observed in human [93] and murine dry eye [94]. Recently, our group has found increased expression of Th17 associated cytokines and IL-17-producing cells in human and experimental murine dry eye [94, 95]. Increased expression of Th17 cytokine IL-17A was observed in corneal and conjunctival epithelia of the dry eye mice.

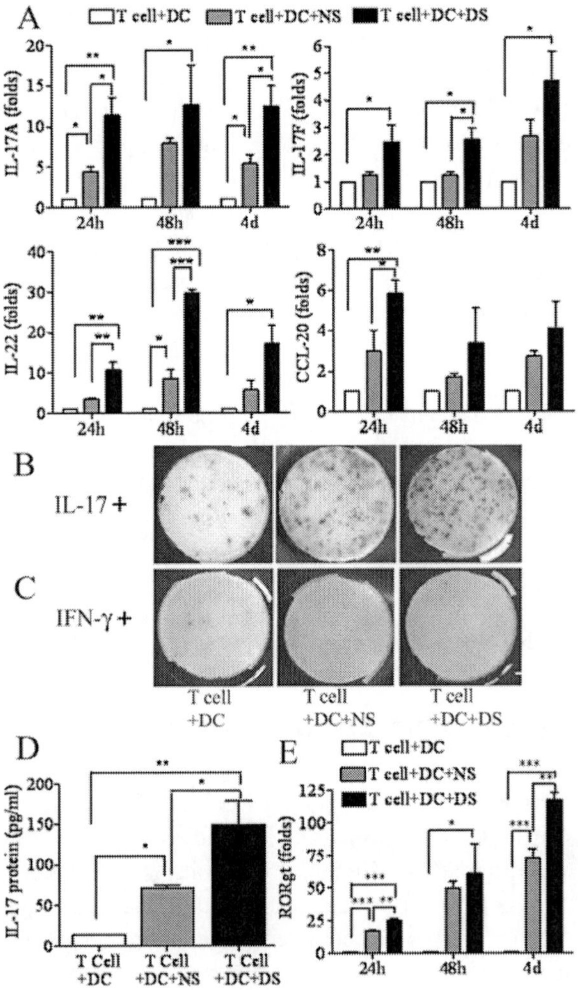

Figure 6. Th17 differentiation of CD4$^+$ T cells induced by the co-culture with dendritic cells (DCs) in the presence of cornea and conjunctival tissues of C57BL/6 mice subjected to desiccating stress. A. Real-time PCR data showing the relative mRNA expression (*x*-fold) of Th17 cytokines (IL-17A, IL-17F, IL-22, CCL-20) in CD4$^+$ T cells (3x10^{5}cells/ well) co-cultured for 1, 2, 4 days with DCs (T cell+DC), or with DCs and cornea and conjunctival explants from non-stressed control mice (T cell+DC+NS) or from mice desiccating stressed for 10-day (T cell+DC+DS), (n=4). B and C. ELISPOT bioassay showing the numbers of IL-17 or IFN-γ- producing cells in these 3 groups. D. IL-17 concentration in the supernatant of CD4$^+$ T cells co-cultured with DCs for 4 days in absence or presence of corneal and conjunctival explants from non-stressed and 10 day desiccating stressed mice. E. Real-time PCR showing the relative mRNA expression of RORγt in CD4$^+$ T cells co-cultured for 1, 2, 4 days in 3 conditions. Data are presented as mean ± SD of 3 or 4 independent experiments. *P<0.05, **P<0.01, ***P<0.001.

Since IL-17A is produced by T cells, not by epithelial cells, the Th17 reaction on the ocular surface is likely due to $CD4^+$ T cells, which have previously been found to infiltrate ocular surface tissues following experimental desiccating stress [96, 97]. Antibody neutralization of IL-17 ameliorated experimental dry eye-induced corneal epithelial barrier dysfunction and decreased the expression of MMP-3 and -9 [94]. These findings provide clear evidence that changes in the ocular surface environment following desiccating stress are capable of inducing Th17 differentiation [98], which plays an important role in dry eye disease.

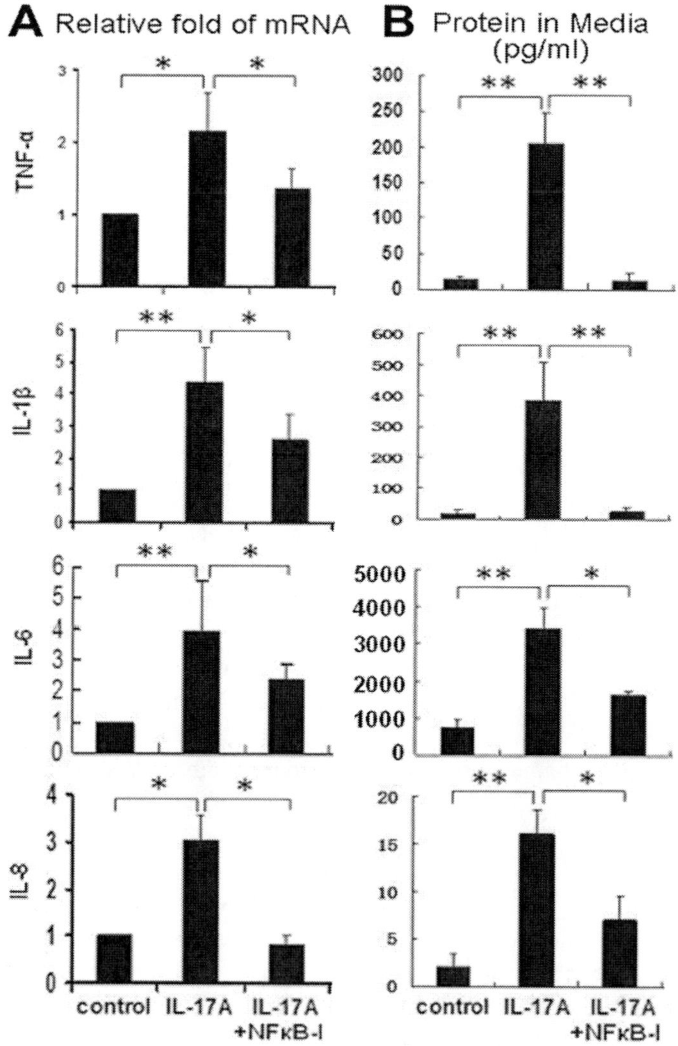

Figure 7. The inflammatory effects of IL-17 on HLECs. Primary HLECs (5×10^5 cells/well) were treated with recombinant human IL-17A at 10 ng/ml for 4-48 hours with or without NF-κB activation inhibitor quinazoline (NFκB-I, 5μM). The pro-inflammatory cytokines (TNF-α, IL-1β and IL-6) and chemokine IL-8 were measured by RT-qPCR for mRNA (A) and by ELISA or Luminex immunobead assays in culture supernatants (B). Results shown are mean ± SD of 4 independent experiments. * $P < 0.05$, ** $P < 0.01$.

We found that efficient differentiation of $CD4^+$ T cells to IL-17 producers required the combination of ocular surface explants from dry eye mice and dendritic cells. $CD4^+$ T cells that were co-cultured with ocular epithelial explants from desiccating stress-induced dry eye mice and dendritic cells were found to express increased mRNA levels of Th17 cytokines (IL-17A, IL-17F, IL-22) and chemokine CCL-20 transcripts (Figure 6A) and to produce and release IL-17A (Figure 6B and 6D), but not Th1 cytokine IFN-γ (Figure 6C). Exposure of dendritic cells to conditioned media from ocular surface explants of dry eye mice did not sufficiently activate these cells to promote T cell differentiation. The possible explanations for these findings include the need for direct contact between these cells, more efficient activation of cytokines such as TGF-β1 produced by the ocular surface epithelial cells or insufficient concentrations of Th17 inducing factors in the explants conditioned media. The transcription factor-RORγt was identified as a candidate master regulator that drives Th17 cell lineage differentiation [81]. Expression of RORγt is induced by TGF-β or IL-6, and overexpression of RORγt was found to promote Th17-cell differentiation when both Th1- and Th2-cell differentiations were blocked. In a model of experimental autoimmune encephalomyelitis, mice with RORγt-deficient T cells were found to have attenuated autoimmune disease and lacked tissue-infiltrating Th17 cells [81]. We found robust up-regulation (up to 100 fold) of the level of Th17 cell transcription factor-RORγt in T cells co-cultured with desiccated ocular surface tissues and dendritic cells (Figure 6E). This provides further evidence of the potent Th17 prone environment induced by desiccation.

IL-17 producing T cells are distinct from Th1 cells [73]. In vitro stimulation of YFP-negative sorted naive $CD4^+$ T cells from Yeti mice indicated that differentiation to IL-17 progresses without opening of the IFN-γ locus. Analysis of the expression of transcription factors showed clearly that IL-17-producing T cells express neither GATA-3 nor T-bet and its target Hlx [73]. In the T cells co-cultured with desiccated ocular surface tissues and dendritic cells, we observed the lower expression of Th1 associated factors (IL-2, T-bet) and Th2 associated factors (IL-4, IL-13, GATA-3). There was no change in production of IFN-γ and IL-12 transcripts as well as in the number of IFN-γ-producing $CD4^+$T cells in this coculture system. Taken together, these findings indicate that desiccating stress may selectively promote the Th17 pathway, a finding that is consistent with the increased level of IL-17 in dry eye disease [94].

IL-17 initiates pro-inflammatory effects by binding to the IL-17 receptor (IL-17R), which is expressed by a variety of cell types including epithelial, endothelial, and fibroblastic stromal cells [66, 99]. As shown in Figure 7, recombinant human IL-17A (10 ng/ml) significantly increased mRNA levels (2-4 fold) of proinflammatory cytokines (TNF-α, IL-1β and IL-6) and chemokine IL-8 expressed by HCECs. These stimulatory responses to IL-17A were confirmed by 4-7 fold increases at protein levels. Interestingly, stimulated production of these inflammatory cytokines and chemokine was significantly suppressed at both mRNA (all $P<0.05$, n=4) and protein levels ($P<0.05$ or 0.01) by NF-κB activation inhibitor quinazoline [100], indicating NF-κB pathway is involved in the inflammatory effect of IL-17 on mucosal epithelium.

These findings demonstrate that desiccating stress stimulates the expression and production of Th17 inducing cytokines by corneal and conjunctival epithelia, and that desiccation creates an environment promoting Th17 differentiation through a dendritic cell-mediated pathway. We hypothesize that Th17 inducing cytokines produced by the ocular epithelium may participate in Th17 differentiation in three ways: (1) activation of immature dendritic cells on the ocular surface; (2) direct transfer to the lymph node in lymphatic liquid; or (3) direct promotion of differentiated Th17 cells that infiltrate the ocular surface.

Conclusion

This chapter focused on recent breakthroughs in mucosal immunology, including the discoveries of TLR signaling in innate immunity, novel epithelial cell-derived pro-allergic cytokines TSLP and IL-33, and a new Th17 cell population in adaptive immunity, which link innate and adaptive immunity. All these advances provide compelling evidence that epithelial cells actively participate, as initiators, mediators and regulators, in innate and adaptive immune responses for host defense, in addition to physical barrier function. Although this new concept has been recognized, the molecular regulation of innate and adaptive immunity by the epithelial cells remains to be elucidated. These novel molecules may be critical for inflammatory, allergic and autoimmune diseases on mucosal epithelia, and may become potential molecular targets for new therapies to treat these conditions.

Abbreviations

DC: dendritic cell; dsRNA: double stranded RNA; IL: interleukin; MyD88: myeloid differentiation primary response gene 88; PCR: polymerase chain reaction; PGN: Peptidoglycan; polyI:C: polyinosinic-polycytidylic acid; PRR: pattern recognition receptor; qPCR: quantitative real-time PCR; R837: imiquimod; RT: reverse transcription; RT-qPCR: reverse transcription and quantitative real-time PCR; ssRNA: single stranded RNA; TGF-β: transforming growth factor; Th1: T helper cell type 1; Th17: T helper cell producing IL-17 family; Th2: T helper cell type 2; TLR: Toll-like receptor; TRIF: TIR-domain-containing adaptor inducing interferon; TSLP: thymic stromal lymphopoietin.

References

[1] Hoebe K, Janssen E, Beutler B. The interface between innate and adaptive immunity. *Nat. Immunol.* 2004;5:971-974.

[2] Meyer T, Stockfleth E, Christophers E. Immune response profiles in human skin. *Br. J. Dermatol.* 2007;157 Suppl 2:1-7.

[3] Buchau AS, Gallo RL. Innate immunity and antimicrobial defense systems in psoriasis. *Clin. Dermatol.* 2007;25:616-624.

[4] Wollenberg A, Klein E. Current aspects of innate and adaptive immunity in atopic dermatitis. *Clin. Rev. Allergy Immunol.* 2007;33:35-44.

[5] Kato A, Schleimer RP. Beyond inflammation: airway epithelial cells are at the interface of innate and adaptive immunity. *Curr. Opin. Immunol.* 2007;19:711-720.

[6] Mayer AK, Dalpke AH. Regulation of local immunity by airway epithelial cells. *Arch. Immunol. Ther. Exp. (Warsz).* 2007;55:353-362.

[7] Schleimer RP, Kato A, Kern R, Kuperman D, Avila PC. Epithelium: at the interface of innate and adaptive immune responses. *J. Allergy Clin. Immunol.* 2007;120:1279-1284.
[8] Muller CA, Autenrieth IB, Peschel A. Innate defenses of the intestinal epithelial barrier. *Cell Mol. Life Sci.* 2005;62:1297-1307.
[9] Ishihara S, Rumi MA, Ortega-Cava CF, Kazumori H, Kadowaki Y, Ishimura N, Kinoshita Y. Therapeutic targeting of toll-like receptors in gastrointestinal inflammation. *Curr. Pharm. Des.* 2006;12:4215-4228.
[10] Sharma R, Young C, Neu J. Molecular modulation of intestinal epithelial barrier: contribution of microbiota. *J. Biomed. Biotechnol.* 2010;2010:305879.
[11] Hazlett LD. Role of innate and adaptive immunity in the pathogenesis of keratitis. *Ocul. Immunol. Inflamm.* 2005;13:133-138.
[12] Kumar A, Yu FS. Toll-like receptors and corneal innate immunity. *Curr. Mol. Med.* 2006;6:327-337.
[13] Chang JH, McCluskey PJ, Wakefield D. Toll-like receptors in ocular immunity and the immunopathogenesis of inflammatory eye disease. *Br. J. Ophthalmol.* 2006;90:103-108.
[14] Akira S, Uematsu S, Takeuchi O. Pathogen recognition and innate immunity. *Cell.* 2006;124:783-801.
[15] Takeda K, Kaisho T, Akira S. Toll-like receptors. *Annu. Rev. Immunol.* 2003;21:335-376.
[16] Trinchieri G, Sher A. Cooperation of Toll-like receptor signals in innate immune defence. *Nat. Rev. Immunol.* 2007;7:179-190.
[17] Gay NJ, Gangloff M. Structure and function of Toll receptors and their ligands. *Annu. Rev. Biochem.* 2007;76:141-165.
[18] Kawai T, Akira S. Signaling to NF-kappaB by Toll-like receptors. *Trends Mol. Med.* 2007;13:460-469.
[19] Miller LS, Modlin RL. Toll-like receptors in the skin. *Semin Immunopathol.* 2007;29:15-26.
[20] Li D-Q, Zhou N, Zhang L, Ma P, Pflugfelder SC. Suppressive effects of azithromycin on zymosan-induced production of proinflammatory mediators by human corneal epithelial cells. *Invest Ophthalmol. Vis. Sci.* 2010;51:5623-5629.
[21] Kumar A, Zhang J, Yu FS. Innate immune response of corneal epithelial cells to Staphylococcus aureus infection: role of peptidoglycan in stimulating proinflammatory cytokine secretion. *Invest Ophthalmol. Vis. Sci.* 2004;45:3513-3522.
[22] Huang X, Barrett RP, McClellan SA, Hazlett LD. Silencing Toll-like receptor-9 in Pseudomonas aeruginosa keratitis. *Invest Ophthalmol. Vis. Sci.* 2005;46:4209-4216.
[23] Kumar A, Zhang J, Yu FS. Toll-like receptor 3 agonist poly(I:C)-induced antiviral response in human corneal epithelial cells. *Immunology.* 2006;117:11-21.
[24] Ueta M, Hamuro J, Kiyono H, Kinoshita S. Triggering of TLR3 by polyI:C in human corneal epithelial cells to induce inflammatory cytokines. *Biochem. Biophys. Res. Commun.* 2005;331:285-294.
[25] Liu YJ. Thymic stromal lymphopoietin and OX40 ligand pathway in the initiation of dendritic cell-mediated allergic inflammation. *J. Allergy Clin. Immunol.* 2007;120:238-244.

[26] Liu YJ, Soumelis V, Watanabe N, Ito T, Wang YH, Malefyt RW, Omori M, Zhou B, Ziegler SF. TSLP: an epithelial cell cytokine that regulates T cell differentiation by conditioning dendritic cell maturation. *Annu. Rev. Immunol.* 2007;25:193-219.
[27] Holgate ST. The epithelium takes centre stage in asthma and atopic dermatitis. *Trends Immunol.* 2007;28:248-251.
[28] Lee HC, Ziegler SF. Inducible expression of the proallergic cytokine thymic stromal lymphopoietin in airway epithelial cells is controlled by NFkappaB. *Proc. Natl. Acad. Sci. U S A.* 2007;104:914-919.
[29] Demehri S, Morimoto M, Holtzman MJ, Kopan R. Skin-derived TSLP triggers progression from epidermal-barrier defects to asthma. *PLoS Biol.* 2009;7:e1000067.
[30] Zhang Z, Hener P, Frossard N, Kato S, Metzger D, Li M, Chambon P. Thymic stromal lymphopoietin overproduced by keratinocytes in mouse skin aggravates experimental asthma. *Proc. Natl. Acad. Sci. U S A.* 2009;106:1536-1541.
[31] Sims JE, Williams DE, Morrissey PJ, Garka K, Foxworthe D, Price V, Friend SL, Farr A, Bedell MA, Jenkins NA, Copeland NG, Grabstein K, Paxton RJ. Molecular cloning and biological characterization of a novel murine lymphoid growth factor. *J. Exp. Med.* 2000;192:671-680.
[32] Quentmeier H, Drexler HG, Fleckenstein D, Zaborski M, Armstrong A, Sims JE, Lyman SD. Cloning of human thymic stromal lymphopoietin (TSLP) and signaling mechanisms leading to proliferation. *Leukemia.* 2001;15:1286-1292.
[33] Reche PA, Soumelis V, Gorman DM, Clifford T, Liu M, Travis M, Zurawski SM, Johnston J, Liu YJ, Spits H, de Waal MR, Kastelein RA, Bazan JF. Human thymic stromal lymphopoietin preferentially stimulates myeloid cells. *J. Immunol.* 2001;167:336-343.
[34] Soumelis V, Reche PA, Kanzler H, Yuan W, Edward G, Homey B, Gilliet M, Ho S, Antonenko S, Lauerma A, Smith K, Gorman D, Zurawski S, Abrams J, Menon S, McClanahan T, de Waal-Malefyt RR, Bazan F, Kastelein RA, Liu YJ. Human epithelial cells trigger dendritic cell mediated allergic inflammation by producing TSLP. *Nat. Immunol.* 2002;3:673-680.
[35] Allakhverdi Z, Comeau MR, Jessup HK, Yoon BR, Brewer A, Chartier S, Paquette N, Ziegler SF, Sarfati M, Delespesse G. Thymic stromal lymphopoietin is released by human epithelial cells in response to microbes, trauma, or inflammation and potently activates mast cells. *J. Exp. Med.* 2007;204:253-258.
[36] Bogiatzi SI, Fernandez I, Bichet JC, Marloie-Provost MA, Volpe E, Sastre X, Soumelis V. Cutting Edge: Proinflammatory and Th2 cytokines synergize to induce thymic stromal lymphopoietin production by human skin keratinocytes. *J. Immunol.* 2007;178:3373-3377.
[37] Kato A, Favoreto S Jr, Avila PC, Schleimer RP. TLR3- and Th2 cytokine-dependent production of thymic stromal lymphopoietin in human airway epithelial cells. *J. Immunol.* 2007;179:1080-1087.
[38] Ma P, Bian F, Wang Z, Zheng X, Chotikavanich S, Pflugfelder SC, Li D-Q. Human corneal epithelium-derived thymic stromal lymphopoietin links the innate and adaptive immune responses via TLRs and Th2 cytokines. *Invest Ophthalmol. Vis. Sci.* 2009;50:2702-2709.

[39] Mohan RR, Mohan RR, Kim WJ, Wilson SE. Modulation of TNF-alpha-induced apoptosis in corneal fibroblasts by transcription factor NF-kappaB. *Invest Ophthalmol. Vis. Sci.* 2000;41:1327-1336.

[40] Ozawa T, Koyama K, Ando T, Ohnuma Y, Hatsushika K, Ohba T, Sugiyama H, Hamada Y, Ogawa H, Okumura K, Nakao A. Thymic stromal lymphopoietin secretion of synovial fibroblasts is positively and negatively regulated by Toll-like receptors/nuclear factor-kappaB pathway and interferon-gamma/dexamethasone. *Mod. Rheumatol.* 2007;17:459-463.

[41] Ying S, O'Connor B, Ratoff J, Meng Q, Fang C, Cousins D, Zhang G, Gu S, Gao Z, Shamji B, Edwards MJ, Lee TH, Corrigan CJ. Expression and cellular provenance of thymic stromal lymphopoietin and chemokines in patients with severe asthma and chronic obstructive pulmonary disease. *J. Immunol.* 2008;181:2790-2798.

[42] Harada M, Hirota T, Jodo AI, Hitomi Y, Sakashita M, Tsunoda T, Miyagawa T, Doi S, Kameda M, Fujita K, Miyatake A, Enomoto T, Noguchi E, Masuko H, Sakamoto T, Hizawa N, Suzuki Y, Yoshihara S, Adachi M, Ebisawa M, Saito H, Matsumoto K, Nakajima T, Mathias RA, Rafaels N, Barnes KC, Himes BE, Duan QL, Tantisira KG, Weiss ST, Nakamura Y, Ziegler SF, Tamari M. TSLP Promoter Polymorphisms are Associated with Susceptibility to Bronchial Asthma. *Am. J. Respir. Cell Mol Biol.* 2010.

[43] Matsuda A, Ebihara N, Yokoi N, Kawasaki S, Tanioka H, Inatomi T, de Waal MR, Hamuro J, Kinoshita S, Murakami A. Functional role of thymic stromal lymphopoietin in chronic allergic keratoconjunctivitis. *Invest Ophthalmol. Vis. Sci.* 2010;51:151-155.

[44] Kay AB. Allergy and allergic diseases. First of two parts. *N. Engl. J. Med.* 2001;344:30-37.

[45] Komine M. Analysis of the mechanism for the development of allergic skin inflammation and the application for its treatment:keratinocytes in atopic dermatitis - their pathogenic involvement. *J. Pharmacol. Sci.* 2009;110:260-264.

[46] Magone MT, Chan CC, Rizzo LV, Kozhich AT, Whitcup SM. A novel murine model of allergic conjunctivitis. *Clin. Immunol. Immunopathol.* 1998;87:75-84.

[47] Stern ME, Siemasko K, Gao J, Duong A, Beauregard C, Calder V, Niederkorn JY. Role of interferon-gamma in a mouse model of allergic conjunctivitis. *Invest Ophthalmol. Vis. Sci.* 2005;46:3239-3246.

[48] Zheng X, Ma P, de Paiva CS, Cunningham MA, Hwang CS, Pflugfelder SC, Li D-Q. TSLP and downstream molecules in experimental mouse allergic conjunctivitis. *Invest Ophthalmol. Vis. Sci.* 2010;51:3076-3082.

[49] Tominaga S. A putative protein of a growth specific cDNA from BALB/c-3T3 cells is highly similar to the extracellular portion of mouse interleukin 1 receptor. *FEBS Lett.* 1989;258:301-304.

[50] Kakkar R, Lee RT. The IL-33/ST2 pathway: therapeutic target and novel biomarker. *Nat. Rev. Drug Discov.* 2008;7:827-840.

[51] Schmitz J, Owyang A, Oldham E, Song Y, Murphy E, McClanahan TK, Zurawski G, Moshrefi M, Qin J, Li X, Gorman DM, Bazan JF, Kastelein RA. IL-33, an interleukin-1-like cytokine that signals via the IL-1 receptor-related protein ST2 and induces T helper type 2-associated cytokines. *Immunity.* 2005;23:479-490.

[52] Schmitz J, Owyang A, Oldham E, Song Y, Murphy E, McClanahan TK, Zurawski G, Moshrefi M, Qin J, Li X, Gorman DM, Bazan JF, Kastelein RA. IL-33, an interleukin-

1-like cytokine that signals via the IL-1 receptor-related protein ST2 and induces T helper type 2-associated cytokines. *Immunity.* 2005;23:479-490.

[53] Moussion C, Ortega N, Girard JP. The IL-1-like cytokine IL-33 is constitutively expressed in the nucleus of endothelial cells and epithelial cells in vivo: a novel 'alarmin'? *PLoS One.* 2008;3:e3331.

[54] Smith DE. IL-33: a tissue derived cytokine pathway involved in allergic inflammation and asthma. *Clin. Exp. Allergy.* 2009.

[55] Allam JP, Bieber T, Novak N. Dendritic cells as potential targets for mucosal immunotherapy. *Curr. Opin. Allergy Clin. Immunol.* 2009;9:554-557.

[56] Haraldsen G, Balogh J, Pollheimer J, Sponheim J, Kuchler AM. Interleukin-33 - cytokine of dual function or novel alarmin? *Trends Immunol.* 2009;30:227-233.

[57] Hayakawa M, Hayakawa H, Matsuyama Y, Tamemoto H, Okazaki H, Tominaga S. Mature interleukin-33 is produced by calpain-mediated cleavage in vivo. *Biochem. Biophys. Res. Commun.* 2009;387:218-222.

[58] Talabot-Ayer D, Lamacchia C, Gabay C, Palmer G. Interleukin-33 is biologically active independently of caspase-1 cleavage. *J. Biol. Chem.* 2009;284:19420-19426.

[59] Krishnan J, Selvarajoo K, Tsuchiya M, Lee G, Choi S. Toll-like receptor signal transduction. *Exp. Mol. Med.* 2007;39:421-438.

[60] Mosmann TR, Coffman RL. TH1 and TH2 cells: different patterns of lymphokine secretion lead to different functional properties. *Annu. Rev. Immunol.* 1989;7:145-173.

[61] Dong C. Diversification of T-helper-cell lineages: finding the family root of IL-17-producing cells. *Nat. Rev. Immunol.* 2006;6:329-333.

[62] Komiyama Y, Nakae S, Matsuki T, Nambu A, Ishigame H, Kakuta S, Sudo K, Iwakura Y. IL-17 plays an important role in the development of experimental autoimmune encephalomyelitis. *J. Immunol.* 2006;177:566-573.

[63] Hwang SY, Kim HY. Expression of IL-17 homologs and their receptors in the synovial cells of rheumatoid arthritis patients. *Mol. Cells.* 2005;19:180-184.

[64] Chang SH, Dong C. A novel heterodimeric cytokine consisting of IL-17 and IL-17F regulates inflammatory responses. *Cell Res.* 2007;17:435-440.

[65] Williams IR. CCR6 and CCL20: partners in intestinal immunity and lymphorganogenesis. *Ann. N Y Acad. Sci.* 2006;1072:52-61.

[66] Moseley TA, Haudenschild DR, Rose L, Reddi AH. Interleukin-17 family and IL-17 receptors. *Cytokine Growth Factor Rev.* 2003;14:155-174.

[67] Gaffen SL, Kramer JM, Yu JJ, Shen F. The IL-17 cytokine family. *Vitam. Horm.* 2006;74:255-282.

[68] Kuestner RE, Taft DW, Haran A, Brandt CS, Brender T, Lum K, Harder B, Okada S, Ostrander CD, Kreindler JL, Aujla SJ, Reardon B, Moore M, Shea P, Schreckhise R, Bukowski TR, Presnell S, Guerra-Lewis P, Parrish-Novak J, Ellsworth JL, Jaspers S, Lewis KE, Appleby M, Kolls JK, Rixon M, West JW, Gao Z, Levin SD. Identification of the IL-17 receptor related molecule IL-17RC as the receptor for IL-17F. *J. Immunol.* 2007;179:5462-5473.

[69] Liang SC, Tan XY, Luxenberg DP, Karim R, Dunussi-Joannopoulos K, Collins M, Fouser LA. Interleukin (IL)-22 and IL-17 are coexpressed by Th17 cells and cooperatively enhance expression of antimicrobial peptides. *J. Exp. Med.* 2006;203:2271-2279.

[70] Zheng Y, Danilenko DM, Valdez P, Kasman I, Eastham-Anderson J, Wu J, Ouyang W. Interleukin-22, a T(H)17 cytokine, mediates IL-23-induced dermal inflammation and acanthosis. *Nature.* 2007;445:648-651.

[71] Hirota K, Yoshitomi H, Hashimoto M, Maeda S, Teradaira S, Sugimoto N, Yamaguchi T, Nomura T, Ito H, Nakamura T, Sakaguchi N, Sakaguchi S. Preferential recruitment of CCR6-expressing Th17 cells to inflamed joints via CCL20 in rheumatoid arthritis and its animal model. *J. Exp. Med.* 2007;204:2803-2812.

[72] Yang XO, Panopoulos AD, Nurieva R, Chang SH, Wang D, Watowich SS, Dong C. STAT3 regulates cytokine-mediated generation of inflammatory helper T cells. *J. Biol. Chem.* 2007;282:9358-9363.

[73] Veldhoen M, Hocking RJ, Atkins CJ, Locksley RM, Stockinger B. TGFbeta in the context of an inflammatory cytokine milieu supports de novo differentiation of IL-17-producing T cells. *Immunity.* 2006;24:179-189.

[74] Mangan PR, Harrington LE, O'Quinn DB, Helms WS, Bullard DC, Elson CO, Hatton RD, Wahl SM, Schoeb TR, Weaver CT. Transforming growth factor-beta induces development of the T(H)17 lineage. *Nature.* 2006;441:231-234.

[75] Murphy CA, Langrish CL, Chen Y, Blumenschein W, McClanahan T, Kastelein RA, Sedgwick JD, Cua DJ. Divergent pro- and antiinflammatory roles for IL-23 and IL-12 in joint autoimmune inflammation. *J. Exp. Med.* 2003;198:1951-1957.

[76] Zhou L, Ivanov II, Spolski R, Min R, Shenderov K, Egawa T, Levy DE, Leonard WJ, Littman DR. IL-6 programs T(H)-17 cell differentiation by promoting sequential engagement of the IL-21 and IL-23 pathways. *Nat. Immunol.* 2007;8:967-974.

[77] Langrish CL, Chen Y, Blumenschein WM, Mattson J, Basham B, Sedgwick JD, McClanahan T, Kastelein RA, Cua DJ. IL-23 drives a pathogenic T cell population that induces autoimmune inflammation. *J. Exp. Med.* 2005;201:233-240.

[78] Leonard WJ, Spolski R. Interleukin-21: a modulator of lymphoid proliferation, apoptosis and differentiation. *Nat. Rev. Immunol.* 2005;5:688-698.

[79] Korn T, Bettelli E, Gao W, Awasthi A, Jager A, Strom TB, Oukka M, Kuchroo VK. IL-21 initiates an alternative pathway to induce proinflammatory T(H)17 cells. *Nature.* 2007;448:484-487.

[80] Nurieva R, Yang XO, Martinez G, Zhang Y, Panopoulos AD, Ma L, Schluns K, Tian Q, Watowich SS, Jetten AM, Dong C. Essential autocrine regulation by IL-21 in the generation of inflammatory T cells. *Nature.* 2007;448:480-483.

[81] Ivanov II, McKenzie BS, Zhou L, Tadokoro CE, Lepelley A, Lafaille JJ, Cua DJ, Littman DR. The orphan nuclear receptor RORgammat directs the differentiation program of proinflammatory IL-17+ T helper cells. *Cell.* 2006;126:1121-1133.

[82] Zheng X, Bian F, Ma P, de Paiva CS, Stern M, Pflugfelder SC, Li D-Q. Induction of Th17 differentiation by corneal epithelial-derived cytokines. *J. Cell Physiol.* 2010;222:95-102.

[83] Aujla SJ, Dubin PJ, Kolls JK. Th17 cells and mucosal host defense. *Semin Immunol.* 2007;19:377-382.

[84] Holtta V, Klemetti P, Sipponen T, Westerholm-Ormio M, Kociubinski G, Salo H, Rasanen L, Kolho KL, Farkkila M, Savilahti E, Vaarala O. IL-23/IL-17 immunity as a hallmark of Crohn's disease. *Inflamm. Bowel Dis.* 2008;14:1175-1184.

[85] Li D-Q, Luo L, Chen Z, Kim HS, Song XJ, Pflugfelder SC. JNK and ERK MAP kinases mediate induction of IL-1beta, TNF-alpha and IL-8 following hyperosmolar stress in human limbal epithelial cells. *Exp. Eye Res.* 2006;82:588-596.

[86] Barton K, Monroy DC, Nava A, Pflugfelder SC. Inflammatory cytokines in the tears of patients with ocular rosacea. *Ophthalmology.* 1997;104:1868-1874.

[87] Shainheit MG, Smith PM, Bazzone LE, Wang AC, Rutitzky LI, Stadecker MJ. Dendritic cell IL-23 and IL-1 production in response to schistosome eggs induces Th17 cells in a mouse strain prone to severe immunopathology. *J. Immunol.* 2008;181:8559-8567.

[88] Corrales RM, Villarreal A, Farley W, Stern ME, Li D-Q, Pflugfelder SC. Strain-related cytokine profiles on the murine ocular surface in response to desiccating stress. *Cornea.* 2007;26:579-584.

[89] Turner K, Pflugfelder SC, Ji Z, Feuer WJ, Stern M, Reis BL. Interleukin-6 levels in the conjunctival epithelium of patients with dry eye disease treated with cyclosporine ophthalmic emulsion. *Cornea.* 2000;19:492-496.

[90] Solomon A, Dursun D, Liu Z, Xie Y, Macri A, Pflugfelder SC. Pro- and anti-inflammatory forms of interleukin-1 in the tear fluid and conjunctiva of patients with dry-eye disease. *Invest Ophthalmol. Vis. Sci.* 2001;42:2283-2292.

[91] Pflugfelder SC, de Paiva CS, Li D-Q, Stern ME. Epithelial-immune cell interaction in dry eye. *Cornea.* 2008;27 Suppl 1:S9-11.

[92] Nguyen CQ, Hu MH, Li Y, Stewart C, Peck AB. Salivary gland tissue expression of interleukin-23 and interleukin-17 in Sjogren's syndrome: findings in humans and mice. *Arthritis Rheum.* 2008;58:734-743.

[93] Chotikavanich S, de Paiva CS, Li D-Q, Chen JJ, Bian F, Farley WJ, Pflugfelder SC. Production and Activity of Matrix Metalloproteinase-9 on the Ocular Surface Increase in Dysfunctional Tear Syndrome. *Invest Ophthalmol. Vis. Sci.* 2009.

[94] de Paiva CS, Chotikavanich S, Pangelinan SB, Pitcher JD, III, Fang B, Zheng X, Ma P, Farley WJ, Siemasko KF, Niederkorn JY, Stern ME, Li D-Q, Pflugfelder SC. IL-17 disrupts corneal barrier following desiccating stress. *Mucosal. Immunol.* 2009;2:243-253.

[95] Chauhan SK, El AJ, Ecoiffier T, Goyal S, Zhang Q, Saban DR, Dana R. Autoimmunity in dry eye is due to resistance of Th17 to Treg suppression. *J. Immunol.* 2009;182:1247-1252.

[96] de Paiva CS, Villarreal AL, Corrales RM, Rahman HT, Chang VY, Farley WJ, Stern ME, Niederkorn JY, Li D-Q, Pflugfelder SC. Dry eye-induced conjunctival epithelial squamous metaplasia is modulated by interferon-gamma. *Invest. Ophthalmol. Vis. Sci.* 2007;48:2553-2560.

[97] Stern ME, Siemasko KF, Gao J, Calonge M, Niederkorn JY, Pflugfelder SC. Evaluation of ocular surface inflammation in the presence of dry eye and allergic conjunctival disease. *Ocul. Surf.* 2005;3:S161-S164.

[98] Zheng X, de Paiva CS, Li D-Q, Farley WJ, Pflugfelder SC. Desiccating stress promotion of Th17 differentiation by ocular surface tissues through a dendritic cell-mediated pathway. *Invest. Ophthalmol. Vis. Sci.* 2010;51:3083-3091.

[99] Molesworth-Kenyon SJ, Yin R, Oakes JE, Lausch RN. IL-17 receptor signaling influences virus-induced corneal inflammation. *J. Leukoc Biol.* 2008;83:401-408.

[100] Bian F, Qi H, Ma P, Zhang L, Yoon KC, Pflugfelder SC, Li D-Q. An immunoprotective privilege of corneal epithelial stem cells against Th17 inflammatory stress by producing glial cell-derived neurotrophic factor. *Stem Cells.* 2010;28:2172-2181.

In: Endothelium and Epithelium
Editors: J. Carrasco and M. Mota, pp. 87-111
ISBN 978-1-61470-874-2

Chapter IV

Morphofunctional Characteristic Features of Static (Endothelium) and Dynamic (Epithelium) Cell Population in Norm and Some Pathology

T. M. Yavisheva* and S. D. Shcherbakov
Blokhin Cancer Research Center, Russian Academy of Medical Science, Kashirskoe sh. 24, Moscow, 115478 Russia

I. Introduction

In biology and medicine, the big attention to structure features, functional reactance, regeneration and other properties of epithelial tissue is paid. The special interest causes the skin epidermis. Recently, there was considerable data about its histostructure complexity (Letuchaya and Ketlinskii, 1980; Savost'anov, 2001; Potten and.Booth, 2002). Thus, in the overwhelming majority of works, morphofunctional rearrangement of epitheliocytes only in vertical direction was investigated.

Besides, available data on quantitative basement cells morphology (quantitative correlations between various cells groups in it, cells distribution according to their area and form, correlations between progenitor patterns and their various derivatives) is obviously insufficient.

Essentially, the objective information on a basal epitheliocytes state represents the special interest from positions of modern view about the organization and reaction of epithelial layer in various situations.

Skin epidermis is intensively proliferating tissue. Therefore, it is difficult to give quantitative and quality value of structural and population basement cells rearrangements at change of vital organism conditions.

*E-mail: Javisheva@rambler.ru

Considering this circumstance, we have carried out relative analysis of two types of epithelial tissue: the corneal endothelium, which has lost proliferation ability during a postnatal ontogenesis and the skin's multi-layered epithelium – actively regenerating in the specified period. It has enabled us to analyze more deeply the structural and functional organization of epithelial tissue in norm and at influence of different factors on epithelial histic structures.

In this article, we summarize our own material related to spatial organization of different epithelia (the posterior corneal epithelium - endothelium and ear skin epidermis), reviewed in 350 mature male BALB/cj mice. We also examined cornea and skin obtained from 60 adult nominally healthy men between 20 and 90 years old who died from cardiovascular disease or from accidental traumas (control to the experimental group –60 persons died from chronic alcohol intoxication); we also examined tumor tissue of two patients groups with breast cancer or fibroadenoma; each group consisted of 15 patients 40 to 60 years old.

II. Age Changes of a Tissue with Low Reparative Abilities on an Example of Human Corneal Endothelium

The study shows that there is a gradual depression of endothelium cell density from 2324 cell/mm² in persons 20-40 years old, to 1410 cell/mm² over 80 years old (Figure 1) (Fyodorov et al., 1993; Yavisheva et al., 2000).

This factor confirms augmentation of cell size with the years. The main part is represented by cells of the average size ("average" cells) which the area fluctuates depending on age, gradually enlarging from 274 mcm² (260-292 mcm²) in persons 20-40 years to 431 mcm² (337-458 mcm²) in persons over 80 years old(Table 1).

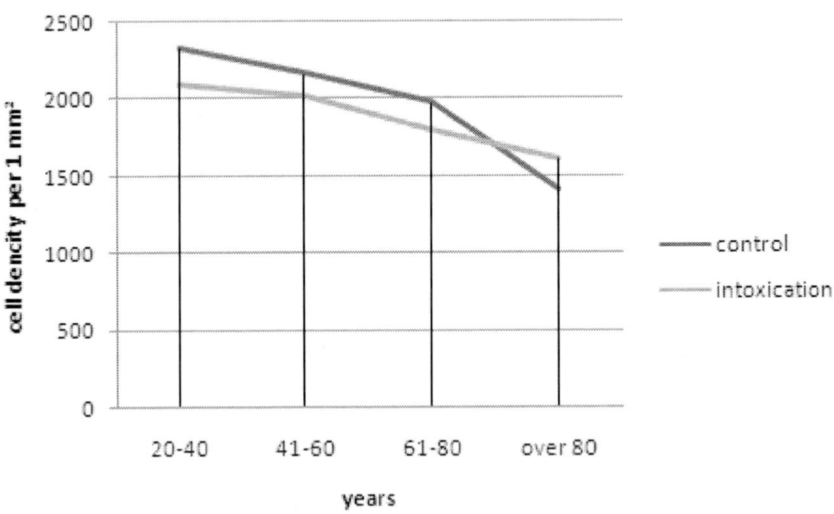

Figure 1. Endothelial cell density in control and alcohol intoxication.

Table 1. Areas of "average", "large" and "small" cells, and nuclei areas and nucleus-cytoplasm ratio in different age groups endothelium in normalcy

groups (years)	eyes quantity	"average cells"			"large cells"			"small cells"		
		cell area, mcm²	nucleus area, mcm²	n/c ratio %	cell area mcm²	nucleus area mcm²	n/c ratio %	cell area mcm²	nucleus area mcm²	n/c ratio %
20-40	30	274 (260-292)	69 (62-79)	21,1	454 (400-560)	68 (57-75)	15,2	152 (120-180)	47 (31-65)	30,8
41-60	37	363 (359-382) p< 0.05	61 (47-75) p> 0.05	22,3	579 (490-640) p> 0.05	68 (48-91) p> 0.05	11,7	164 (158-169) p> 0.05	48 (33-59) p> 0.05	29,8
61-80	40	399 (350-485) p< 0.05	54 (42-64) p< 0.05	13,5	594 (492-746) p> 0.05	57 (48-73) p> 0.05	9,6	160 (154-169) p 0.05	43 (34-58) p> 0.05	25,4
Over 80	25	431 (337-458) p< 0.05	45 (40-55) p< 0.05	10,4	566 (445-650) p> 0.05	51 (43-63) p> 0.05	9,4	153 (132-180) p> 0.05	36 (27-41) p> 0.05	23,7

The data was statistically processed with the use of Wilcoxon-Mann-Whitney (U) test. P was defined in relation to 20-40 year-old persons.
n/c – nucleus-cytoplasm ratio.

The maintenance of "average" cells in persons between 20 and 80 years old remains stable and the greatest in cell population of endothelium, compounding 61-62% and only in persons over 80 years it decreases to 49.6%. Thus, among 'average' cells, there are "prevailing" cells which form the greatest part of this population of the cells in each age group. Their quantity is stable in 20-80 year old men and compounds 9.68-10.12%, and over 80, it really decreases to 7.72%.

The area of the "average" cell nucleus decreases with the years and the nucleus-cytoplasm ratio also decreases. The quantity of "large" cells in 20-80 years is stable, and over 80 years is increased. The area of "large" cell tends to augmentate with the years, and nucleus–to depression, as a result the nucleus-cytoplasm ratio has decreased from 15.2% in 20-40 year-old men to 9.4% over 80 years. The number of "small" cells tends with the years to reduction. The area and nucleus of the "small" cell are stable as a rule and the nucleus-cytoplasm ratio is high in comparison with the "large" and "average" cells. Cells with right 6-side form, often meet in cornea endothelium of persons of any age and compound 16% from endothelium population in persons of 20-90 years. In each age group, prevalence of 4-5-side form cells over 7-8-side form cells is observed (Table 2).

Thus, the study showed that in human corneal endothelium (a tissue with limited regenerative abilities), cell density decreases with aging, thus the area of the"average" cell is enlarged, and the nucleus area is decreased in it. It leads to a depression of the nucleus-cytoplasm ratio and confirms the attenuation of synthetic processes in endothelium with aging. However, in endothelium monolayers, there are the processes of intracellular regeneration, and the increasing of polyploidy cells in 61-80 year-old and over 80 year old persons testifies to this (Table 2).

The number of cells with tetraploidy nuclei in a group of 61-80 yearolds and persons over 80 year-old forms 43-45, 0 on a cornea while in 20-40 and 41-60 year-old patients, this parameter significantly low and fluctuates from 4.6 to 9.67. Because polyploidy cells possess the high functional activity and prolonged life duration , a rising number of polyploidy cells with age in the conditions of physiological decrease of endothelial cells total quantity may be considered as the compensatory process referred on maintenance of cellular layer vital activity.

Endothelium is a tissue with a low level of cell regeneration, therefore aging processes here proceed slowly. In populations of endothelium up to 80 years old, the maintenance of "average" and "prevailing" cells (it is the most functional active), and also cells with right hexagonal form (optimal for endothelium monolayer) remains stable. Besides, there is a stable relationship between the number of "average" and "prevailing" cells (from 5.4 to 6.4) in all age groups. It obviously argues for enough high level of regulative mechanisms providing processes of homeostasis in endothelial cells. Heterogeneity of endothelial cells is associated on the whole with the increasing of size variability of the "average" cells during aging and appearance of cells of the larger area, instead of to the form. This process is clearly visible in 61-80 year-old persons and considerably expressed over 80 years. Thus, aging processes in endothelium monolayers start to be revealed distinctly in persons 61-80 years-old and is appreciably found only in persons over 80 years when organism adaptic possibilities decrease.

Table 2. Endothelial cell polymorphism and polyploidy cell number with aging

Groups (years)	Eyes quantity	Cell polymorphism per 1 mm² (cell/mm²)		Number of polyploidy cells (on the whole cornea)	
		4-5-side form cells	7-8-side form cells	Two nuclei cells	Tetraploidy nucleus cells
20-40	30	269 (230-319)	109 (5-195)	2,75 (0-4)	4,6 (1-10)
41-60	37	200 (130-386) P> 0.05	65 (20-154) P> 0.05	4,57 (2-8) P> 0.05	9,6 (1-30) P> 0.05
61-80	40	212 (140-264) P> 0.05	43 (34-55) P> 0.05	6,4 (1-22) P< 0.05	43 (5-71) P< 0.05
Over 80	25	305 (180-311) P> 0.05	93 (50-150) P> 0.05	11,25 (3-35) P< 0.05	45,0 (6-70) P< 0.05

The data was statistically processed with the use of Wilcoxon-Mann-Whitney (U) test. P was defined in relation to 20-40 year-old persons.

III. Morphofunctional Organization of Human Corneal Endothelium at Chronic Intoxication

It is known that the studying of pathological processes allows us to understand more deeply the natural laws existing in norm. The carried out researches have shown that populations of the endothelial cells inpeople of 20-40 years most actively reacts to pathological influence in the conditions of organism chronic alcohol intoxication. Considerable depression of cell density in comparison with other age groups argues this (Figure 1). Besides, the change of the nucleus-cytoplasm ratio in endothelial "average", "small" and "large" cells as a result of nuclei reduction obviously causes the greatest loss of endothelium cells at chronic pathology in 20-40 year-old persons in comparison with other age groups (Figure 2, 3,4). It leads to the augmentation of "average" cell size in comparison with the same age control. Considerable changes in corneal endothelium at chronic intoxication occur as well in persons over 80 yearsold. In "average", "large" and "small" cells, the augmentation of the nucleus area that leads to an increase of nucleus-cytoplasm ratio is noticed.

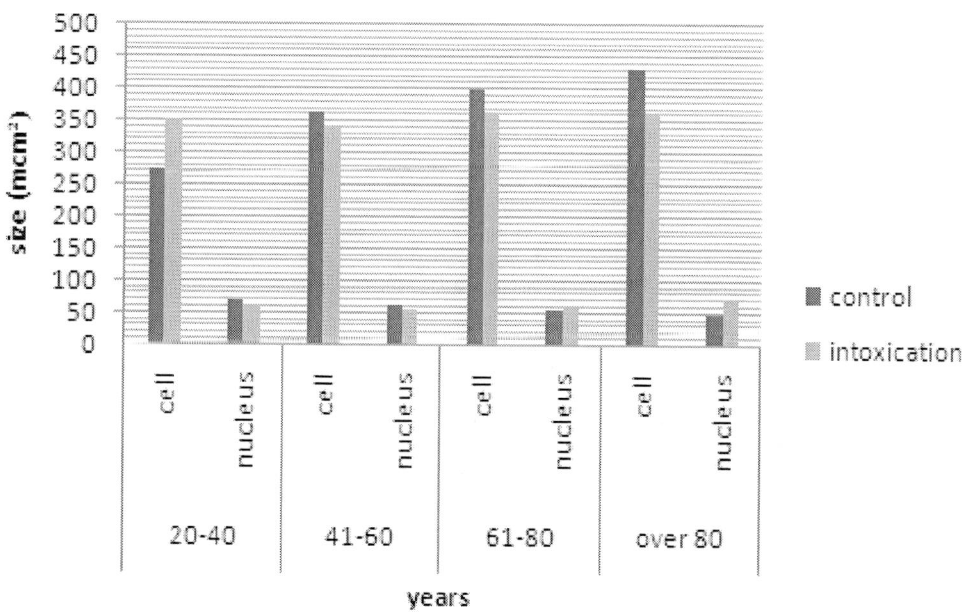

Figure 2. Endothelial "average" cell and nucleus area in control and intoxication.

Besides, persons over 60 years old and especially over 80 years old in comparison with the control, the quantity of polyploidy cells essentially decreases, i.e. ability for intracellular regeneration of endothelium cells is depressed (Table 3).

On one hand, these changes at persons over 60 years old and especially over 80 years old are associated with toxic influence on cell nucleus, and on the other hand, with depression of reaction and adaptic possibilities of older persons organism. This means that chronic alcohol intoxication influences considerably on corneal endothelium. Thus, in 20-40 year-old persons and over 80 years old, there is an active though differently directed reaction to action of the unfavorable factor. The most stable, i.e. least reactive is corneal endothelium in 41-60 year-old persons.

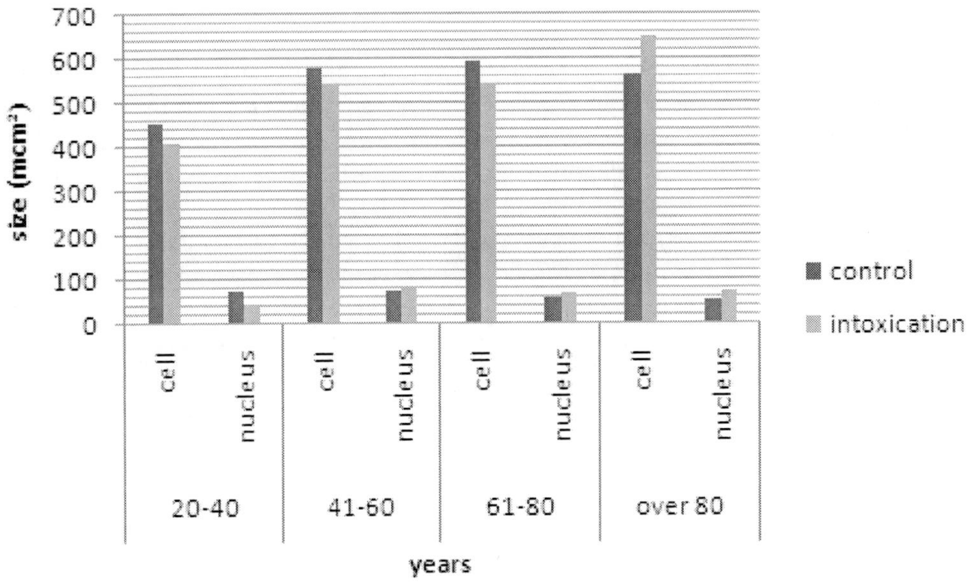

Figure 3. Endothelial "large" cell and nucleus area in control and intoxication.

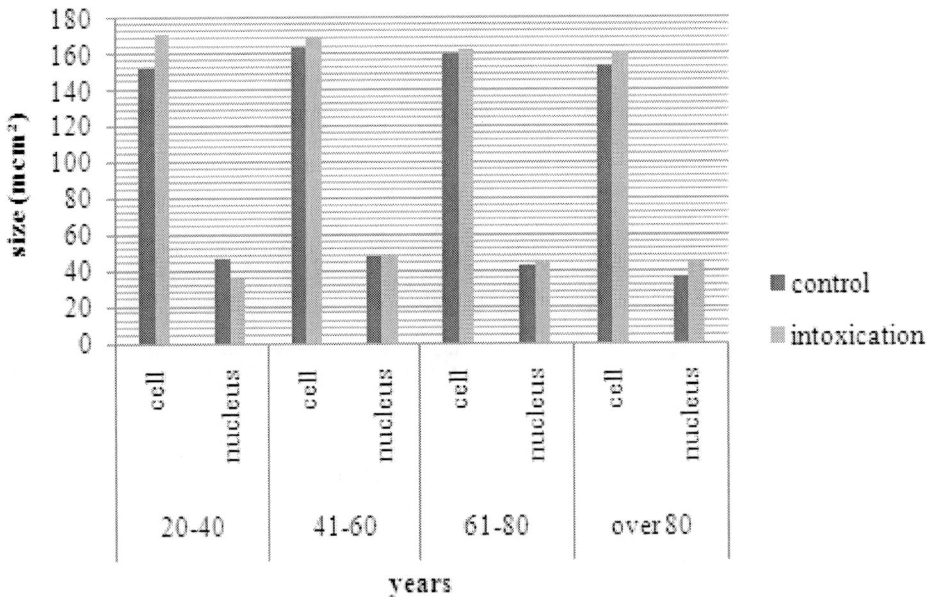

Figure 4. Endothelial "small" cell and nucleus area in control and intoxication.

Table 3. Endothelial polyploidy cell number during alcohol intoxication

Groups (years)	Eyes quantity	Binuclear cells		P (U)	Tetraploidy nucleus cells		P (U)
		intoxication	norma-lcy		intoxication	normalcy	
20-40	10	1.0 (0-4)	2.75 (0-4)	>0.05	2.0 (0-10	46 (1-10)	>0.05
41-60	10	1.2 (0-4)	4.57 (2-8)	>0.05	2.6 (0-13)	9.67 (1-30)	>0.05
61-80	10	1.0 (0-4)	6.4 (1-22)	<0.05	3.0 (1-9)	43 (5-71)	<0.05
Over 80	10	3.3 (0-9)	11.25 (3-35)	<0.05	5.0 (1-13)	45 (6-76)	<0.05

The data was statistically processed with the use of Wilcoxon-Mann-Whitney (U) test.

IV. Quantitative Morphology of Mouse Epidermal Basement Cells and Human in Norm and at Chronic Intoxication with Advancing Age

Because the principle of form cells distribution is identical in mouse and human skin epidermis, we compared not only cells distribution features between human age groups, but also animals as the maximum and minimum proliferative activity can be defined only in experiments.

While studying theproliferative activity of mouse skin epidermis basal cells, it has been established that in the morning, average mitotic activity has been 10% at 6 a.m. and 8%, at 9 a.m., significantly increasing as compared with evening : 2.7%, at 9 p.m., 2% at 12 p.m.

Table 4. Quantitative characterization of mouse epidermal basement cells

Conventional names of cell series	Degree of ellipticity DE	Percentage of cells with a certain DE in the population	Number of cell classes within a cell series	Cell area, mcm^2
Peak 1 cells	0.58	20-30	1	43-67
Peak 2 cells	0.61	8-10	1	52-41
“Narrow” cells	0.11-0.26	9-10	17	43-99
“Elongated” cells	0.27-0.41	16	10	-
“Transitional” cells	0.42-0.57	16	11	-
“Oval” cells	0.62-0.78	10	11	51.69-81.9
“Round” cells	0.79-1.0	6	14	60.3-64.6

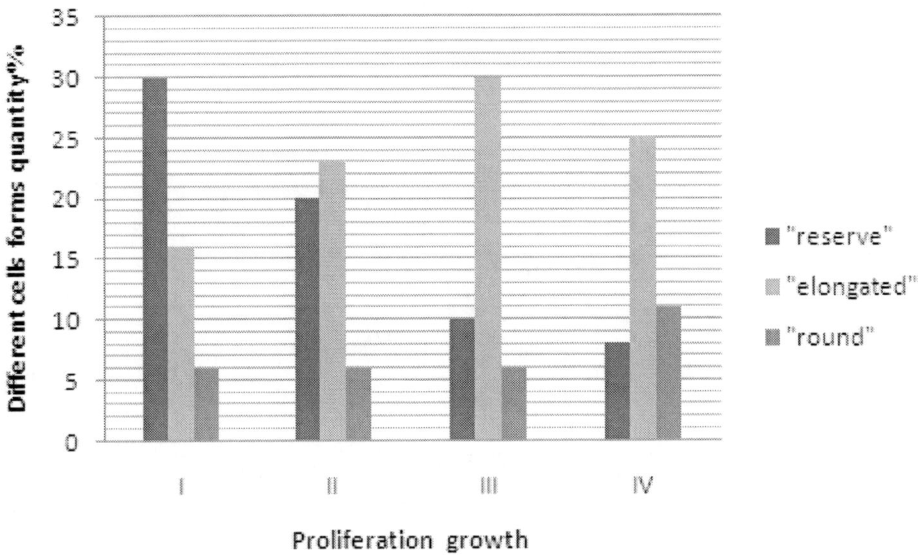

Figure 5. Mouse epidermal basement population composition during proliferation growth.

Comparative analysis of cell distribution histograms with respect to cell area, cell form DE (degree of ellipticity was defined as ratio of ellipsis shortest to longest axes) on the one hand, and mitotic activity on the other hand, in the evening when epithelial cell proliferation was decreased, showed that cells of peak 1 (20-30%) were the most frequent in the whole basement cell population. These cells form a dynamic reserve – the depot of the basal cells which functions consist of the maintenance of basal cells activity during all proliferative processes in the epidermis. So, peak 1cells have been conventionally called "reserve" cells. Study of basal layer revealed, that the of "reserve' cells number at different patients fluctuated both in control and at intoxication from 25-30% to 8-6%.

It is shown, that at intensifying of proliferative processes in human and mice epidermis, there is a gradual decrease of this depot cells from 25-30% to 8%, i.e. these cells will be transformed into their derivative forms – "transitional" and then into "elongated" cells (Figure 5).

Hence, "reserve" cells (peak1) and their derivative forms ("transitional" and 'elongated" cells) represent proliferating group of cells and constitute 62% of basal cells (30, 16 and 16% accordingly). When the "elongated" cells reach 30% and "reserve" 8%, there comes an exit (jump) of 5% of "elongated" cells immediately into mitosis. As a result, the quantity of "round" cells which represent a division product of "elongated' cells is increased. Thus, to human epidermal "elongated" cells, as well as mouse, synchronization and a jumping exit into mitosis are inherent. Transition of "elongated" cells into "round", i.e. their exit into mitosis takes place only until the "elongated" will reach primary quantity – 16%, and the number of "round" will be enlarged to 30%. Hence, regulation of proliferative processes (correlation of "elongated" cells and "round") is fulfilled by a principle of negative feedback. The second common in the basal layer population are the cells of peak 2 – 10%. Earlier, it has been shown, that cells of peak 2 are represented by the earliest descendants of epidermal cambial cells which gradually maturing, will be transformed into " narrow" cells, and then into "oval". As a result, the number of "narrow" cells at their transition into "oval" gradually decreases from 10% to 2-3%, and the number of "oval" – increases accordingly. When the number of "oval" cells reach 25-30%, some part of them (15-16%) simultaneously pass into "reserve" cells, filling them up to primary value. Thus, in the population of epidermis basal layer there is a following order of transition of one cells into others: the earliest descendants of cambial cells (cells of peak 2) pass into their derivative form – "narrow" cells which will, in turn, be transformed into "oval", and the latter intermittently pass into "reserve" cells which participate in layer regeneration. Hence, in population of a basal layer are available two streams of proliferating cells: I stream: the earliest descendants of cambial cells (cells of peak 2, "narrow", "oval"), that are gradually transformed into "reserve" cells; II stream: "reserve" cells and their derivative forms (" transitional" and "elongated" cells), maturing into the final cells. Proliferative activity of these two streams cells descends not simultaneously. So, proliferation of cambial cells and their descendants realizes after the depression of "reserve" cells and their derivates proliferation.

It is shown above, that in normal human corneal endothelium, the correlation between the "average" cells quantity and "prevailing" cells in age groups in 20-90 year-old persons fluctuates from 5.4 to 6.4. In the epidermis of the same age groups, the correlation of proliferating cells number to the peak 2 cells number changes from 6.2 to 6.9. These correlations in corneal endothelium and in human epidermis are very close. Probably, "average" cells in endothelial static population is the proliferating group in epidermis, which reached definitive development stage, and peak 2 cells are the "prevailing" cells. Drawing an analogy between peak 2 cells and endothelial "prevailing" cells, it is possible to assume that peak 2 cells (the earliest descendants of cambial cells) together with cambial cells, also carry out regulatory functions in the epidermal basal layer. There is a decrease of cambial cells number from 12 cells in 20-40 years-old persons to 7 in older than 75 years (Yavisheva and Shcherbakov, 2009).

At toxic influence in all age groups, there is a depression of 'reserve' cells number in comparison with a skin of control persons. However, to the greatest degree, this depression is revealed at young persons 20-40 years-old and over 80 years-old. In two other age groups, this parameter is rather stable. Thus, it is necessary to notice that depression of "reserve" cells quantity in young men descends at the expense of active transition into their derivative form – "elongated" cells and the number of the latter considerably increases – to 26% (in control of 22.5%) (Tables 5, 6).

Table 5. "Elongated" and "round" cell number and their correlation in human epidermal basement cell population in control with advancing age

Groups (years)	Patients number	"elongated" cells %	"round" cells %	proportion of "elongated" cells to "round" cells %
20-40	10	22.5 (20-30)	8.3 (7.8-11)	2.7
41-60	10	22.6 (20-27)	7.6 (7.0-9.0)	3.0
61-80	10	22.9 (21-25)	7.9 (7.5-10)	2.9
Over 80	10	20.0 (18-22)	8.8 (7.6-11.5)	2.3

Table 6. "Elongated" and "round" cell number and their correlation in human epidermal basement cell population at intoxication with advancing age

Groups (years)	Patients number	"elongated" cells %	"round" cells %	proportion of "elongated" cells to "round" cells %
20-40	10	26.2 (25-30)	7.1 (6.2-8.0)	3.7
41-60	10	23.7 (21-28)	7.9 (7.2-9)	3.0
61-80	10	25.4 (23-30)	8.6 (7.9-11)	2.9
Over 80	10	19.9 (18-22)	5.3 (4.6-6.8)	3.8

In older people, reduction of "reserve" cells quantity descends in connection with their destruction since the portion of "elongated" cells (19.9%) is practically not changed in comparison with control (20%). It has been shown above that as a result of mitotic activity of "elongated" cells, "round" cells are formed which number increases during proliferation process. Hence, the correlation between the number of "elongated" cells and "round" cells can serve as an indicator of proliferative processes activity in the epidermal basal layer. The greatest depression of "round" cells in comparison with other age groups and with the control is revealed in patients in two groups – 20-40 year-olds and over 80 years-old. In 20-40 year-old persons and persons over 80 years-old, the "round" cells number in comparison with control decreases 16.9%, in age groups of 41-60 and for 61-80 years old, this parameter is on the contrary- enlarged by 3.9% and 8.;9% accordingly. This data argues that at toxic influence, the number of "round" cells is increased only in patients at the age from 41-60 years-old and 61-80 years-old, and this augmentation is accompanied by equivalent ascending of "elongated" cells number. From here, the proportion of "elongated" cells to "round" cells is practically not changed in comparison with the control (2.9-3.0). In patients 20 – 40 years-old and patients over 80 years-old on the contrary, the quantity of "round" cells decreases. Hence, the ratio of "elongated" cells to "round" cells is enlarged to 3.8 % in these two age groups (in control – 2.3, 2.7). It has been shown above that "round" cells are the product of "elongated" cells division. Depression of their quantity in the epidermis basal layer in 20–40 year-old patients and over 80 year-old patients at toxic influence can beevidence that at this pathology in the given age groups, there is depression of basal layer proliferative activity.

Thus, the carried out research has shown that both corneal endothelium and epidermal basal layer cells react to changes of vital organism conditions. However, the endothelium (static population) responses to unfavorable factors by the changing of cell size and epidermal basal layer (dynamic population) – cell form. By means of these parameters (despite their differences) in both types of epithelium, identical changes in tissue reaction at persons of certain age groups have been revealed. It was thus revealed, that for young people (20-40 years-old) and senile people(over 80 years-old), hyperactive response to unfavorable factors action is characteristic. In 41-60 and 61-80 year-old persons, reactance depression in response to intoxication takes place.

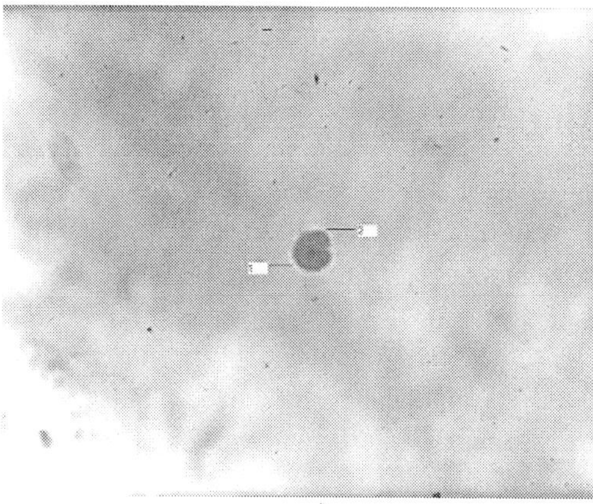

a)

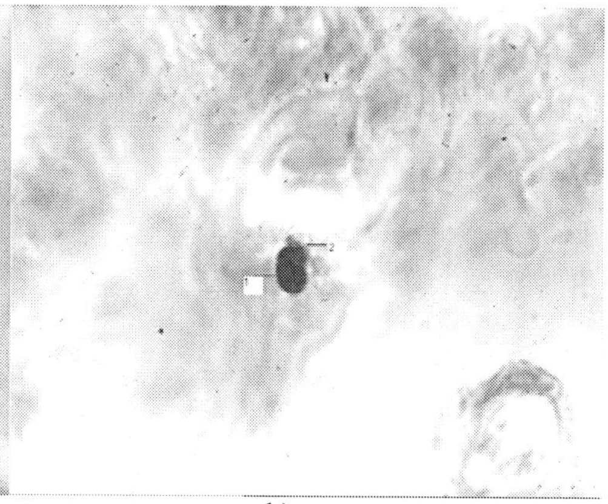

b)

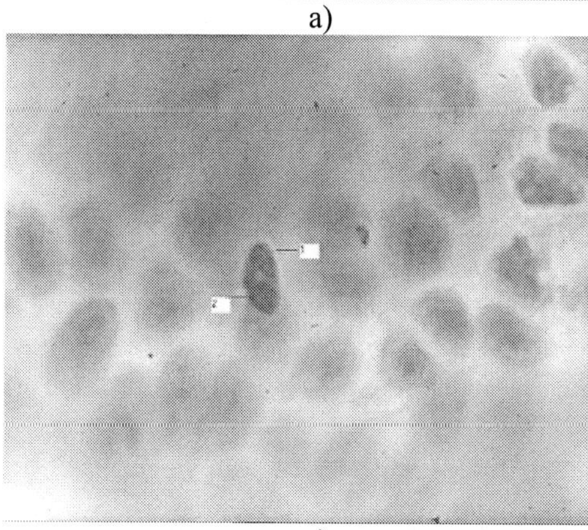

c)

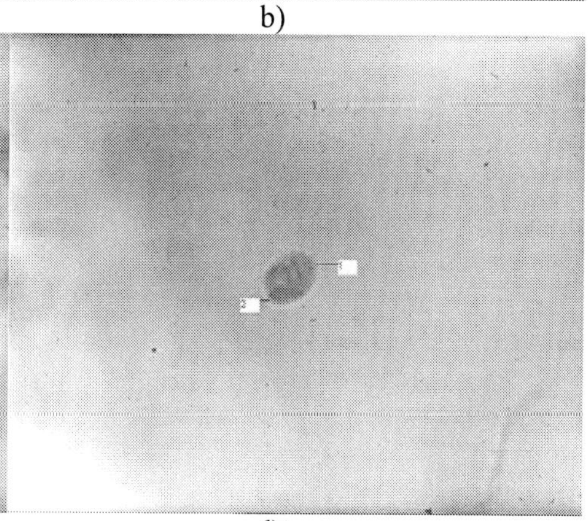

d)

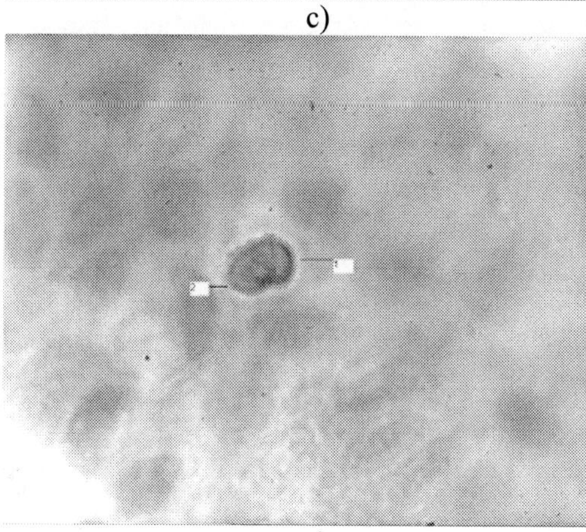

e)

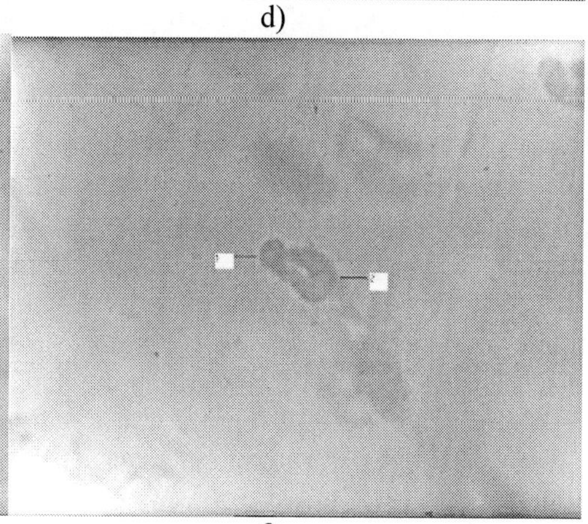

f)

Figure 6. Some stages of transformation of cambial cell during cell division. Staining with hematoxylin, x1000. a). cells resulting from division of the central cells are not separated but remain in close contact with each other; b, c, d).gradual daughter cell polarization in parallel to mother cell; e). mother and daughter cells form the circular structure; f). after disintegration of this structure, the daughter cell transforms itself into the cell of other form. 1 – mother cell; 1 – daughter cell.

V. Structural and Functional Organization of Normal Epidermal Basement Cells

The architecture of the epidermal basal layer and corneal endothelium are basically identical. The differences established consist only that in human corneal endothelium "average", the most active cells are definitive structures, and in the epidermal basal layer during proliferation and differentiation, "average" cells will be transformed into various morphofunctional groups of cells: in so-called "reserve", "elongated", "round". The "prevailing" cells in endothelium and cambial cells with their earliest descendants in the epidermis carry out the regulatory functions in cell population. "Round" cells are gradually transformed into the final cells and eliminated from population. Thus, the physiologically normal desquamation of final cells leads to the activation of the "reserve" ones, i.e., feedback relationships exist between these cells. A decrease in the population density of the "reserve" cells leads, in turn, to the proliferation of cambial cells; the number of the latter is reduced, whereas that of the mother and daughter cells (the product of their division) is increased, i.e. the feedback effect is also characteristic of the cambial and "reserve" cells. The proliferation of cambial cells has some specific features and closely related to the layer structural organization. Each cambial cell is located in the center of special structure, rosette, consisting of six or seven cells at the periphery and three or four cells in the center. Small round or oval cells with large round dark-stained nuclei surrounded by a narrow cytoplasmic rim were detected in all experimental mice on days 2 and 3 after neupogen injection. Each cell occupied the center of a rosette.

These cells were seen in some rosettes (about 20-24) and were absent in others (also about 24), that is, alternating zones with or without central dark cells were observed over the entire surface of the sample. Unlike other cells, the long axis of the cambial cell is square with a basal membrane, not in a line. Therefore, the mitosis of central cells was characterized by certain specific features: two cells formed as a result of division did not part, but closely adhered to each other (Figure 6, a). One of these cells located closer to the basal membrane (the mother cell), the other at some distance from it (the daughter cell). This leads to the axial polarization block of the mother cell because it is mechanically compressed by both the daughter cell and the basal membrane. Thus, the mother cell becomes planar, and the negatively charged chromatin density rises sharply, compared to the daughter cell, and this brings a charge redistribution between the two cells and causes electric field formation (Yavisheva et al., 2004; Yavisheva et al., 2007). While remaining in contact with each other, the mother and daughter cells form a circular structure (Figure 6b,c,d,e). In this structure, there is a field-like effect on the daughter cells. This hypothesis is supported by the detection of the circular structures in the center of epidermal rosettes (Figure 7a,b,c). The changes in circular structures are indicated by the deformation of mother cells only, because the daughter cells are situated in the other plane preventing their analysis. When polarity of the mother cell coincided with the direction of the external electric field applied for 2 min, the surface of the mother cell facing the negative pole was flattened. When the direction of the electric field was inverted, the flattened surface of the mother cell resumed its elongation towards the positive pole. Identical orientation of the circular structures is an important factor for positive correlation of electrical activity of the cells and generation of endogenous electricity. After disintegration of these structures , the mother cells were chaotically distributed, while the daughter cells transformed themselves into the cells of other forms (Figure 6f). Thus, the electric field affects differentiation of the daughter cells only at the stage of circular structures. Nevertheless, performance of daughter cells and maturation take place not in one rosette but in the morphofunctional zone (region) composed of 24 rosettes. This region is functionally divided into two equal subunits (containing 12 rosettes with cambial cells inside each other), acting in turn. Primarily, one subunit of cambial cells undergoes division and forms 12 pairs of mother and daughter cells, then other subunit also forms 12 pairs. When the amount of daughter cells in both subunits reaches a total of 24 items, 12 of them, formerly born, and 12 others are then gradually transformed into "reserve" cells (30%), providing physiological regeneration in the morphofunctional region (Yavisheva and Yagubov, 1996). Interestingly, each subunit undergoing proliferation does not have 12 cells entering mitosis together. At first, one subunit of 6 cells divide and only then do the other 6 cells because only up to half of all cells can divide simultaneously. Daughter cells from the first six pairs of mother and daughter cells become slightly drawn along their axis; this means that the field excited by these pairs is not enough for differentiation. Moreover, only because the amount of these pairs rises up to 12, due to other six cambial cells of the same subunit division, the daughter cells lengthen along the field gradient and gravitate towards negatively charged mother cells at the same time, resulting in daughter cells stretching along the basal membrane. Then, unlike charged mother and daughter cells conjoin forming a circular structure, which disbanding leads to the daughter cell transforming to other forms (i.e., differentiation (Yavisheva et al., 2004; Yavisheva et al., 2007).

Therefore, differentiation of the epithelium daughter cell can be passed in an electric field excited by 12 mother and daughter cell pairs that are disposed in the same subunit. Six pairs of these cells form a field with a limit value when daughter cells differentiation does not happen. However, epidermal daughter cells cannot differentiate without derma cells. It emerged that derma has morphofunctional regions identical with the epidermal regions (Yavisheva et al., 2007). The derma papillary layer fibroblast cambial cells function simultaneously with epidermal cells over the basal membrane, entering mitosis together and passing equal transformation stages. Because the number of mother and daughter cells in one derma subunit reaches 12 items, their daughter cells undergo differentiation. However, dermal circular structures have a more elongated form, and the daughter cells long axis is directed normally to the basal membrane (Figure 8).

Consequently, epidermal and dermal regions function principles are identical. Acting together, epidermal and dermal regions form a unified region. Thus, there appear 12 pairs of epidermal and dermal mother and daughter cells in one subunit of a unified zone with synchronous cambial epidermis and derma cell division. Dermal daughter cells first leave the unified zone because of their long axis, unlike epidermal daughter cells which are directed normally to the basal membrane. After this, there is a light constriction of other cells in 1.3-1.5 times and their mitosis begins. Consequently, epidermal and dermal daughter cells exert unlike effects to each other due to the production of their regulatory factors: epidermal factors cause the sensitive cells to constrict, and dermal factors lead to the lengthening of cells. Thus, 12 formed epidermal daughter cells constricted by their growth factors are to lengthen out in an electric field, meaning they are to differentiate only when the 12 dermal daughter cells growth factors relax their cortexes.

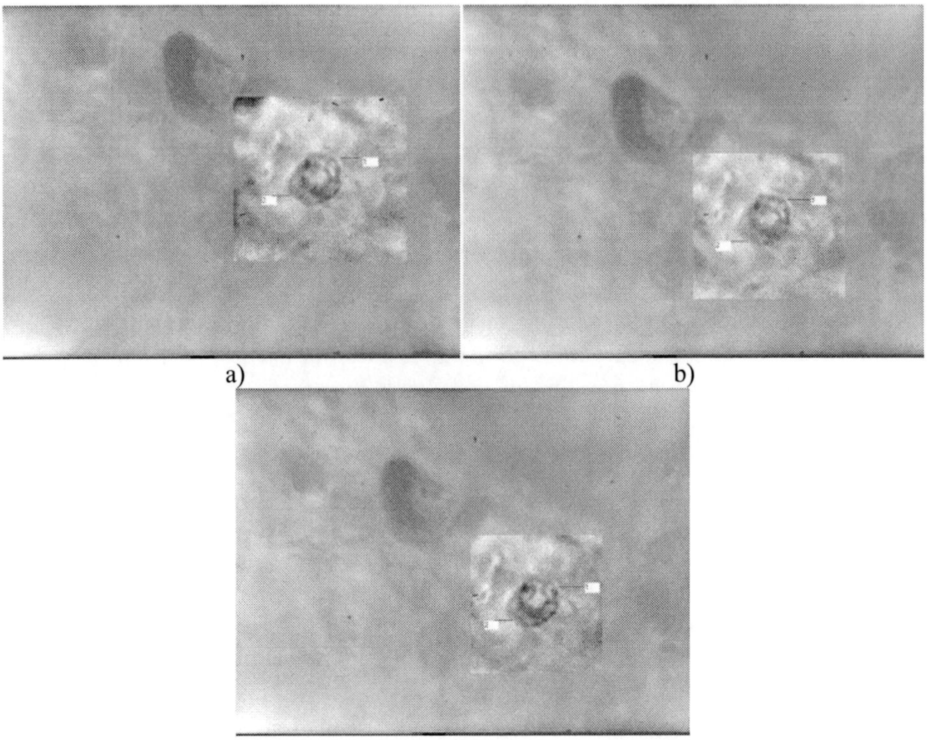

Figure 7. Effect of electric field (300V, 2 min) on a mother cell. Staining with methylene blue, x1000. a) mother and daughter cells incorporated in the circular structure before electrical stimulation; b) if the polarities of electric field and mother cell coincided, the cell was deformed, and its surface facing the negative electric pole was flattened; c) after changing the polarity of the electric field, the flattened surface of the mother cell was elongated towards the positive pole of electric generator. 1 – mother cell; 2 – daughter cell.

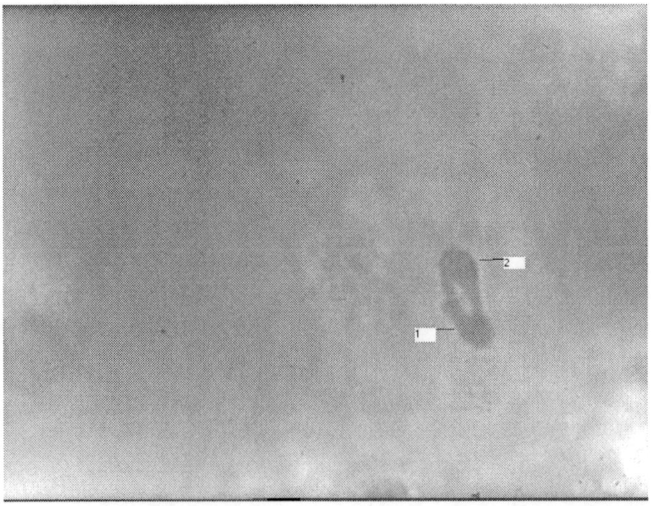

Figure 8. Dermal circular structure has more elongated form, and the daughter cells long axis is directed normally to basal membrane.Staining with hematoxylin,x1000. 1 – mother cell; 2 – daughter cell.

Vi. Characteristic Features of Function of Morphofunctional Zones in Breast Fibroadenoma and Breast Cancer

Because histological structures in benign and malignant tumors represent the same cell systems that the epidermal proliferative units, functional units, and rosettes that are characteristic of normal epithelium, it can be concluded that these tumors occur as a result of certain rearrangements of functional units of normal tissue. In this study, we found that breast fibroadenoma develops from a single morphofunctional zone of normal epithelium in which the cambial cells number decreased 1.5 times (Yavisheva et al., 2005). The disturbance occurs in one of epithelium subunits while the underlaying stroma is not changed. The epithelial cambial cells number in this subunit drops to eight, and in the corresponding stromal subunit remains in normalcy of twelve cells. Unlike the other subunit where there is a balance between the number of cambial epithelial and stromal cells, here the number of stromal cambial cells on four more than epithelial ones is primary. Therefore, in the first subunit, the stromal mother and daughter cells pairs also are formed on four more than epithelial. Then, at entrance into proliferation of second subunit cambial cells and during the mother and daughter cells formation, four epithelial mother and daughter cells pairs of this subunit will be exposed at once to influence already available regulative factors of four "odd" stromal daughter cells of the first subunit that will cause a cortex relaxation of these second subunit four epithelial cells and their possibility to be polarized under the influence of electric field. In spite of the fact that the electric field in the second subunit was created by four epithelial mother and daughter cells pairs that is below threshold (it is created by six pairs), the stretching of daughter cells does take place, since this field will co-operate with the first subunit field formed by eight epithelial mother and daughter cell pairs, which value is above the threshold. Thus, the epithelial daughter cells with a positive electrical charge in one subunit will deviate and will be polarized in the direction towards the negatively charged mother cells in the other subunit and vice versa. And, the deviation of the second subunit four daughter cells in the electric field (as more weak) will be more appreciable than of the first subunit eight cells. Hence, in each proliferative period, the daughter cells growth vectors number is increased by four until it becomes twenty. This is connected with the fact that the epithelial daughter cells quantity in both subunits is twenty (eight + twelve), and each daughter cell has the constant angle of deviation (voltage lines according to which they are polarized, are always definitely distributed in space). Therefore, such daughter cells and their derivatives will be not in parallel to the basal membrane as in normalcy, but under an angle to it, that leads to the development of nodus formation. As a result, the epithelial focus of non-cancer growth consists of four, eight, twelve, sixteen and twenty radiate sectors, each of which is presented by neogenic histological structures. Hence, such structures in radiate sector are derivatives of one cambial cell division.

We discovered that a malignant tumor as well as a benign tumor develops from a single morphofunctional zone of normal epithelium, and stroma hasn't been changed. But the cambial cells number (and, therefore, the mother and daughter cells number) decreased at least two times. For example, during functioning of the first subunit, only six daughter cells are formed in the tumor (instead of twelve cells in normal tissue). However, these cells will not undergo differentiation: as shown above, six mother and daughter cells pairs is the critical level when no electric field is created; as a result, no differentiating effect on daughter cells will be exerted. Unlike the benign tumor, there is no interaction between two subunits, because of the electric field threshold level in each subunit.

Thus, the cambial cells quantity regulates the processes of proliferation and differentiation in epithelium. During twelve cambial cells division, the electric field is created in which the daughter cells are stretched and differentiated. Six cambial cells cannot create an electric field and stretch the daughter cells, therefore there is no daughter cells differentiation.

VII. Inter-relation of the Described Processes of Proliferation and Differentiation with Tissue Transduction Paths

All of the processes may be expressed biochemically. So, on the basis of literature data, we analyzed the above described proliferation and differentiation processes with the help of proteins.

We know that cell constriction happens due to the contraction cortical actin-myosin complex (Riddley, 1996). The critical importance is the RhoA protein, one of the small G_proteins (Cdc42, Rac, RhoA) regulating actin cytoskeleton maintenance. The RhoA protein activates Rho kinase, which phosphorylates the myosin light chain, resulting in increasing myosin actin binding. Thus, a decrease in RhoA activity is needed for the epidermal as daughter cells relax their cortexes.

The central role there belongs to tyrosine kinase Src, a basic proliferation, differentiation, and cell migration regulator. In the active state, Src phosphorylates p190 RhoGAP protein, which inactivates RhoA (Arthur et al., 2000; Brunton et al., 2004). Src has one catalytic and two other domains located in the N terminus: SH2 domain binds to specialized phosphotyrosine sites and SH3 to polyproline sites. The C terminus contains a specific tyrosine residue, which C_terminal Src kinase (Csk)_dependent phosphorylation leads to its SH2 domain interaction and consequential Src inactivation. Src activation can be bypassed by this tyrosine residue dephosphorylation or ligand binding of the SH2 and SH3 domains. Therefore, factors in dermal daughter cell growth seem to activate the Src kinase epidermal cells. This can be achieved as follows. Epidermal daughter cells in a unified region are affected by epidermal growth factor (EGF) produced by them and dermal daughter cells growth factor (HGF/SF or FGF type); these factors stimulate results in activating the appropriate epidermal cells membrane receptors and their autophosphorylation. This leads to p85 protein activation–3-phosphoinosite-kinase (PI3K) adaptor subunit–cell proliferation and migration regulator of great importance (Cuevas et al., 2001; Jimenes et al., 2002) acting not only at receptor areas but throughout some mediators located in the central parts of the cell (Walker et al., 2007). The protein p85 is activated through its module domains CH3, BCR, polyproline PRD, and two SH2 domains (cSH2 and nSH2) and in between iSH2 located ad interim and connected to the catalytic subunit of protein p110. The inactive p85 nSH2 domain is aligned toward the iSH2 domain the way that nSH2 inhibits p110 activity (Shekar et al., 2005). Despite the fact that the nSH2 domain plays a primary role in inhibiting p110, both the nSH2 and cSH2 domains stimulation only leads to complete p110 inhibition deprivation. Moreover, epidermal receptors activated by EGF bind mainly to the nSH2 domain, which directly modulate p110 activity; HGF/SF Met receptors primarily bind to the cSH2 domain and induce conformational alteration in the SH3, BCR, and PRD domains (Suenaga et al., 2005). This brings us to a break in p85 intramolecular SH3 and PRD interactions resulting in possible PRD interactions with other SH3 domain proteins. One of these proteins is inactive Src located at the perinuclear area in and around centrosome membranes. Src joins the active p85 subunit PRD with its SH3 resulting in Src SH3 domain and its kinase activation. Consequently, activation of Met receptors, or activation of other receptors having a high affinity to the PI3K cSH2 domain, furthers the expression of abundant Src kinase SH3 domain in cells. Thus, in PI3K, activation of both the cSH2 and nSH2 domains disrupts the inhibition of the p110 protein and makes RAS joining possible—joining of G protein plays a great role in ERK (extracellular signal_regulated kinase) activation. Furthermore, PI3K kinase activity is increased.

Thus, inactive Src primary activation taking place in the center of the cell occurs through its SH3 domain, joining the active PI3K. Then, activated Src kinase decreases RhoA expression phosphorylating p190 protein RhoGap. Yet RhoA protein has two effectors, Rho kinase and formins—big multi-domain proteins acting in forming non-branching actin filaments (Brunton et al., 2004). Therefore, a decrease in RhoA activity lowers Rho kinase function, resulting in the attenuation of the actin_myosin interaction and cortex relaxation in epidermal daughter cells. Furthermore, a decrease in RhoA reduces formins, another RhoA effector level. At the same time, actin redistributes from the cortical cell layer to the central layer. Indeed, formins are in an inactive state due to the inhibitory N_terminal domain (DID, diaphanous inhibitory domain) joining the autoregulatory C_terminal DAD domain (diaphanous autoregulatory domain). There are FH1 and FH2 domains in between. RhoA activates formins joining GBD (GTPase_binding domain) and DID domains of the formin molecule and promotes DID from DAD dissociation resulting in activation of the FH2 domain, which is necessary and sufficient for the nucleation of actin filaments. Proline_rich FH1 domain adjunction to FH2 leads to more effective actin nucleation. However, formation of the FH1–FH2 complex is not sufficient for generating rapid non-branching filaments growing from the barbed end. It is necessary that FH1 domain binds to the profilin:actin complex because FH1–FH2 construction without profilin acts in a classic capping protein mode or nucleates actin from a new fixed point (Kovar et al., 2003; Michelot et al., 2005). On average, more FH1 domain active profilin:actin binding sites means a bigger velocity of actin filament elongation from the barbed end. FH1 has low affinity to profilin, however, that is why it needs to be activated, securing by Src expressed in the cell center where lots of formins are found. Binding to polyproline FH1 domain sites by its SH3 domain Src kinase phosphorylates them, providing profilin:actin adjunction. In respect that the active number of formins decreases with lowering of the amount of RhoA, and the active Src kinase level is kept the same, more FH1 domain polyproline sites are to be activated. As a result, binding the profilin:actin complex to the active FH1 domain increases the concentration of actin and promotes its orientation correctly toward the barbed end, that leads to rapid, long non-branching filaments at the center of the cell. Their RhoA_kinase interaction leads to the formation of stress fibers.

Thus, the action of 12 dermal daughter cells growth factors to 12 epidermal cells results in the relaxation of epidermal cells cortex due to the activation of the Src kinase SH3 domain, which decreases RhoA expression and leads to the beginning of actin redistribution from the cortical layers to the body of the cell and provides primary cell polarization. Only now can epidermal daughter cells and their nuclei stretch out in the electric field. However, stretching of the nucleus is possible only with microtubule development, which, interacting with intermediate filaments, stretches the nucleus (Alberts et al., 1994). Consequently, with relaxed epidermal cells cortex stretching in the electric field, microtubule system development takes place, and primary filament system growth occurs contemporarily, of which assembly and formation is provided by microtubules (Chang et al., 2006; Godsel et al., 2008). Thus, stretching of an electric field cell leads to increased exocytose bubbles splitting off from the Golgi complex, which are to be inserted into the cell membrane, and to active outflowing of the new dose of the Src, which is localized in the membranes. Src, PI3K, and γ_tubulin in centrosome membranes form large protein complexes with nucleating microtubules (Sulimenko et al., 2006). However, inactive Src interacts with γ_tubulin weakly because it has a low affinity to it. Active PI3k can interact with γ_tubulin via its SH2 domain, which binds to certain γ_tubulin tyrosine residues and phosphorylates them, resulting in γ_tubulin activation (Macurek et al., 2008). Just after that, inactive Src can associate with the phosphotyrosine γ_tubulin site via its SH2 domain, which is needed for steady tubulin dimmer SH2 joining. At the same time, the SH2 domain kinase site is activated. Consequently, localized Src in the inactive cell center is going to be activated again but with help from its own SH2 domain, not SH3, which plays an important role in the nucleation and polymerization of the microtubule. Moreover, Src kinase binds up γ_tubulin to membranes that do not have any membrane association sites. This leads to microtubules minus ends fixing at the centrosome preventing dissociation of tubulin dimmers and promoting nucleation. Therefore, the epidermal daughter cell stretching in the electric field conduces to its Src SH2 domain activation, providing development of microtubules, which facilitate formation of intermediate filaments and nucleus stretching.

On entering and exiting, the mitosis cambial cell axis is normal to the basal membrane. This occurs due to the lack of Src SH2 domain expression—as this cell stretching mechanistically has been disabled by a daughter cell—and to sharp RhoA increasing versus the norm. Hence, contraction of cambial cells increases and stress fibers, microtubules;, intermediate filaments do not appear, preventing cell spreading along the basal membrane and determining the normal axis of the cell. It is known that chromosome-ordered orientation occurs at the end of mitosis because chromosomes are being separated by microtubules attached to the centromeres (anatelophase orientation): thus, in daughter cells with a vertical division line, centromeres are located in the upper cell pole and telomeres in the lower pole and conversely in mother cells. Therefore, the axis of the cambial cell and the axis of the pre-stretched daughter cell coincide in a certain chromosome direction. The most effective stretching force produced by the electric field affects lower cell areas, bordering mother cells; that is why formed microtubules and intermediate filaments shift the lower poles of the daughter cells nucleus and stretch them in parallel with the basal membrane. Consequently, the centrosome and telomeres fixed at the nucleus membrane turn from a vertical to a horizontal direction and gather before the leading edge of the epidermal cell. Therefore, chromosomes having certain nucleus periphery, binding sites and with rosette-like glomerular packing are going to stretch nearby telomeres resulting in loops unwinding and chromatin decondensation at these sites making transcription possible. RhoA activity in dermal daughter cells is decreased better than in epidermal cells due to an increased SH3 domain and kinase site activity of Src. Thus, the amount of active formins and stress fibers is lowered, leading to significant relaxation of the cortex of these cells and to their electric field-mediated stretching boost; the SH2 domain expression also rises and, therefore, microtubules amount rises as well. Increased microtubule polymerization stretches the dermal daughter cell nucleus along its axis in a vertical direction, driving the centrosome there (i.e., closer by centromeres unlike in epidermal cells with its telomere-close manner). Thus, an electric field-differentiated cells centrosome, with Src kinase located in its membrane, localizes not only along the cell axis but at its leading edge and at the nucleus pole with centromeres or telomeres where chromosome looping and epitheliocytes or fibroblast-like cells formation occurs. At the same time, the active edge of a cell with a Src kinase concentration is formed there, and the Src to RhoA expression ratio provides cell differentiation.

After stretching results in differentiation, epidermal daughter and mother cells come apart and the daughter cells spread along the basal membrane according to the polarity gained. In this cell, due to Src SH2 and SH3 domain activation, stress fibers, microtubules, and intermediate filaments are preformed and can provide this process. Activated through its SH3 domain, Src descends to focal contacts from the centrosome area (i.e., from the leading edge of the cell). However, focal contacts should be prepared. Indeed, with stimulation of cell receptors by growth factor, phospholypase C (PLC) is activated immediately in the focal contact area that finally leads to phosphatidylinosit-4,5-diphosphate (PIP2), cofilin, Cdc42, and WASP (Wiskott_Aldrich syndrome protein) protein activation regulating actin polymerization (Mouneimne et al., 2004; 2006). WASP proteins expressed with the help of PIP2 and Cdc42 communicate with Src, bringing it to this area. Then, different signal molecules are stimulated through Src, with Rac among them, which is involved with lamellopodium formation. With the cell spreading, actin redistribution from cortical layers to stress fibers area is still in progress. Cell spreading stretches the cell surface and increases Src new-dose appearance and its SH2 activation, resulting in its total kinase activity increasing and temperate RhoA and active formins decreasing. Enhanced Src activation leads to the amplification of active FH1 domain profilin:actin binding sites and to the formation of long actin strands at the cell center. Their interaction with Rho kinase results in stress fiber formation, which is due to tractive force, stretches focal contacts transforming them to focal adhesion areas. Thus, activated Src primarily stimulates different ways of transduction including the RAS/ERK cascade and the PI3K/AKT pathway (serine/threonine protein kinase B). Entering the nucleus, ERK enhances many genes and results in two events: cyclin D transcription, with ERK-activated Myc protein involvement, and yclin-dependent kinase inhibitors induction (p21/p27), which prevents progression of the cell cycle. First, ERK phosphorylates Myc and increases its activity but then provides conditions for Myc proteasome-mediated degradation. The other signal cascade, PI3K/AKT, promotes Myc stabilization leading to preventing p27 inhibitory action on cyclin-dependent kinases and to catalytic telomerase subunit activation (Aubert, 2008). Therefore, AKT induces p21 to exit from the nucleus to the cytoplasm where proteasome degradation occurs (Chambard et al., 2007). Yet, with increasing Src/RAS/ERK and inhibitor activity, AKT inactivates them incompletely due to promotion of both inhibitors inactivation. This also leads to low-level Myc expression and its action on telomerase reduction.

The DNA synthesis process cannot start in the presence of proliferation inhibitors and needs to be under other conditions. Thus, there are diverged epidermal and dermal mother and daughter cells in a unified region, staying in the interphase. When dermal daughter cells leave the region, light, 1.3–1.5 fold, constriction of all the surfaces of the cells occurs, whereupon DNA synthesis starts. Primarily, dermal daughter cells exit the region because their axis is directed inward to the tissue not along basal membrane, unlike epidermal cells axis. Thanks to epidermal daughter cells located in place, and dermal cells, with activated Src SH3 domain, drift inward to the tissue, epidermal daughter cells with expressed SH2 domain solely begin to prevail in a unified zone. This leads to a decrease in Src expression and the prevalence of RhoA in these cells. RhoA prevails even more in cambial mother cells, as the SH2 domain is low- expressed in them. Therefore, cell stretching and the decrease in RAS/ERK cascade expressions happens thanks to the lowering of Src and consequent inhibitor synthesis reduction. However, because RAS activates not only ERK but also PI3K, decrease in RAS activity can lead to a lowering of its PI3K binding. At this time, PI3K is going to be activated primarily through expressed RhoA, which can interact with p110 (Reuveny et al., 2004), thus AKT activity is not going to decrease. AKT-mediated stabilized Myc protein can activate telomerase, and consequently, the prevalence of RhoA over Src expression promotes its synthesis. Owing to inhibitor synthesis reduction and to AKT activity keeping the same level, all inhibitors are going to be inactivated. As a result, the cyclin/cdk2 complex will pass over its inactivation, and DNA synthesis will proceed unlike RNA. Indeed, cell contraction brings to DNA loops arrangement density increasing in nuclei, which may lead, in turn, to a blocking of transcription factors (Georgiev, 1989). Thereby, RNA synthesis begins to lower in epithelial daughter cells, which brings a slowed down use of ribonucleotides that is permanently formed in the organism. Because unused ribonucleotides are gradually transformed into desoxyribonucleotides, the latter is going to accumulate in the cell, resulting in potentially active genes of their induction of ferment complex action. Thus, synthesis of the DNA-matrix-based nucleic acids will commence but with a desoxyribonucleotide intake. Active chromatin is going to be replicated first because the mechanisms of DNA-matrix-based RNA and DNA nucleic acids formation are the same. Thanks to desoxyribonucleotides involved in the synthesis, which is, in principle, different chemically from ribonucleotides, DNA molecules will have a double-stranded structure (Mushkambarov and Kuznetsov, 2003).

The aforesaid cambial cell proliferation and differentiation scheme is significant when all 12 cambial cells work in each subunit of unified zone, forming an electric field sufficient for differentiating daughter cells (i.e., for their conventional stretching and for necessary Src kinase SH2 domain activation). Such an amount of cells was detected in not only 20–40 year-old men but also in 41–59 year-old men. Men older than 60 years have 11 cells and men older than 75 have 7 cells (Yavisheva and Shcherbakov, 2009). Therefore, the amount of human skin cambial cells specifying differentiation lessens as life progresses. Because the number of human skin cambial cells stays at one level during a long lifetime, however, and the cell differentiation rate changes and senescence marks accumulate, the function of epidermal cambial cells is up-regulated by other factors. Indeed, these cells act in a unified epidermal-dermal region where they are markedly affected by dermal fibroblast growth factors. We know that fibroblasts do not have an active telomerase, this leads to incomplete replication of DNA end regions and telomere shortening, which slowly accumulates as life progresses. The lack of telomerase is a consequence of a fibroblast prevailing of Src over RhoA expression of its activity block. Every division telomere shortening leads to greater chromosome stretching between their nucleus membrane fix points and an increase in Src SH2 domain expression. Thus, with the fibroblast division number growing, the SH2 domain expression and total kinase activity of Src also increases, and that of RhoA decreases. However, as the RhoA protein promotes proliferation, inhibitors exit from the cells and DNA synthesis, RhoA lowering with time will lead to fibroblasts growth factors and their proliferation decreasing. With critical Src increasing, the level of cell proliferation stops as Src kinase inactivates RhoA protein that is essential for cell division. Consequently, the Src to RhoA expression ratio determines the number of cell divisions. Because fibroblast growth factors activate epidermal cambial and daughter cells in the Src kinase SH3 domain, factors cut gradually leads to this domain expression decreasing and boosting RhoA. Consequently, rigidity of the cortical layers increases in epidermal cambial cells that results in chromatin density and lowering electric activity. Such an electric field cannot stretch daughter cells sufficiently with high contraction, which means that SH2 domain expression is lowered. Therefore, despite 12 cambial cells acting in each subunit in 20–40 year-old and 41–59 year-old men, SH2 domain expression is gradually decreased during one's lifetime because the imbalance of Src and RhoA affects the way these cells behave. Thus, chromosome stretching near telomeres decreases resulting in their incomplete untwisting and looping along with alteration in protein transcription. In the future, with incremental RhoA expression and lowering of the Src level, contraction of epidermal cambial cells will increase derating transcription processes and synthetic activity in these cells and being able to lead to cell death with straightforward occurrence. This process can probably cause the reduction in the number of these cells with advancing age. After 60 years of living, the number of these cells becomes 11 resulting in a significant decrease in the electric field and reduction in the Src SH2 domain expression and altering protein transcription. Thanks to the reduction of the Src level and boost in RhoA, stress formation of fibers and microtubules declines, stretching out the cell at an angle with the basal membrane (i.e., cells migration and maturation are going to be passed along this axis not in parallel to the membrane). This provides quick cell desquamation and skin layer diminution and skin thinning. A reduction in the number of epidermis cells leads to a reduction in the formation of the epidermal growth factor, resulting in total RhoA expression and decrease in cell proliferation. Therefore, the RhoA to Src expression ratio is kept the same due to corresponding Src activity reduction in these cells. Over 75 years, the number of cambial cells number is reduced to 7, causing their death. Stretching epidermal daughter cells and Src SH2 domain expression fall to a greater degree according to 60 year-old men and over, this enhances RhoA toward the expression of Src. Therefore, stress fibers and microtubules formation is decreased, bringing more vertical stretching (i.e., closer to centromeres) to the daughter cell where chromosome looping of fibroblast-like cells occurs. Cells desquamation also increases due to a shifting of the direction of the main axis resulting in a reduction in epidermal growth factor, in RhoA expression in layer, and a decrease in cambial cells and proliferation of their offspring. Consequently, epidermis parenchymal cells from people older than 75 years reduces, thanks not only in increased desquamation but to probable formation of connective tissue cells. Interestingly, obtained from 75+ years men, cambial cells number is close to a critical level (6 cells) when daughter cells differentiation doesn't occur or does weakly.

Therefore, these cells number lowering down to 6 in one subunit of region can lead to malignant tumor formation. Thus, among senior and old people oncologic disease occurrence is high. But tumor can develop in junior age, when total cambial cells number is still high. Moreover, first changes occur in a one epithelium subunit, and the stroma hasn't been changed (Yavisheva et al., 2007). Cambial cells number there lowers to 6 but in corresponding stromal, the subunit is still normal—12. Thus, with simultaneous epithelium and stromal cambial cells proliferation in one subunit, stromal mother and daughter cells 6 pairs prevail to epidermal pairs. Then, as pathologically unchanged second subunit cambial cells enter mitosis, growth factors formed by 6 "odd" stromal daughter cells of one subunit start to affect the newly formed 6 pairs of epithelium mother and daughter cells in another subunit, i.e., activate their Src kinase SH3 domain inducing cortex relaxation. But each epithelial subunit field is formed by only six mother and daughter cells pairs and has critical value, at which different subunits fields interaction doesn't occur and, therefore, daughter cells stretching is insignificant or lacks. As epithelium subunit cambial cells number becomes less (less 6) and, consequently, mother and daughter cells pairs as well, formed field and stretching force decrease. Therefore, along with Src SH3 domain expression corresponding to age, epithelial daughter cells SH2 domain is weakly expressed resulting in total kinase Src activity reduction and in notable RhoA activity and active formins amount increasing. This leads to FH1 domain profilin: actin binding sites number lowering. FH1–FH2 complex is going to nucleate actin from new points, as is shown above, what will bring to short actin cords number increasing in cell cortex and to its certain spasm. Therefore, long actin filaments formation leading to stress fibers production are going to be sharply decreased. Critical stress fibers, microtubules and intermediate filaments number lowering occurring due to very low SH2 domain activity, brings to focal adhesion size and number lowering and cell spreading failure. RhoA expression increasing leads to turn cell and nucleus axis vertical. Weak chromosome looping will occur at not only telomeres, but at centromeres. Consequently, with process aggravation, the tumor cell might have not only epithelial but fibroblast phenotype. Indeed, cytokeratin and vimentin microfilaments, and growth factors with spastic and relaxing crucial for cell growth and division, can be expressed together resulting in independence of the cell from the presence of serum growth factors. The cell axis turns at an angle with the basal membrane directing migration of tumor cells in the same direction, which provides tumor growth in tissue place or on its surface in the form of nodus. Very weak Src SH2 domain expression does not lower RhoA back to normal; therefore, the RAS/ERK and RhoA/PI3K/AKT paths are stimulated at the same time in interphase. Their overexpression, due to stress fibers and the reduction in the number of microtubules, which slows down the transport of endocytosis vesicles, increases the formation of proliferation inhibitors, but high RhoA activity stimulates p110 PI3K/AKT path, which results in their inactivation. This allows the cell to grow irrespective to substrate adhesion.

Conclusion

Both corneal endothelium, and epidermis basal layer cells react to changes of vital organism conditions. However, the endothelium (static population) responds to unfavorable factors by changing cell size, and epidermal basal layer (dynamic population) cell form. By means of these parameters (despite their differences) in both types of epithelium, identical changes in tissue reaction at persons of certain age groups have been revealed. It was thus revealed, that for young people (20-40 years-old) and senile people (over 80s year-old), hyperactive response to unfavorable factors action is characteristic. In 41-60 and 61-80 year-old persons, reaction depression in response to intoxication takes place.

Proliferation and differentiation of cambial epithelial cells take place in morphofunctional zones consisting of two subunits on 12 rosettes in each and working in turn. Both subunits produce 30 % of the "reserve" cells serving for physiological regeneration of the layer. At division of the 12 cambial cells, 12 mother and daughter cells pairs are formed which create a sufficient electric field in which there is a distention of daughter cells that leads to their differentiation. But the stretching of daughter cells may happen only if their cortex will be preliminary relaxed by the stromal cells growth factors. Hence, the epithelial and the stromal cambial cells form the uniform morphofunctional zone. Depression of epithelial cambial cells number not less than in 2 times leads to the creation of threshold electric field in which there is no daughter cells distention, and, hence their differentiation. With the years, there is a depression of epithelial cambial cells number, therefore the risk of malignant tumor occurrence appears.

References

[1] Alberts, B., Bray, D., Lewis, L., et al. (1994). *Molekulyarnaya biologiya kletki* (Molecular Biology of the Cell). Moscow: Mir.,

[2] Arthur, W., Petch, L., and Burridge, K. .(2000). Integrin Engagement Suppresses RhoA Activity via a c_Src-dependent Mechanism, *Curr. Biol, 10.* 719–722.

[3] Aubert, G. and Lansdorp, P. (2008). Telomeres and Aging. *Physiol. Rev 88.* 557–579.

[4] Booth, C. and Potten, C.(2000). Gut Instincts: Thoughts on Intestinal Epithelial Stem Cells, *J. Clin. Invest.105.* 1493–1499.

[5] Brunton, V., MacPherson, Y., and Frame, M.(2004). Cell Adhesion Receptors, Tyrosine Kinases and Actin Modulators: A Complex Three_way Circuitry. *BBA_Molec.Cell. Res.1692,* 121–144.

[6] Chambard, J.C., Lefloch, R., Pouyssegur, J., and Lenormand, P. (2007), ERK Implication in Cell Cycle Regu_lation, *BBA_Mol. Cell Res. 1773.* 1299–1310.

[7] Chang, L., Shav_Tal, Y., and Trcek, T.(2006). Assembling an Intermediate Filament Network by Dynamic Contratranslation, *J. Cell Biol. 172.* 747–758.

[8] Cuevas, B., Lu, Y., and Mao, M. (2001). Tyrosine Phosphorylation of p85 Relieves Its Inhibitory Activity on Phosphatidylinositol-3-kinase, *J. Biol. Chem. 276.* 27455–27461.

[9] Fyodorov S.N., Ronkina T.I., Tavisheva T.M.(1993). *Human corneal endothelium.* Moscow: "Eye Microsurgery".

[10] Georgiev, G.P. (1989). *Geny vysshikh organizmov i ikh ekspressiya* (Genes of Higher Organisms and Their Expression). Moscow: Nauka.

[11] Godsel, L., Hobbs, R., and Green, K. (2008). Intermediate Filament Assembly: Dynamics to Disease, *Trends Cell Biol. 18,* 28–37.

[12] Jimenes, C., Hernandez, C., and Pimentel, B.(2002). The p85 Regulatory Subunit Controls Sequential Activation of Phosphoinisitide_3_kinase by Tyr Kinases and RAS. *J. Biol. Chem. 277,* . 41556–41562.

[13] Kovar, D., Kuhn, R., and Tichy, A.(2003). The Fission Yeast Cytokinesis Formin Cdc12p Is Barbed End Actin Filament Capping Protein Gated by Profilin, *J. Cell. Biol.. 161,* . 875–887.

[14] Letuchaya, F.M. and Ketlinskii, S.A.,(1980). Rat Epidermis Functional Units Development During Its Histogenesis. *Tsitologiya.,* 178–180.

[15] Macurek, L., Draberova, E., and Richterova, V. (2008). Regulation of Microtubule Nucleation from Membranes by Complexes of Membrane-bound Gamma_tubulin with Fyn_kinase and Phosphoinositide_3_kinase, *J. Biochem. (Cell).416.* 421–430.

[16] Michelot, A., Huang, S., and Ingouff, M. (2005), The Formin Homoly 1 Domain Modulates the Actin Nucleation and Bundling Activity of Arabidopsis FORMIN1, *Plant Cell. 17.* 2296–2313.

[17] Mouneimne G., Soon, L., and DesMarais, V. (2004). Phospholipase C and Cofilin are required for carcinoma cell directionality in response to EGF stimulation, *J.Cell Biol.166.* 697–708.

[18] Mouneimne G., DesMarais, V., and Sidani, M.(2006). Spatial and Temporal Control of Cofilin Activity Is Required for Directional Sensing During Chemotaxis, *Curr. Biol.16.* 2193–2205.

[19] Mushkambarov, N.N. and Kuznetsov, S.A. 2003., *Molekulyarnaya biologiya* (Molecular Biology), Moscow: MIA.

[20] Potten, C., Booth, C. (2002). Keratinocyte Stem Cells: A Commentary, *J. Invest. Derm.19.*889–899.

[21] Reuveny, M., Helter, H., and Bengal, E. (2004). RhoA Controls Myoblast Survival by Inducing the Phosphatidylinositol_3_kinase_Akt Signaling Pathway, *FEBS Lett.569.*129–134.

[22] Riddley, A. (1996). RhoA: Theme and Variations, *Curr. Biol.6.* 1256–1264.

[23] Savost'yanov, G.A., (2001). Epithelial Places Spatial Organization Theory by the Example of Neuroepithelium, *Morfologiya. 120,* 18–30.

[24] Semkin, V.I. (2001). Morphofunctional Alterations of Epithelial Differons with Aging, *Klinicheskaya Gerontologiya.* 27–31.

[25] Shekar, S., Wu, H., and Fu, Z. (2005). Mechanism of Constitutive Phosphoinositide3_kinase Activation by Oncogenic Mutants of the p85 Regulatory Subunit, *J. Biol.Chem.280.* 27850–27855.

[26] Suenaga, A., Takada, N., Hatakeyama, M. (2005). Novel mechanism of Interaction of p85 Subunit of Phosphatidylinositol3_kinase and ErbB3 Receptorderived Phosphotyrosil Peptides, *J. Biol. Chem.280.* 1321–1326.

[27] Sulimenko, V., Draberova, E., and Sulimenko, T. (2006). Regulation of Microtubule Formation in Activated Mast Cells by Complexes of Gamma_tubuline with Fyn and Syk Kinases, *J. Immunol.15.* 7243–7253.

[28] Walker, V., Ammer, A., and Gao, Z. (2007). PI3K Activation IsRequired for PMA_directed Activation of Src by AFAP_110, *Amer. J. Physiol. Cell Physiol. 293.* 119–132.

[29] Yavisheva, T.M. and Yagubov, A.S. (1996), Quantitive Morphology of Basal Layer Cells from Multi-layer Epithelium of Mouse Cornea and from Human Lungs Malignant Tumor, *Ontogenez.. 27,* 95–99.

[30] Yavisheva T.M., Yagubov A.S., Rukavishnikov S.G. (2000). *Epithelial tissue in norm and cancer*. Samara: Samarskii meditsinskii universitet.

[31] Yavisheva, T.M., Shcherbakov, S.D., Golubeva, I.S.,et al. (2004). Structural Organization of the Basal Layer and Morphofunctional Characteristics of Mouse Epidermal Cambial Cells, *Bull. Exp. Biol. Med. 13.* 584–588.

[32] Yavisheva, T.M., Shcherbakov, S.D., and Dubinkin I.V.(2005).Mouse Skin Epidermal Cambial Cells Proliferation Peculiarities, *Kletochnye Technol. Biol. Med. 3.* 136–139.

[33] Yavisheva, T.M., Shcherbakov, S.D., and Savluchinskaya, L.A.(2005). Morphofunctional Zones Action in Normal Epithelium, Fibro_adenoma and Breast Cancer Cells, *Dokl. Acad. Sci. 404.* 125–128.

[34] Yavisheva, T.M., Shcherbakov, S.D., Sharafutdinov, G.Z.,(2005). About Epithelial Cells Differentiation in Tissue Units System, Dokl. Acad. Sci. *401.* 833–836.

[35] Yavisheva, T.M., Shcherbakov, S.D., and Sharafutdinov, G.Z. (2007). Dermal Cambial Cells (Fibroblasts) Inteaction with Epidermis in Mouse Skin Morphofunctional Zone, *Bull. Exp. Biol. Med.11..* 594–599.

[36] Yavisheva, T.M. and Shcherbakov, S.D. (2009). Morhpofunctional changes of human skin cambial cells and their derivates during aging. *Bull.Exp.Biol.Med. 9,*326-329.

In: Endothelium and Epithelium
Editors: J. Carrasco and M. Mota, pp. 113-135
ISBN 978-1-61470-874-2

Chapter V

Changes of Respiratory Epithelium in Bronchial Asthma

Adriana Grigoraş and Irina-Draga Căruntu
Department of Histology, "Gr. T. Popa" University of Medicine and Pharmacy, Iaşi, Romania

Abstract

Airway epithelium lines the upper part of the respiratory tract and functions to maintain airway integrity by providing both a physical barrier to inhaled noxious agents and a mechanism for their clearance *via* the mucociliary escalator. Ciliated columnar cells and goblet cells are the most important cell types in the normal respiratory epithelium. The other types of cells are represented by basal cells, brush cells and Kulchitsky cells (small granule cells). The normal architecture and function of airway epithelium is disturbed in chronic airway inflammatory diseases such as bronchial asthma. Airway remodeling is the general description for the thickening and restructuring of the airways seen in asthma patients. The characteristics of airway remodeling include inflammatory cell infiltration in *lamina propria*, subepithelial fibrosis, muscular and glandular hypertrophy, thickening of basement membrane and increased vascularity. Damage to the airway epithelium is one prominent feature of bronchial asthma. Epithelium damage includes gap openings, partial or total denudation, goblet cell metaplasia. The increased production of mucus can lead to the formation of plugs that contribute to airway obstruction. Mucin glycoproteins (MUC) are the major macromolecular component of mucus in normal airway epithelium. An increased expression of MUC2, MUC4 and MUC5B has been reported in asthmatic bronchial biopsies. At the epithelium level, in patients with bronchial asthma, appears an increased expression of Ki-67, p21 (WAF1) and decreased expression of FasL, Bcl-2 or Bax, too. Apoptosis of the airway epithelium is increasingly recognized as a potential mechanism by which damage may occur. More work is now needed to comprehend all aspects of this disease, including minimizing the ongoing damage to airway epithelium and consequent airway remodeling.

Keywords: Airway epithelium, bronchial asthma, epithelial lesions

1. Normal Respiratory Epithelium

The respiratory epithelium is a ciliated pseudostratified epithelium with goblet cells which line the airways, from the nasal cavities (except for the olfactory mucosa) to the intralobular bronchi (all the way to the entry in the pulmonary lobule) [1]. Scientifically, clinical terminology frequently uses the term “airway epithelium” while histology and histopathology use the term “respiratory epithelium”. Because this text is founded on microscopic morphological information, characteristic for histology and histopathology, we opted to use “respiratory epithelium”.

1.1. Characterization of The Respiratory Epithelium in Standard Light Microscopy

In standard light microscopy it appears as a heterogeneous mass of cells placed on a basement membrane, often thicker than for other epithelia with nuclei aligned at various levels. This characteristic led to the classification of the epithelium as a pseudostratified epithelium. In the structure of the respiratory epithelium one can easily notice, even in standard Hematoxylin-Eosin (HE) staining, the ciliated and goblet cells together with regenerative cells [2, 3].

Ciliated cells are tall columnar cells with acidophil cytoplasm and an ellipsoidal nucleus situated centrally or towards the basal domain. At the apical pole one can generally recognize cilia, which are numerous tapered expansions 7-10 μm in length and 0.25 μm in diameter (Figure 1). They normally float in the mucus on the surface of the epithelium produced by the goblet cells and the mucous glands in the wall of the airways [2, 4].

Goblet cells dispersed throughout the entire thickness of the respiratory epithelium are also tall columnar cells but smaller than the ciliated cells. In the apical 2/3 of the cell cytoplasm contains mucinogen granules with a merging tendency, making the staining with eosin very faint (Figure 1). The irregular, often triangular nucleus is pushed towards the basal domain of the cell. This microscopic aspect led to the comparison of the goblet cell with a “champagne glass” or with a “chalice flower” [2, 5]. Special stains such as mucicarmin, Alcian Blue or Periodic Acid Schiff (PAS) allow for easy identification of this type of cell by targeting the cytoplasmic mucus [4].

Goblet cells are not present in the structure of the respiratory epithelium at birth, appearing when the newborn is exposed to the external environment. Goblet cells are less numerous than ciliated cells, the ratio in the structure of the general respiratory epithelum being 1 goblet cell to 5 ciliated cells. Variations of this proportion can occur; the number of goblet cells may decrease towards the intralobular territory [5]. The exposure to a polluted environment (including second-hard smoke) leads to an increase in the number of goblet cells throughout the respiratory epithelium.

Renewal cells are cuboidal and can be identified by the alignment of rounded nuclei above the basement membrane [1, 2] (Figure 1).

1.2. Characterization of The Respiratory Epithelium in Electron Microscopy

Electron microscopy uncovers 6 types of cells in the structure of the respiratory epithelium. Alongside the three cell types depicted previously, there are cells with microvilli at their apical pole (type I and II) and endocrine cells (Kulchitsky cells) [2, 5].

In electron microscopy, ciliated cells have cell organelles common for any other cell, the only specificity being the presence at the apical pole of hundreds of vibrating cilia placed in concentric circles.

The cilia ensure the unidirectional displacement, by quick and synchronized movements, of mucus on the on the surface of the cell by which they remove large particles (3-5 μm) from inhaled air. This mechanism of filtration makes them a component of the muco-ciliated barrier. In the nasal cavities, the movement of the cilia is orientated towards the pharynx, while in the trachea it is orientated towards the larynx [6, 7].

Every cilium has a peripheral plasma membrane and an interior cytoskeleton named the axoneme. The axoneme consists of a complex of microtubules: a central pair (C1 and C2) around which 9 doublets of peripheral microtubules are uniformly arranged. The central pair is separated from the peripheral doublets by a discontinuous central sheath projection at 14-nm intervals along the length of the cilium. Every peripheral pair consists of an A-microtubule and a B-microtubule. Every microtubule has 13 protofilaments arranged helicoidally due to alternating polymerization of α tubulin and β tubulin. In the peripheral doublets the complete A-microtubule, consisting of 13 protofilaments, attaches an incomplete B-microtubule which consists of 10 protofilaments. On each A-microtubule there are dynein molecules (a protein with ATP activity) arranged as arms radiating towards the B-microtubule of the neighboring doublet, spaced at 24 nm intervals. Molecules of nexin, present at 86 nm intervals, play an important role in the attachment of an A-microtubule of one doublet to the B-microtubule of the following doublet. The connection between the central doublet and the peripheral one is achieved by radial spokes expanding from each peripheral doublet towards the central pair spaced at 29 nm intervals. At the base of the cilium, in the cell's apical domain, there is a basal body which is a structure similar to the centrosome and is considered as a modified centrosome [2, 4].

Goblet cells have at their enlarged apical pole a series of short microvilli; the basal pole is tapered and fixated on the basement membrane. In the apical two thirds of cell, the cytoplasm is occupied by secretory vacuoles containing mucus. In the basal third of cell we can find the nucleus along with the rest of the cytoplasmic organelles (SER, Golgi complex) [4, 8].

In electron microscopy the renewal cells in the structure of the respiratory epithelium are cuboidal cells which do not touch the surface of the epithelium. Their cytoplasm contains a small number of cell organelles [3].

Type I brush-border cells are tall columnar cells, impossible to differentiate in the epithelial cellular mass in light microscopy. In electron microscopy they have microvilli at their apical pole but in the cytoplasm there can be noted a small number of cell organelles except for mitochondria. At the basal pole, type I brush-border cells come into close proximity with sensitive axonal terminations and thus are considered as cells with a role in chemoreception in the respiratory epithelium [2, 5].

Type II brush-border cells have, like the type I cells, microvilli at their apical pole but they can be differentiated from the latter by better represented organelles, numerous basal corpuscles and the absence of contact with nervous terminations. Due to these characteristics they are believed to be immature ciliated cells [5, 8].

Kulchitsky cells, which appear either isolated or organized into small groups, are cuboidal cells. They are similar to renewal cells but in the cytoplasm there are a series of secretory granules, thus being called small granule cells. The granules have between 100 and 300 nm diameter, are lined by a membrane and have an electron-dense core surrounded by a more transparent halo. The immunocytochemistry techniques proved that these cells produce various polypeptides with endocrine and paracrine activity (serotonin, encephalin, endorphins, calcitonin, bombesin). Consequently, they were classified as enteroendocrine cells. Their number is highest at birth and decreases progressively with aging. Their presence explains the possibility of carcinoid tumors occurring at this level [2].

The cells are consolidated in the epithelium by a series of cell junctions called desmosomes. The desmosomes (*macula adherens*) are junctions between two epithelial cells with the shape of oval discoid structures with a long axis of 400-500 nm and a short axis of 100 nm.

In electron microscopy, the structure of the desmosome is centered by an intercellular space of approximately 30 nm, in the middle of which appears a discrete electron-dense line. On either side of the electron-dense line, the cytoplasm adjacent to the inner face of the plasmalema contains a dense area called an "attachment plaque" or a "desmosomal plaque" (400 x 250 x 10 nm).

Towards these dense areas, intermediate filaments converge. The main molecules in the intercellular space are represented by desmoglein 1, desmocolin 1 and desmocolin 2, which is a part of the cadherin superfamily. In the desmosome plaque there are unglycosylated attachment proteins: desmoplakin 1, desmoplakin 2, plakoglobin, desmiokin or a basic polypeptide. These molecules establish links with intermediate cytokeratin filaments within the cytoplasm of the epithelial cells which insert into the plaque and come out again in a "hairpin-like" trajectory [4, 7, 8].

Alongside adherent junctions there are also cummunication junctions (gap junctions), which allow selective intercellular diffusion of ions and small molecules. In electron microscopy, in the territory of communication junctions a small space (3 nm) is noticed because the cells are not joined directly with their plasma membranes. In this space there are described severeal bridge-like structures passing from one cell to the other, called connexons. Connexons have a tubular arrangement (with a central space) through which ions and small molecules such as cyclic AMP, some hormones or aminoacids can pass. This is similar to two cells which are joined and form a channel with a 2 nm diameter [4, 8].

In association with these cell types, characteristic for respiratory epithelium, in some areas of the respiratory airways there can be found M cells which are epithelial cells associated with lymphoid tissue which differentiate during organogenesis (in the digestive epithelium as well).

The M cell is also called an "en dome" cell due to its peculiar bell shape, with short irregular microvilli at the apical pole. This cell functions as a transport pathway for antigen structures (and not as an antigen-presenting cell). Without intracellular modification, they are released through the basal pole into the space created by the bell-shape of the cell. After transport, antigens are engulfed by macrophages or intraepithelial lymphocytes or can completely cross the surface epithelium reaching the *lamina propria* [5].

1.3. The Relationship between the Respiratory Epithelium and the Basement Membrane

Between the surface epithelium and the subjacent connective tissue there is a special structure called the basement membrane. For the respiratory epithelium the basement membrane is thicker than in other places. It can be readily observed in standard light microscopy due to special stains: PAS (staining burgundy red) or argentic impregnation (it colors in brownish-black) (Figure 2) [3, 8].

In electron microscopy, the basement membrane of the respiratory epithelium respects the complex organization of all basement membranes, consisting of three laminas: *lamina lucida*, *lamina densa* and *lamina reticularis*. *Lamina lucida* and *lamina densa* together form the basal lamina. The basal lamina is synthetized by the surface epithelium. *Lamina reticularis* is produced by the connective tissue [4]. *Lamina lucida* is transparent in electron microscopy, being 45 nm thick. One can observe moderate condensations in hemidesmosome territories.

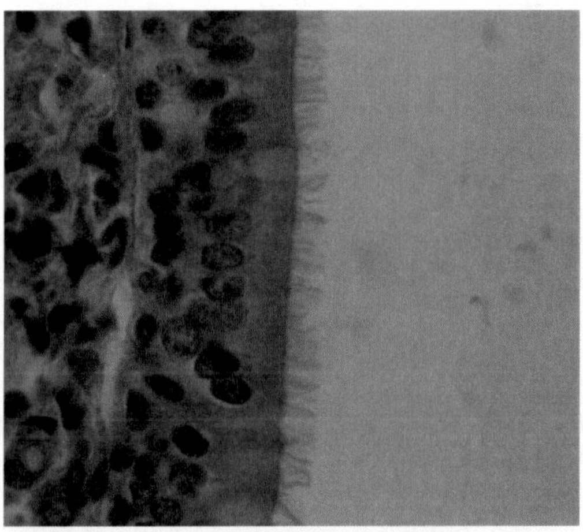

Figure 1. Normal respiratory epithelium (trichrome Szekelly, x 400).

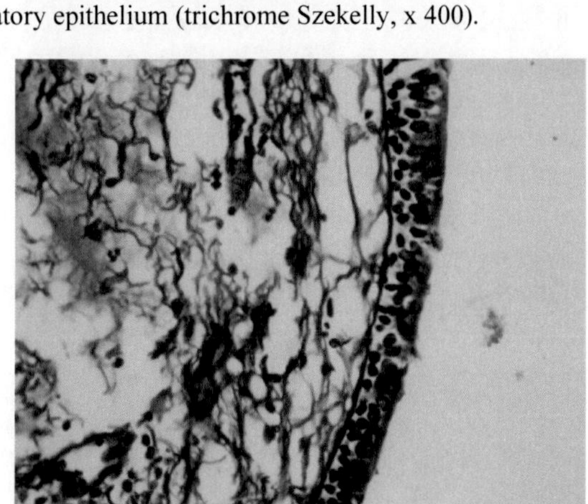

Figure 2. Normal basement membrane (Gordon Sweet, x 200).

Through biochemistry, at this level, a series of glycoproteins were identified, such as: laminin, entactin, basement membrane glycoprotein as well as transmembranary proteins from the family of the integrins which project from the epithelial cell membrane towards the basal lamina [2, 4]. *Lamina densa* is an electron-dense area, 50 nm thick. Through biochemistry it was determined that it consists of protein molecules: laminin and type IV collagen arranged in a "wire fence" pattern surrounded by proteoglycans such as perlecan, fibronectin, type V collagen, adhesion molecules from the family of integrins [2, 4]. *Lamina reticularis* is formed by type I and III collagen. It is placed between the basal lamina and the subjacent connective tissue, its thickness being variable according to the forces exerted on the surface epithelium [2, 4]. Electron microscopy showed that all the cells of the respiratory epithelium, even if of various heights, come in direct contact with the basement membrane. Consequently, the respiratory epithelium can be considered a simple epithelium. If the fixation of the cells within the epithelium is achieved by intercellular desmosomal type junctions, their anchorage on the basement membrane is realized through hemidesmosomal junctions [3, 4]. In electron microscopy, the hemidesmosome has a similar structure to the above described desmosome having an intracellular attachment plaque in the cytoplasm adjacent to the cell membrane of the basal pole. Biochemically, the attachment plaque consists of proteins from the desmoplakin-like family (plectin, BP 230 and erbin), able to secure cytokeratin intermediate filaments which do not reenter the cytoplasm after entering the plaque. Thus, the U-shape trajectory characteristic for the desmosomes is absent. Transmembranar molecules such as calcium-dependent cadherins, present in the structure of the desmosomes, are replaced by integrins ($\alpha 4\beta 6$ integrin) or similar molecules (type XVII collagen, CD151 glycoprotein) acting as matrix-cells receptors which ensure the attachment to the laminin and type IV collagen in the structure of the basement membrane [4, 7, 8].

1.4. Lamina Propria

The respiratory epithelium is placed on a connective tissue sheet called the *lamina propria* [1]. This is a loose connective tissue with a rich blood and nerve supply, with many elastic fibers, along with fixed (fibroblasts, fibrocytes) and free cells (macrophages, mast cells, plasma cells, lymphocytes). An increased number of migrated cells determine the classification of this type of connective tissue as a "defense chorion" together with the one in the digestive tube [5]. The increased number of lymphocytes in the wall of the big respiratory airways explains the possibility of lymphoid infiltrations or even of lymphoid nodules occurrence – components of BALT (bronchus-associated lymphoid tissue) [5]. *Lamina propria* also contains tubulo-acinous sero-mucous glands (Bowman's glands) consisting of acini and short excretion ducts lined by a simple cuboidal epithelium allowing the release of secretion products on the surface of the epithelium [5].

2. Fibrobronchoscopy in Asthma: Pros and Cons

Fibrobronchoscopy represents the most adequate evaluation method of airway modifications in asthma because it allows a direct visualization of the large airways and a bronchoalveolar lavage and/or a bronchial biopsy. Nevertheless, the procedure is invasive and therefore rarely accepted by asthma patients with respiratory insufficiency. Although the scientific impact of its results, leading to further research in the pathogeny of asthma, is uncontestable, fibrobronchoscopy accompanied or not by a biopsy is not a routine exam in asthma. Within the context of asthmatic pathology, the fibronchoscopy investigation is practiced mainly with research purposes, fact supported by a relatively small number of publications in literature (approximately 200 in 20 years) [9]. According to the literature, microscopic exams of bronchial mucosa fragments harvested through fibrobronchoscopy obviously opens new perspectives for deciphering physiopathology in asthma. In order to minimalize the risk of an incomplete harvest it is recommended to obtain 5 to 7 fragments of bronchial mucosa *via* a biopsy for each patient examined through fibrobronchoscopy [9].

Because the investigated patients may have bronchial obstruction in varying degrees, fibrobronchoscopy calls for a series of supplementary safety measures (careful supervision of patients, the administration of bronchodilators in the same day as the investigation, supplementation of oxygen therapy during the exam) in order to prevent the risk of complications (decrease of oxygen saturation [10, 11], cough, wheezing [12].

A pioneering study in fibrobronchoscopy [13] states that a biopsy combined with lavage can be performed on asthma patients but it demands supplementary caution on patients with increased bronchial reactivity. The explanation is the significant decrease in Forced Expiratory Volume 1 (FEV1) recorded in asthmatic as well as in healthy patients [13]. The decrease in FEV1 is inversely correlated with the quantity of methacholine necessary to reduce FEV1 by 20% compared to values previous to fibrobronchoscopy but it cannot be correlated with the score for the symptoms, the usage of inhaled beta2-agonists and with the variations of Peak Expiratory Flow (PEF) recorded in the 2 weeks preceding the exam. Moreover, the authors report a significant decrease in oxygen saturation during the biopsy (an average of 3% with variations between -1 and 17%) without correlation with severity indexes in asthma [13].

Our research focused on the pathology of asthma which was made possible by the close cooperation with the staff of The Clinical Pneumology Hospital in Iaşi. The experience of the staff in the performance of fibrobronchoscopy with bronchial biopsy in asthma patients with varying degrees of the disease's severity confirmed the presence of desaturation in severe forms of the illness and its absence in moderate and milder forms (intermittent and slightly persistent asthma). For the patients investigated on our joint research the most important decrease of oxygen saturation was 10% (from 90% to 80%) the average recorded value being 1.3%.

The evaluation of the clinical (score of the symptoms), paraclinical (arterial saturation, PEF) and therapeutic (usage of inhaled beta2-agonists) parameters consequent to the fibrobronchoscopy reveals the absence of a direct relationship between the invasiveness of the procedure and the evolution of the disease [14]. The degree of arterial O_2 saturation and the decrease in PEF were similar for asthmatic and healthy patients. Nevertheless, the decrease in PEF and bronchial hyperreactivity were significantly reduced in the bronchial biopsy by comparison with the situation where the biopsy was associated with bronchoalveolar lavage [14].

The research dedicated to the advantages and disadvantages, to the limits and the risks of this diagnosis technique allowed the revision of the initial guidelines, recommending the fibrobronchoscopy with biopsy and lavage including for the patients with more severe obstruction (FEV1 < 60%) [15]. There is data supporting the statement that bronchial biopsy is well tolerated even by asthma patients who did not receive bronchodilating medication [11]. This conduct should be avoided due to the risk of a much more severe obstruction occurring after bronchoscopy. According to a meta-analysis [16], including 6 independent studies on the safety of the technique bronchoscopy, endobronchial biopsy and bronchoalveolar lavage, these can be applied on asthma patients with different degrees of severity. The most complications arise when both harvesting procedures are practiced, lavage as well as biopsy, which leads to the conclusion that bronchoalveolar lavage is responsible in most cases for the reported adverse effects.

The results of our interdisciplinary research team are based on a bronchoscopic examination performed in complete safety, in a controlled environment, without unpredicted events or secondary effects leading to a possible interruption of the exam. None of the patients presented an exacerbation of the disease in the days following the fibrobronchoscopy and did not require systemic corticotherapy or the increase of the dosage in corticotherapy.

3. Epithelial Lesions in Asthma

The bronchial mucosa fragments, harvested through fibrobronchoscopy, were routinely processed for HE and other special stains.

The microscopic exam in HE stain is mandatory as the first stage towards the evaluation of the quality of the harvested fragments in order to confirm the possibility of interpretation of the lesions. Thus, it is necessary the existence of the epithelium and of a subepithelial connective sheet of minimum 0.3 mm and the quality of the fixation and coloration has to be also appraised. The analysis of a bronchial biopsy is affected by the presence of tissue fragmentation, pieces of hyaline cartilage, extended hemorrhage or large quantities of mucus.

In the bronchial biopsies harvested from asthma patients the most important modifications are in the epithelium such as epithelial desquamation, hyperplasia of goblet cells and squamous epithelial metaplasia.

3.1. Epithelial Desquamation

Frequently, the respiratory epithelium presents partial epithelial desquamation lesions with the preservation of a "basal" layer of cells on the basement membrane or areas of complete desquamation with complete exposure of the basement membrane. In favor of the idea that epithelial desquamation (complete or partial) could represent a lesion marker in asthma, the following can be observed:

- bronchial epithelial desquamation lesions were noticed both from tissue fragments harvested postmortem from patients deceased after an acute asthma attack and from those obtained through endoscopy from patients with stable chronic asthma [17, 18];
- the presence of "Creola" bodies in the sputum of asthma patients [19];

- the presence of an increased number of epithelial cells in the bronchoalveolar lavage liquid in patients with asthma [20], in which the number of shed cells were directly correlated with the degree of bronchial hyperreactivity [21].

Nevertheless, numerous studies in literature support the idea that bronchial epithelial desquamation in asthma represents in fact an artifact of harvest, processing and inclusion of tissue fragments obtained through endoscopy and not a characteristic morphologic lesion [22]. The idea is founded on the conclusions of various research performed on large groups of volunteer patients as opposed to groups of asthma patients.

Thus, partial or complete epithelial desquamation was also identified in fragments of bronchial mucosa harvested from healthy patients [23, 24]. Cell composition of sputum and of the bronchoalveolar lavage liquid was identical for the control and asthma patients [23]. The degree and spread of epithelial desquamation are not statistically significant correlated with values of FEV1 [24]. Some specialists explain the presence of an increased number of epithelial cells in the bronchoalveolar lavage liquid in patients with asthma by the fact that the harvesting procedure represents a direct trauma on an already fragile epithelium [21].

Data in literature [25] reconfirms the role of the epithelial junctions by indicating the preservation of only one cell layer depleted of ciliated or goblet cells (interpreted as partial denudation) on 50% of the length of the analyzed fragments and complete denudation of the basement membrane on 10 to 15% of the length of the same fragments. Partial denudation on larger areas versus complete denudation on smaller areas can also be explained by the fact that desmosome junctions are easier to damage than junctions between epithelial cells and the basement membrane such as hemidesmosomes which are associated with adhesion molecules like integrins and anchorage filaments [25]. Within the same context, recent research shows that beta 4 integrin, a molecule ensuring the adhesion of epithelial cells in normal respiratory epithelium, has a decreased expression in asthma patients. This molecular modification could explain the perturbation of epithelial cell homeostasis in asthma patients [26].

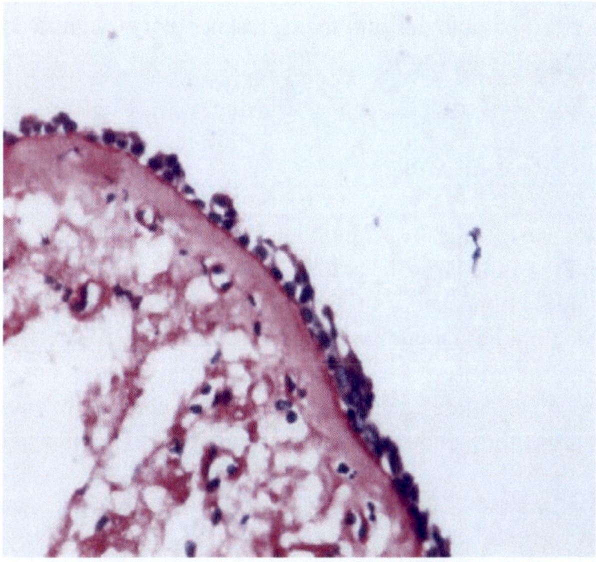

Figure 3. Partial epithelial desquamation – mild persistent asthma (HE, x 100).

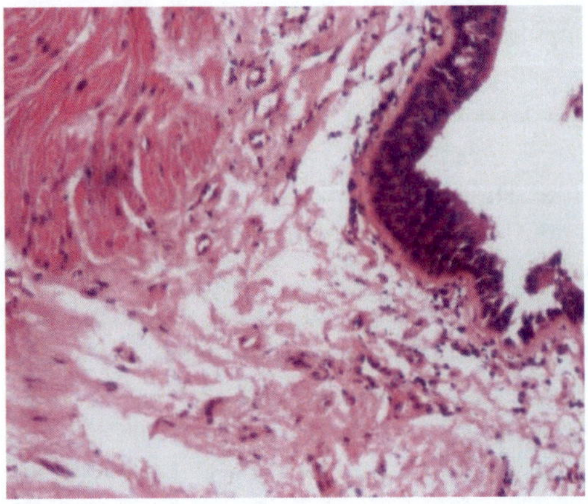

Figure 4. Partial epithelial desquamation associated with a complete epithelial desquamation area – moderate persistent asthma (HE, x 100).

These controversial opinions on epithelial desquamation have an important reflection on the theory according to which the loss of epithelial barrier may favor bronchial hyperreactivity in asthmatic patients. The classic theory is replaced by a new one, focused on the intact epithelium acting as a "proinflammatory organ" directly intervening in reshaping of bronchial wall [27, 28].

The research of our team on epithelial modifications associated with asthma revealed partial or complete epithelial desquamation of the bronchial epithelium. The areas of partial epithelial desquamation were present both in healthy and in almost all asthma patients (94.73%). Complete epithelial desquamation was not noticed on bronchial biopsies harvested from control patients but it was present in 55.26% of the asthma patients regardless of the severity of the disease (Figure 3, 4).

Our results support the statement that although these lesions are frequently observed in the microscopic exam they are not specific for asthma. Epithelial desquamation may also occur in other respiratory pathologies and their presence is not significantly correlated with the severity of the disease.

3.2. Hyperplasia of Goblet Cells

In the bronchial mucosa, mucus secretion has two points of origin: one from goblet cells of the epithelium and the other from glandular mucous cells. Normal mucus secretion vary between 10 and 100 ml/24 hours and is 96% water, the rest represented by electrolytes, lipids, amino acids, immunoglobulins, albumins and enzymes [8, 29]. The mucus layer [29, 30] consists of two layers: an inferior one, 5 μm thick, and a superior one, 1-2 μm thick. The inferior layer has a relatively more watery consistency and ensures the movement of the cilia. The upper layer has a greater mucous consistency and comes in contact only with the tips of the cilia.

Normally the mucus layer continuously moves towards the pharynx where it is swallowed with a speed of 1 cm per minute due to the effect of the cilia's movement. The cilia's movement is coordinated, and this phenomenon is called metachronism. Coughs and sneezes associated with muco-ciliary activity ensure the elimination of mucous secretions through air and digestive pathways (by deglutition) [29, 30].

Literature does not mention the development of the glandular area in bronchial mucosa fragments harvested from patients with intermittent and slightly persistent forms of asthma. Consequently, the increased mucus secretion is caused by goblet cell hyperplasia [31], which is a lesion with a constant frequency and a variable severity [31, 32].

One must note that the occurrence of a large number of goblet cells in the respiratory epithelial structure is not characteristic in asthma. This occurrence is a "response" of the epithelium to toxins such as tobacco smoke, to irritating particles in the environment and microorganisms [33, 34].

Although hyperplasia of goblet cells is reported even in the less severe stages of the disease, there are opposing opinions [35] based on studies with a work methodology which does not solidly support the conclusions (the identification of goblet cells being performed in standard coloration HE and the quantification through a semi-quantitative method). An extremely interesting study [36] indicates an increased number of goblet cells in asthma patients (93.043 ± 15.82) as opposed to healthy patients (41.95 ± 9.23) without size variations from one group to another. The level of mucin deposition was three times bigger in asthma patients as opposed to healthy patients. There were no differences in the levels of mucin deposition between patients with slightly persistent asthma and those with moderately persistent asthma, except that the levels of mucin secretion were significantly lower. According to this data, slightly persistent asthma is associated with hyperplasia of goblet cells with an increase in mucus deposits from the epithelium while in moderate cases of the disease there is an increase in the deposits as well as mucus secretion. These elements suggest the fact that the acute degranulation phenomenon of the goblet cells could represent the mechanism of symptom exacerbation in mild and moderate forms of asthma and chronic degranulation could contribute to the thickening of the bronchial wall in moderately persistent asthma [36].

The research of our team revealed the presence of hyperplasia lesions in goblet cells even on standard HE coloration. In order to quantify more precisely the degree of these lesions we performed an assessment of the extension of these modifications in the structure of the respiratory epithelium in asthma patients with different degrees of severity through the microscopic examination of specimens stained using the PAS technique (Figure 5, 6). Thus, we could observe the hyperplasic lesions of the goblet cells even in the early stages of the disease (intermittent asthma), with an increasing intensity showing a decrease in the FEV1 values.

Current research on the physiopathology of bronchial hypersecretion of mucus as well as on the mechanism of goblet cell hyperplasia is focused on several molecules involved either in regulation of mucus exocytosis (such as MARCKS – Myristoylated Alanine-Rich C Kinase) or in the increase of goblet cells numbers (such as EGF - Epidermal Growth Factor and CLCA - Calcium Activated Chloride Channels) or in the preservation of a large number of goblet cells (apoptotic factors Bcl-2 and Bax) [33]. Within the same context of the pathogenic mechanisms of asthma, experimental studies proved that there is an increase in the expression of some mucigen genes [22, 32, 37].

In humans there were described 20 MUC genes (genes that produce membrane-associated mucins), of which 9 are also expressed in the respiratory tract (MUC1, MUC2, MUC4, MUC5AC, MUC5B, MUC7, MUC8, MUC11 and MUC13). Recent research showed that normally, the genes coding MUC1, MUC2 and MUC8 are expressed in the respiratory epithelium as well as in submucosal glands, while MUC5AC and MUC13 are expressed only in the respiratory epithelium [38, 39]. Later, through immunohistochemistry (IHC), MUC5AC was identified in goblet cells and MUC5B and MUC7 were exclusively at the glandular level [40]. Interestingly, ciliated cells of the respiratory epithelium express MUC4 [40].

This distribution of MUC gene expression in the respiratory tract changes in pathological conditions characterized by mucus hypersecretion such as asthma. Most studies in literature support an increase in MUC5AC, MUC 2 and MUC4 gene expression in asthma patients with varying degrees of severity of the disease [36, 37, 41].

On the other hand, deceased patients with asthma present a significant expression of the MUC5B gene [42, 43]. This sustains the implication of submucosa glands in the hypersecretion of mucus in these patients [42, 43]. This observation raises the importance of the primary involvement of goblet cells.

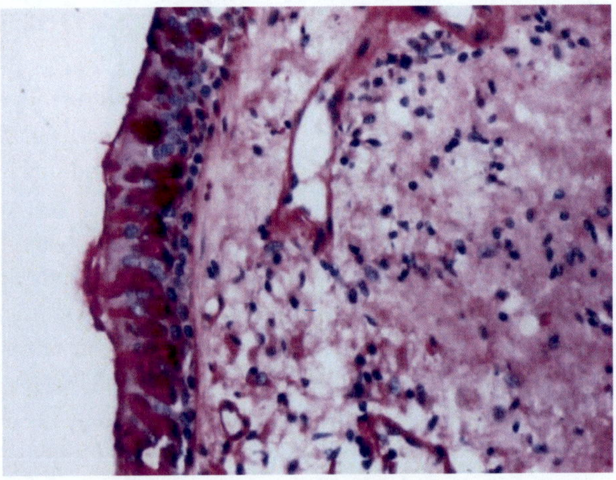

Figure 5. Moderately hyperplasia of goblet cells – mild persistent asthma (PAS, x 200).

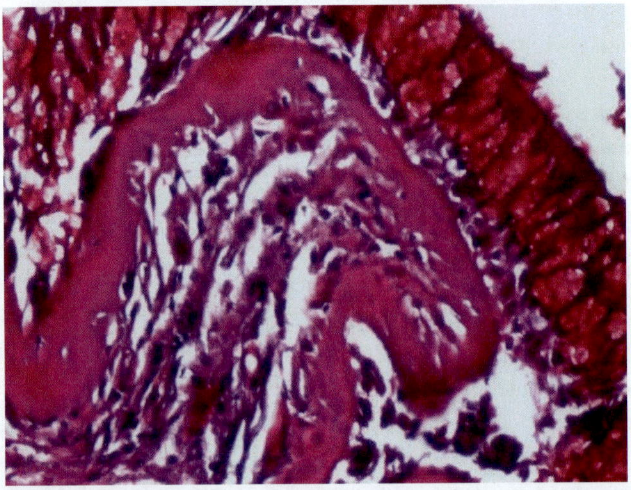

Figure 6. Important hyperplasia of goblet cells – moderate persistent asthma (PAS, x 200).

Mucus hypersecretion remains an important morbidity and mortality factor in pulmonary pathology (including asthma). The possibility of controlling the mucus quantity in the respiratory pathways focuses on the inhibition of nerve activity (by tachikinic receptor antagonists) or the inhibition of mucin exocytosis (through Munc-18B blockers). Other possible therapeutic methods are based on mucin synthesis and goblet cell hyperplasia inhibition (through EGF inhibitors and tyrosin-kinase receptor inhibitors) or the control of goblet cells apoptosis (inductor Bax molecules or Bcl-2 inhibitors) [44]. The latter aspect is also supported by recent research on lab animals, according to which the resolution of goblet cell allergen induced hyperplasia is controlled by γ-interferon through Bax [45].

3.3. Epithelial Metaplasia

Epithelial metaplasia, described in literature as a possible manifestation of the respiratory epithelium subjected to a chronic inflammatory aggression [31, 46], was identified in our research performed on 38 asthmatic patients in 16 cases, 4 with a mild form, 7 with moderate persistent form and 3 diagnosed with severe persistent asthma. We must emphasis the fact that we excluded from our study groups smokers, because they constantly present epithelial transformations secondary to the chronic exposure to substances in cigarette smoke.

3.4. Epithelial Proliferative Activity

In response to the chronic aggression expressed in the morphology by an increased fragility, leading to cell desquamation the respiratory epithelium, will react in order to reestablish its integrity through two fundamental mechanisms: either an increase in cell proliferation in order to replace the injured cells or an extension of cell survival through the inhibition of the factors controlling programmed cellular death (apoptosis). It appears that both processes coexist in the respiratory epithelium affected by asthma.

Consequently, numerous studies in literature focused on the transformations in the expression of cell proliferation factors (Ki-67, PCNA) as well as that of apoptosis factors in the respiratory epithelium in patients with varying degrees of severity and also by comparison with normal respiratory epithelium.

Data in literature reveals the fact that the aggression exerted by several allergens leads to an increase in the proliferative activity of the respiratory epithelium in intermittent and slightly persistent asthma [46] and also that there appears to be a direct relationship between the levels of PCNA expression and the thickness of the respiratory epithelium [47].

The comparative examination [48] of three groups (control, asthma patients with or without corticosteroid treatment) revealed that the respiratory epithelium of patients under treatment express at the same time Bcl-2, Fas, FasL and PCNA. The authors conclude that steroid treatment applied to patients with asthma leads to a decrease in the inflammation of the bronchial wall (especially the eosinophil one) and it promotes cell survival and proliferation, possibly due to the influence on the expression of Bcl-2 and PCNA [48]. However, corticosteroid treatment applied chronically to asthma patients does not succeed to control the balance between the proliferation and cell death in the respiratory epithelium. As a result, the attempt to reestablish epithelial integrity fails. The alteration of epithelial structural integrity constitutes a key element in the pathologic mechanism by which the disease maintains itself [49]. Consequently, epithelial dysfunction contributes on a morphological level to the structural reshaping of the bronchial wall and on a functional level to the decline of pulmonary functions. Therefore, recent opinions support the necessity of new therapeutic targets in asthma focusing on the control of the epithelial proliferation even more so due to the fact that the steroid treatment proved to be inefficient [49].

In our research, directed to the evaluation of proliferative activity, we determined an increase in the level of Ki-67 expression (corresponding to the proliferation rate) in the respiratory epithelium in asthma patients as compared to the control patients. Nevertheless, the bronchial epithelium did not prove to be hyperproliferative [50]. Ki-67-positive cells were present sporadically near the basement membrane along the epithelium in milder forms and reached a maximum as number and a clustering tendency in severely persistent forms of asthma (Figure 7, 8) [50].

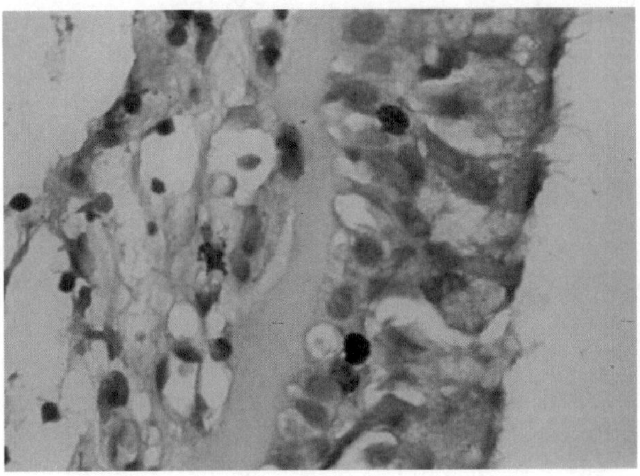

Figure 7. A small number of Ki-67-positive cells in bronchial epithelium – intermittent asthma (IHC, anti-Ki-67, x 400).

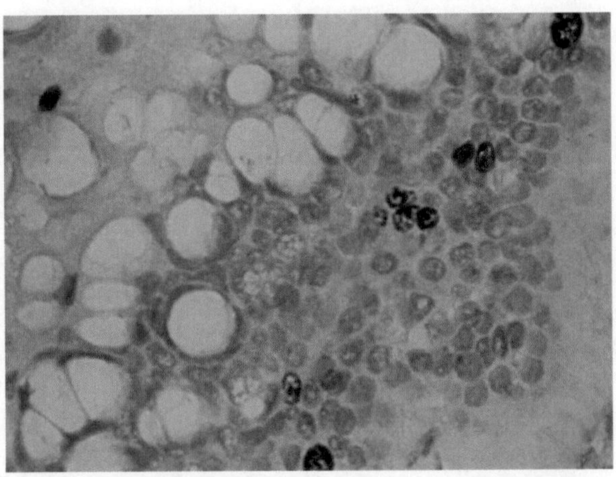

Figure 8. Ki-67-positive cells forming small clusters in bronchial epithelium – severe persistent asthma (IHC, anti-Ki-67, x 400).

The statistical analysis revealed the existence of a statistically significant correlation between the number of Ki-67 positive cells and the severity of the disease ($r=0.887$, $p<< 0.01$, 95% CI) [50].

3.5. Epithelial Apoptosis

Apoptosis, defined as programmed cell death, constitutes a normal sequence in tissue development, morphogenesis and equilibrium. In the context of regulation deficiencies, causing perturbations in the DNA repair mechanism and the modification of the cell cycle, apoptosis contributes to the establishment of pathologic conditions. Apoptosis is a multiple-factor phenomenon produced by the specific activation of an intracellular cysteine proteinase family named caspases [51]. The apoptotic process can be achieved two different ways. The intrinsic path (mitochondrial path, type I) is regulated by apoptotic protein members of the Bcl-2 family and by specific caspases through the implication of the mitochondria which release proteins activating the caspases in the cytosol thus triggering apoptosis [52]. The extrinsic path (transmembranar path, type II) is regulated by Fas proteins and by the proteins of the tumor necrosis factor receptor (TNF-R), involving specific caspases inducing apoptosis by linking the ligand to the cell surface [53]. The molecular family Bcl-2 includes members which by promoting cell survival are believed to be anti-apoptotic (Bcl-2, Bcl-X_L), together with members coordinating cell death, considered as pro-apoptotic (Bax, Bak). The Bax molecule, initially identified by immunoprecipitation as the link partner for Bcl-2 [54] has a physiological role in the embryonic development as it is overexpressed as a response to the presence of apoptotic stimuli.

Most studies in the literature support the existence of a growing proliferation rate in injured respiratory epithelium even in the early stages of the disease. Not the same can be stated about the aspects regarding the level and control of apoptosis as the second mechanism through which epithelial integrity can be restored. This is mainly due to the fact that, currently, there is little information with regard to the level and the control mechanism behind programmed cell death in normal respiratory epithelium [55]; obtaining fragments of normal bronchial mucosa through a fibrobronchoscopy examination is difficult. The apoptotic process in the respiratory epithelial cells is absolutely normal and necessary because the cell number can be controlled and lesions secondary to viral infections or allergen induced can be limited. In literature there are data referring to the apoptosis of ciliated and goblet cells in normal bronchial epithelium [56] but none regarding the apoptosis of renewal cells. Research on respiratory epithelium apoptosis in a normal and a pathologic context (including asthma) is oriented on the anti and pro-apoptotic intervention of members of the Bcl-2 family [49, 55, 57, 58]. Studies on the respiratory epithelium in asthma reveal an increase of epithelial cell apoptosis through the different interventions of anti and pro-apoptotic factors. In the category of inductors we include, besides allergen substances, the elements in cigarette smoke, some cytokines (TGF-1), infectious agents (viruses and bacteria) and hypoxia [55, 57]. Sadly, due to the large variability of asthma, the extreme diversity of the study groups and the differences in the interpretations of the immunohistochemistry exams, data regarding the expression of the anti-apoptotic factor Bcl-2 in the respiratory epithelium in asthma are controversial. Moreover, current studies in literature, referring to the level of the pro-apoptotic expression of Bax in the respiratory epithelium in asthma patients, are very scarce.

- Anti-apoptosis activity – the expression of Bcl-2

Some studies demonstrate an increase of Bcl-2 expression in the bronchial epithelium in asthma patients in comparison with control patients [48, 59], mainly in the renewal basal layer and directly correlated with steroid treatment [48], suggesting that anti-apoptosis represents a self-protection mechanism in chronic inflammatory aggression [59]. However, these studies are contradicted by more recent investigations noting a decrease in the expression of the Bcl-2 factor in the respiratory epithelium especially for the patients with severely persistent forms by comparison with the milder forms of asthma [49, 55].

Our results [60] support these latest findings because in the group we investigated we noted a significant decrease in the expression of the Bcl-2 factor in patients with severely persistent asthma as opposed to patients diagnosed with milder forms of the disease and in control patients (Figure 9, 10).

The statistical analysis revealed the existence of a significant inverse correlation between the severity of asthma (expressed through the FEV1) and the levels of Bcl-2 expression ($r = -0.7349$, $p \ll 0.01$, 95%CI). Pertaining to location, the immunohistochemical positive cells were mostly observed in the basal layer of the respiratory epithelium, except for the squamous metaplasia areas, where Bcl-2 positive cells could be found in the superficial layers too.

- Pro-apoptotic activity – the expression of Bax

The investigation of Bax expression in the respiratory epithelium, harvested post-mortem from asthma patients with varying degrees of severity by comparison with the normal respiratory epithelium, revealed a decrease in the expression of Bax in all asthma patients (even in milder stages of the disease) as opposed to control patients [61]. These observations are also supported by several studies showing a decrease in the number of goblet cells under the influence of γ-interferon, a substance which contributes to the activation of the pro-apoptosis factor Bax [62, 63].

Our research [60], confirmed by statistical analysis, showed a decrease of Bax expression in the bronchial epithelium of all asthma patients in comparison with the control patients even in the intermittent asthma stage ($r=-0.617$, $p=0.0265$, 95%CI). The statistical study performed outlined the absence of a significant association between the degree of severity in asthma and the level of Bax expression ($\chi^2=1.77$, $p=0.6206$, 95%CI) [60]. We must stress the fact that Bax-positive cells were observed "focally" in the structure of the respiratory epithelium, these being present in territories with various lesions such as epithelium denudation or squamous metaplasia (Figure 11, 12). Thus, today it is believed that the decrease in Bax expression levels at the level of the respiratory epithelium in asthma patients even in the early stages of the disease is an element supporting the prolongation of goblet cell life in its structure and the hypersecretion of mucus [44].

Consequently, the identification of a controlled "regulation" for the Bax expression in the respiratory epithelium represents a new research direction, with a benefic effect in the treatment of mucus hypersecretion in asthma. The issue of apoptosis in respiratory epithelium cells in asthma patients still remains open especially because more and more authors, on the basis of their own research, bring into discussion once again the theory according to which the steroid treatment itself (especially in the larger dosage used in patients with severe forms of the disease) would constitute a factor in increased apoptosis of respiratory epithelium in asthma patients [55, 64, 65, 66, 67].

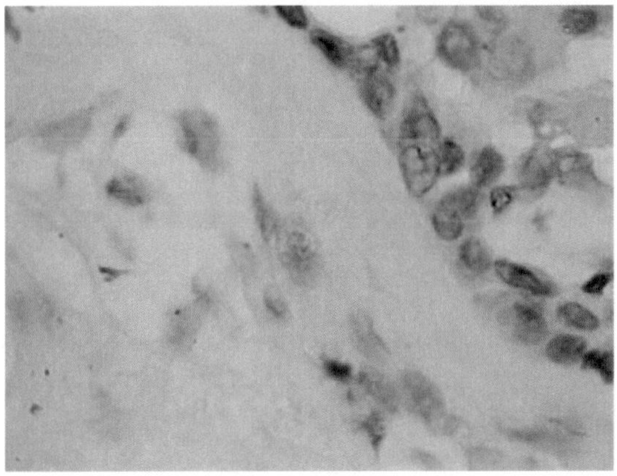

Figure 9. Moderately positive reaction (++) at the bronchial epithelium level – mild ersistent asthma (IHC, anti-Bcl-2, x 600).

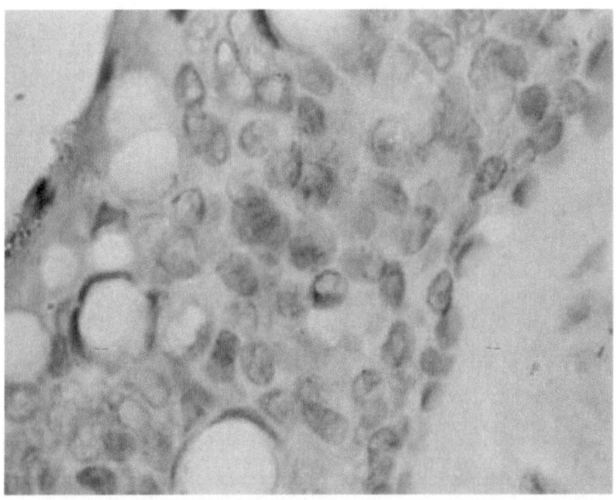

Figure 10. Slightly positive reaction (+) at the bronchial epithelium level – severe persistent asthma (IHC, anti-Bcl-2, x 600).

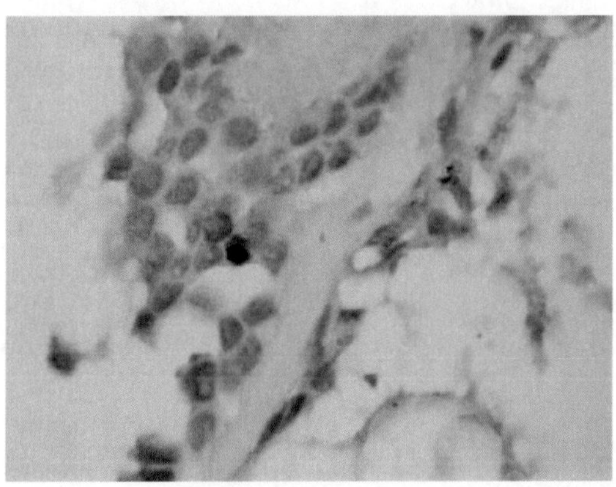

Figure 11. Slightly positive reaction (+) at the bronchial epithelium level – intermittent asthma (IHC, anti-Bax, x 600).

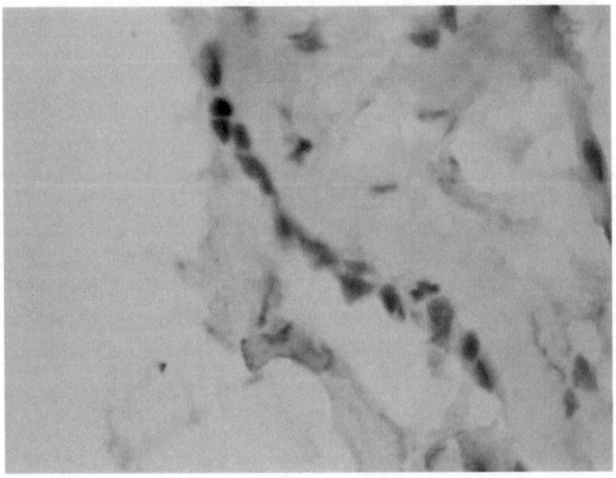

Figure 12. Slightly positive reaction (+) at the bronchial epithelium level – severe persistent asthma (IHC, anti-Bax, x 600).

- Other factors involved in apoptosis

Recent studies showed that Zn has an anti-apoptotic role, intervening through the inhibition of caspases associated with antioxidant activity and cell membrane stabilization [68]. The correlation of this observation with the decreased levels of Zn in chronic inflammatory diseases of the respiratory tract [69] lead to promoting and administrating Zn supplements as a protection factor toward the occurrence of lesions in respiratory epithelial cells [70].

The molecule p21 (Cip1/WAF1), kinase-dependent cyclin inhibitor, intervenes in the regulation of cell proliferation and death [71, 72]. Studies performed on human subjects [73] along with experimental studies (using genetically modified rats with the absence of p21 gene) [73] showed an increase of p21 expression in the respiratory epithelium of asthma patients and consequently an intensification of apoptosis unlike in the control group. The manipulation of p21 expression could represent another new therapeutic target in asthma even more so because the expression of p21 is not influenced by corticosteroid treatment [74].

Another two investigated molecules in the context of the relationship between cigarette smoke and apoptosis are peroxiredoxin V (PRXV), a protein with antioxidant potential [75], and pirin, intensely expressed and associated with the stimulation of apoptosis [76].

4. Final Remarks

Although fibrobronchoscopy is not a current diagnostic method, it allowed the acquirement of bronchial mucosa fragments from patients diagnosed with different stages of asthma. Thus, we were able to describe the epithelial lesions and those of the bronchial wall even from the stages of intermittent and slightly persistent asthma, fact which was not possible until now because bronchial mucosa fragments were harvested only during autopsies in deceased patients due to severe asthma crises. The possibility of analyzing bronchial mucosa fragments represented the "starting point" for research on morphologic, physio-pathological and clinical correlations in asthma.

Current existing data is still insufficient mainly because of the difficulties in acquiring bronchial mucosa fragments. The use of fibrobronchoscopy in asthma is limited. Furthermore, the great variability of the disease and different methodologies used according to different laboratories may influence the interpretation of the final results of microscopic and immunohistochemical exams.

Important steps were recorded in the deciphering of the lesions in the respiratory epithelium injured by asthma. Besides the descriptive characterization of microscopic lesions in the respiratory epithelium, the increase in the Ki-67 cell proliferation factor expression and the decrease in the expression of anti-apoptosis factor Bcl-2 support the idea of the so-called "reparatory phenotype" of the epithelium. We do not have all the answers for all the questions regarding the factors or the mechanisms leading to the collapse of these structural repair processes in the epithelium. Most opinions support the role of the chronic injury of the epithelium sustaining the inflammatory process in the bronchial wall which exerts a new aggressive action on the already injured epithelium. Current research in asthma involve molecular and cellular biology aspects, with a direct orientation on the intervention and action mechanism of growth and repair factors and also of pro and anti-apoptotic factors.

The final purpose is not just to reach an understanding of the intimate pathologic mechanisms taking place in the epithelium and in the entire bronchial wall in asthma, but to also discover new therapeutic modalities in order to control asthmatic phenomenon and to improve the life quality in asthma patients.

References

[1] Fawcett, DW; Jensh, RP. *Bloom and Fawcett's Concise Histology*, 2nd ed. London: Arnold; 2002.

[2] Ross, MH; Pawlina, W. *Histology. A Text and Atlas with Correlated Cell and Molecular Biology*, 6th ed. Philadelphia: Lippincott Williams and Wilkins; 2010.

[3] Ross, MH; Kaye, GI; Pawlina W. *Histology. A Text and Atlas*, 4th ed. Philadelphia: Lippincott Williams and Wilkins; 2003.

[4] Căruntu, ID; Cotuțiu, C. *Histologie. Țesuturi fundamentale*. Iaşi: Ed. Gr. T. Popa; 2004.

[5] Grigoraş, A; Floarea-Strat, A. *Histologie ilustrată*. Iaşi: Ed. Junimea; 2009.

[6] Gartner, LP; Hiatt, JL. *Color Texbook of Histology*, 3th ed. Philadelphia: Saunders; 2007.

[7] Junqueira, LC; Carneiro, J. *Basic Histology: Text and Atlas,* 11th ed. Connecticut: McGraw-Hill/Appleton and Lange; 2005.

[8] Eroschenko, VP. *di Fiore' s Atlas of Histology with Functional Correlations*, 11th ed. Philadelphia: Lippincott Williams and Wilkins; 2007.

[9] Ulmeanu, R; Mihălțan, FD; Cordoş, I; Ulmeanu, DI. *Actualități în endoscopia toracică diagnostică şi terapeutică*, Bucureşti: Ed. Universitară Carol Davila; 2009.

[10] Ciobanu, L. *Ghiduri de practică şi protocoale în fibrobronhos*copie. Iaşi: Ed. Fides; 2001.

[11] Van Vyve, V; Chanez, P; Bousquet, J; Lacoste, JY; Michel, FB; Godard, P. Safety of bronchoalveolar lavage and bronchial biopsies in patients with asthma of variable severity. *Am. Rev. Respir. Dis.* 1992; 146:116-121.

[12] Jarjour, NN; Calhoun, WJ. Bronchoalveolar lavage in stable asthmatics does not cause pulmonary inflammation. *Am. Rev. Respir. Dis.* 1990; 142:100-103.

[13] Djukanovic, R; Wilson, JW; Lai, CK; Holgate, ST; Howarth, PH. The safety aspects of fiberoptic bronchoscopy, bronchoalveolar lavage, and endobronchial biopsy in asthma. *Am. Rev. Respir. Dis.* 1991; 143:772–777.

[14] Humbert, M, Robinson DS, Assoufi B, Kay AB, Durham SR. Safety of fiberoptic bronchoscopy in asthmatic and control subjects and effect on asthma control over two weeks. *Thorax,* 1996; 51:664-669.

[15] Workshop summary and guidelines: investigative use of bronchoscopy, lavage, and bronchial biopsies in asthma and other airway diseases. *J. Allergy Clin. Immunol.* 1991; 88(5):808-814.

[16] Elston, WJ; Whittaker, AJ; Khan, LN; Flood-Page, P; Ramsay, C; Jeffery, PK; Barnes, NC. Safety of research bronchoscopy, biopsy and bronchoalveolar lavage in asthma. *Eur. Respir. J.* 2004; 24(3):375-377.

[17] Laitinen, LA; Heino, M; Laitinen, A; Kava, T; Haahtela, T. Damage of the airway epithelium and bronchial reactivity in patients with asthma. *Am. Rev. Respir. Dis.* 1985; 131:599-606.

[18] Jeffery, PK; Wardlaw, AJ; Nelson, FC; Collins, JV; Kay, AB. Bronchial biopsies in asthma: an ultrastructural quantitative study and correlation with hyperreactivity. *Am. Rev. Respir. Dis.* 1989; 140:1745-1753.

[19] Naylor, B. The shedding of the mucosa of the bronchial tree in asthma. *Thorax,* 1962; 17:69-72.

[20] Kirby, JG; Hargreave, FE; Gleich, GJ; O'Byrne, PM. Bronchoalveolar cell profiles of asthmatic and non-asthmatic subjects. *Am. Rev. Respir. Dis.* 1987; 136:379-383.

[21] Ordonez, C; Ferrando, R; Hyde, MD; Wong, HH; Fahy, VJ. Epithelial Desquamation in Asthma Artifact or Pathology? *Am. J. Respir. Crit. Care Med.* 2000; 162(6): 2324-2329.

[22] Fahy, VJ. Remodeling of the airway epithelium in asthma. *Am. J. Respir. Crit. Care Med.* 2001; 164:S46-51.

[23] Fahy, JV; Liu, J; Wong, H; Boushey, HA. Analysis of cellular and biochemical constituents of induced sputum after allergen challenge: a method for studying allergic airway inflammation. *J. Allergy Clin. Immunol.* 1994; 93:1031-1039.

[24] Redington, A; Springall, D; Ghatel, M; Lau, LC; Bloom, S; Holgate, S; Polak, J; Howarth, P. Endothelin in bronchoalveolar lavage fluid and its relation to airflow obstruction in asthma. *Am. J. Respir. Crit. Care Med.* 1995; 151:1034-1039.

[25] Wegner, CD; Rothlein, R; Gundel, RH. Adhesion molecules in the pathogenesis of asthma. *Agents Actions Suppl.* 1991; 34:529-544.

[26] Liu, C; Xiang, Y; Liu, H; Li, Y; Tan, Y; Zhu, X; Zeng, D; Li, M; Zhang, L; Qin X. Integrin beta4 was downregulated on the airway epithelia of asthma patients. *Acta Biochim. Biophys. Sin. (Shanghai)* 2010; 42(8):538-547.

[27] Holgate, ST. The airway epithelium is central to the pathogenesis of asthma. *Allergol. Int.* 2008; 57(1):1-10.

[28] Holgate, ST. Epithelium dysfunction in asthma. *J. Allergy Clin. Immunol.* 2007; 120(6):1233-1244.

[29] Guyton, AC. *Tratat de fiziologie a omului,* 11th ed. Bucureşti: Ed. Medicală Calisto; 2007.

[30] Patton, KT; Thibodeau, GA; Douglas, MM. *Essentials of Anatomy and Physiology*. Amsterdam: Elsevier; 2011.
[31] Sumi, Y; Hamid, Q. Airway remodeling in asthma. *Allergol. Int.* 2007; 56(4):341-348.
[32] Fahy, JV. Goblet cell and mucin gene abnormalities in asthma. *Chest,* 2002; 122(6):320-326.
[33] Rogers, DF. The airway goblet cell. *Int. J. Biochem. Cell Biol.* 2003; 35(1):1-6.
[34] Rogers, DF. Airway goblet cells: responsive and adaptable front-line defenders. *Eur. Respir. J.* 1994; 7(9):1690-1706.
[35] Groneberg, DA; Eynott, PR; Lim, S; Oates, T; Wu, R; Carlstedt, I; Roberts, P; McCann, B; Nicholson, AG; Harrison, BD; Chung, KF. Expression of respiratory mucins in fatal status asthmaticus and mild asthma. *Histopathology,* 2002; 40(4):367-373.
[36] Ordoñez, C; Khashayar, R; Wong, H; Ferrando, R; Wu, R; Hyde, DM; Hotchkiss, JA; Zhang, Y; Novikov, A; Dolganov, G; Fahy, JV. Mild and moderate asthma is associated with airway goblet cell hyperplasia and abnormalities in mucin gene expression. *Am. J. Respir. Crit. Care Med.* 2001; 163(2):517-523.
[37] Morcillo, EJ; Cortijo, J. Mucus and MUC in asthma. *Curr. Opin. Pulm. Med.* 2006; 12(1):1-6.
[38] Rose, MC; Voynow, JA. Respiratory tract mucin genes and mucin glycoproteins in health and disease. *Physiol. Rev.* 2006; 86(1):245–278.
[39] Hovenberg, HW; Davies, JR; Herrmann, A; Linden, CJ; Carlstedt, I. MUC5AC, but not MUC2, is a prominent mucin in respiratory secretions. *Glycoconj J.* 1996; 13(5):839–847.
[40] Davies, JR; Herrmann, A; Russell, W; Svitacheva, N; Wickstrom, C; Carlstedt, I. Respiratory tract mucins: structure and expression patterns. *Novartis Found Symp.* 2002; 248:76–88.
[41] Rubin, BK. Physiology of airway mucus clearance. *Respir. Care,* 2002; 47(7):761-768.
[42] Thornton, DJ; Howard, M; Khan, N; Sheehan, JK. Identification of two glycoforms of the MUC5B mucin in human respiratory mucus. Evidence for a cysteine-rich sequence repeated within the molecule. *J. Biol. Chem.* 1997; 272(14):9561–9566.
[43] Sheehan, JK; Richardson, PS; Fung, DC; Howard, M; Thornton, DJ. Analysis of respiratory mucus glycoproteins in asthma: a detailed study from a patient who died in status asthmaticus. *Am. J. Respir. Cell Mol. Biol.* 1995; 13(6):748–756.
[44] Rogers, DF; Barnes, PJ. Treatment of airway mucus hypersecretion. *Ann. Med.* 2006; 38(2):116-125.
[45] Xiang, J; Rir-Sim-Ah, J; Tesfaigzi, Y. IL-9 and IL-13 induce mucous metaplasia that is reduced by IFN-gamma in a Bax-mediated pathway. *Am. J. Respir. Cell Mol. Biol.* 2008; 38(3):310-317.
[46] Ricciardolo, F; Di Stefano, A; van Krieken, J; Sont, J; van Schadewijk, A; Rabe, K; Donner, C; Hiemstra, P; Sterk, P; Mauad, T. Proliferation and inflammation in bronchial epithelium after allergen in atopic asthmatics. *Clin. Exp. Allergy,* 2003; 33:905–911.
[47] Vignola, AM; Chiappara, G; Siene, L; Bruno, A; Gagliardo, R; Merendino, AM; Polla, BS; Arrigo, AP; Bonsignore, G; Bousquet, J; Chanez, P. Proliferation and activation of bronchial epithelial cells in corticosteroid-dependent asthma. *J. Allergy Clin. Immunol.* 2001; 108(5):738-746.

[48] Druilhe, A; Wallaert, B; Tsicopoulos, A; Lapa e Silva, JR; Tillie-Leblond, I; Tonnel, AB; Pretolani, M. Apoptosis, proliferation and expression of Bcl-2, Fas, and Fas Ligand in bronchial biopsies from asthmatics. *Am. J. Respir. Cell Mol. Biol.* 1998; 19(5):747-745.
[49] Cohen, L; Tarsi, J; Ramkumar, T ; Horiuchi, TK; Cochran, R; DeMartino, S. Epithelial cell proliferation contributes to airway remodeling in severe asthma. *Am. J. Respir. Crit. Care Med.* 2007; 176(2):138-145.
[50] Grigoraş, CC; Grigoraş, A; Mihăescu, T; Vereş, L; Carasevici, E. Nivelul expresiei factorului de proliferare Ki-67 în astmul bronşic. *Rev. Med. Chir. Soc. Med. Nat. Iaşi.* 2009; 113(4):1034-1039.
[51] Alnemri, ES; Livingston, DJ; Nicholson, DW; Salvesen, G; Thornberry, NA; Wong, WW; Yuan, J. Human ICE/CED-3 protease nomenclature. *Cell,* 1996; 87:171.
[52] Green, DR; Reed, JC. Mitochondria and apoptosis. *Science,* 1998; 281:1309–1312.
[53] Wallach, D; Varfolomeev, EE; Malinin, NL; Goltsev, YV; Kovalenko, AV; Boldin, MP. Tumor necrosis factor receptor and Fas signaling mechanisms. *Annu. Rev. Immunol.* 1999; 17:331–367.
[54] Oltvai, ZN; Milliman, CL; Korsmeyer, SJ. Bcl-2 hetero-dimerizes in vivo with a conserved homolog, Bax, that accelerates programmed cell death. *Cell,* 1993; 74:609–619.
[55] Zalewski, PD; Ruffin, RE. Apoptosis-regulatory factors as potential drug targets in the epithelium of normal and inflamed airways. *Curr. Mol. Pharmacol.* 2008; 1:38-49.
[56] Hackett, TL; Knight, DA. The role of epithelial injury and repair in the origins of asthma. *Curr. Opin. Allergy Clin. Immunol.* 2007; 7:63-68.
[57] Tesfaigzi, Y. Roles of apoptosis in airway epithelia. *Am. J. Respir. Cell Mol. Biol.* 2006; 34: 537-547.
[58] Jyonouchi, H. Airway epithelium and apoptosis. *Apoptosis,* 1999; 4:407-417.
[59] Vignola, MA; Mirabella, F; Costanzo, G; Di Giorgo, R; Gjomarkaj, M; Bellia, V; Bonsignore, G. Airway remodeling in asthma. *Chest,* 2003; 123:417-422.
[60] Grigoras, A; Grigoras, CC; Mihăescu, T; Veres, L; Cotutiu, C; Floarea-Strat, A. Apoptosis quantification at the respiratory epithelium level in asthma. *Cent. Eur. J. Med.* 2010; 5(5):556-564.
[61] Schwalm, K; Stevens, JF; Jiang, Z; Schuyler, MR; Schrader, R; Randell, SH; Green, FH; Tesfaigzi, Y. Expression of the proapoptotic protein Bax is reduced in bronchial mucous cells of asthmatic subjects. *Am. J. Physiol. Lung Cell Mol. Physiol.* 2008; 294(6):1102-1129.
[62] Shi, ZQ; Feng, Y; Hou, YK; Liu, T; Xiu, QY. A study of interferon-gamma induced airway mucous cell apoptosis and its mechanisms. *Zhonghua Jie He He Hu Xi Za Zhi* 2005; 28(3):160-163.
[63] Xiang, J; Rir-Sim-Ah, J; Tesfaigzi, Y. IL-9 and IL-13 induce mucous cell metaplasia that is reduced by IFN-gamma in a Bax-mediated pathway. *Am. J. Respir. Cell Mol. Biol.* 2008; 38(3):310-7.
[64] Fahy, JV; Corry, DB; Boushey, HA. Airway inflammation and remodeling in asthma. *Curr. Opin. Pulm. Med.* 2000; 6(1):15-20.
[65] Dorscheid, D; Wojcik, K; Sun, S; Marroquin, B; Steven, R. Apoptosis of Airway Epithelial Cells Induced by Corticosteroids. *Am. J. Respir. Crit. Care Med.* 2001; 164(10):1939-1947.

[66] White, S; Delbert, R; Dorscheid, M. Corticosteroid-Induced Apoptosis of Airway Epithelium A Potential Mechanism for Chronic Airway Epithelial Damage in Asthma. *Chest* 2002; 122:278-284.

[67] Esmailpour, N; Högger, P; Rabe, K; Heitmann, U; Nakashima, M; Rohdewald, P. Distribution of inhaled fluticasone propionate between human lung tissue and serum in vivo. *Eur. Respir. J.* 1997; 10:1496-1499.

[68] Truong-Tran, AQ; Grosser, D; Ruffin, RE; Murgia, C; Zalewski, PD. Apoptosis in the normal and inflamed airway epithelium: role of zinc in epithelial protection and procaspase-3 regulation. *Biochem. Pharmacol.* 2003; 66:1459-1468.

[69] Zalewski, PD. Zinc metabolism in the airway: basic mechanisms and drug targets. *Curr. Opin. Pharmacol.* 2006; 6:237-243.

[70] Bao, S; Knoell, DL. Zinc modulates airway epithelium susceptibility to death receptor-mediated apoptosis. *Am. J. Physiol. Lung Cell Mol. Physiol.* 2006; 290:L433-L441.

[71] Maddika, S; Ande, SR; Panigrahi, S; Paranjothy, T; Weglarczyk, K; Zuse, A; Eshraghi, M; Manda, KD; Wiechec, E; Los, M. Cell survival, cell death and cell cycle pathways are interconnected: implications for cancer therapy. *Drug Resist. Updat.* 2007; 10:13-29.

[72] Blundell, R; Harrison, DJ; O'Dea, S. p21(Wafl/Cip1) regulates proliferation and apoptosis in airway epithelial cells and alternative forms have altered binding activities. *Exp. Lung Res.* 2004; 30:447-464.

[73] Fedorov, IA; Wilson, SJ; Davies, ED; Holgate, ST. Epithelial stress and structural remodelling in childhood asthma. *Thorax* 2005; 60:389-394.

[74] Puddicombe, SM; Torres-Lozano, C; Richter, A; Bucchieri, F; Lordan, JL; Howarth, PH; Vrugt, B; Albers, R; Djukanovic, R; Holgate, ST; Wilson, SJ; Davies, DE. Increased expression of p21(waf) cyclin-dependent kinase inhibitor in asthmatic bronchial epithelium. *Am. J. Respir. Cell Mol. Biol.* 2003; 28: 61-68.

[75] Serikov, VB; Leutenegger, C; Krutilina, R; Kropotov, A; Pleskach, N; Suh, JH; Tomilin, NV. Cigarette smoke extract inhibits expression of peroxiredoxin V and increases airway epithelial permeability. *Inhal. Toxicol.* 2006; 18:79-92.

[76] Gelbman, BD; Heguy, A; O'Connor, TP; Zabner, J; Crystal, RG. Up-regulation of pirin expression by chronic cigarette smoking is associated with bronchial epithelial cell apoptosis. *Respir. Res.* 2007; 8:10.

In: Endothelium and Epithelium
Editors: J. Carrasco and M. Mota, pp. 137-155

ISBN 978-1-61470-874-2

Chapter VI

Blood Testis Barrier: How does the Seminiferous Epithelium Feeds the Developing Germ Cells?

Luís Rato, Marco G. Alves, Sílvia Socorro, José E. Cavaco and Pedro F. Oliveira*

CICS-UBI - Health Sciences Research Centre, University of Beira Interior, Av. Infante D. Henrique, 6200-506 Covilhã, Portugal

Abstract

The testes of mammals are paired organs that essentially perform two functions: steroid biosynthesis and production of spermatozoa. They consist in two compartments encased by the tunica albuginea: the seminiferous tubules, the functional units of the testes, and the intervening interstitial space. Within the seminiferous tubules, the seminiferous epithelium is compartmentalized by adjacent Sertoli cells(SCs) via specialized junctions creating the blood-testis barrier (BTB) where post-meiotic germ cell development takes place in a specialized microenvironment. The fully functional BTB consists in three components: an anatomical/physical barrier to restrict entry of molecules into the adluminal compartment of the seminiferous tubules; an immunological barrier (inside and outside the tubules), that limits the movement of immune cells and regulates the level of cytokines in the seminiferous epithelium; a physiological barrier (transporters and channels at the basolateral and apical membranes), that are highly dynamic to encounter the needs of germ and SCs. Together, these components are essential to the function of BTB in testes and the special microenvironment created is responsible for proper development of the germ cells into fully functional sperm.

Developing germ cells are unable to use glucose for their energy metabolism and preferentially use lactate, used by spermatocytes and spermatids, as a substrate for ATP production. Sertoli cells on the seminiferous epithelium produce the lactate utilized by the developing germ cells. Carbohydrate metabolism of SCs has been shown to present

*E-mail: pfobox@gmail.com

some unique features. These cells can metabolize various substrates (e.g.: glucose, fatty acids, ketone bodies), but preferentially metabolize glucose being the majority of it converted to lactate and not oxidized via the citric acid cycle.The reasons why SCs preferentially export lactate for germ cells are not entirely understood. However, there are evidences of an anti-apoptotic effect of lactate on germ cells suggesting that it plays a crucial role in spermatogenesis.

Several biochemical mechanisms might contribute to the modulation of lactate secretion by SCs, among which (a) the transport of glucose through the plasma membrane, mediated by a set of homologous glycoprotein molecules (GLUTs); (b) the interconversion of pyruvate to lactate by lactate dehydrogenase (LDH); and (c) the release of lactate from SCs, mediated by the action of the monocarboxylate transporters family (MCTs).

The understanding of the functioning and regulation of these processes are crucial steps to identify key mechanisms associated with seminiferous epithelium (dys)function, and their influence over male (in)fertility.

Introduction

The blood testis barrier (BTB) is a central structural element in testicular physiology. This term has been developed since 1970, and the first studies were performed based in the following evidences: radioactive tracers and dyes when injected in animals, stained most tissues, with the exceptions of the brain and seminiferous tubules of the testis [1-3]. On other hand, it was demonstrated that there were significant differences in fluids compositionand proteins obtained from the rete testis, seminiferous tubule lumen, and the testicular lymph and blood plasma [4, 5].

Blood testis barrier is one of the tightest blood-tissue barriers in mammalian tissues [6-8], creating an adequate microenvironment for proper development of germ cells and conferring cell polarity [8]. Blood testis barrierdivides the seminiferous epithelium in two compartments: outside the barrier is the basal compartment where spermatogonial renewal occurs and inside the barrier is the apical or adluminal compartment where meiosis, spermiogenesis and spermiation take place (Figure 1). When BTB is dysfunctional, germ cell differentiation is arrested [9]. The generation and maintenance of BTB is assured by somatic Sertoli cells (SCs) [10] and its molecular composition has long been a matter of debate [5, 8, 11-13]. Theultrastructure of BTB is composed by co-existing specialized junctions between adjacent SCs near the basement membrane, which include tight junctions, basal ectoplasmic specializations, basal tubulobulbar complex gap junctions and desmosome-like junctions [12, 13], and anchoring junctions, such as apical ectoplasmic specializations, apical tubulobulbar complex gap junctions, desmosome-like junctions and gap junctions between Sertoli cells and germ cells [14].

It is noteworthythat BTB is not closed all time and must allow the developing germ cell migration through the seminiferous epithelium. Despite the complex composition of BTB, this barrier shows a highly dynamic restructuring at specific stages of the spermatogenic cycle, since developing germ cells must cross BTB towards adluminal compartment [15]. Throughout this process, germ cells continue to be tightly anchored to SCs via anchoring junctions [16]. It is a well-coordinated process so that even during BTB "opening and closing" the immunological privilege is maintained. Recently, Cheng and colleagues[13] have suggested that probably gap and desmosome-like junctionsprovide the cross-talk to coordinate the function of multiple junctions at the BTB to effectively safeguard the immunological barrier function. All events that comprise junction disassembly and reassembly are likely involving a complex network of signaling cascade and rapid turnover of junction-associated molecules [8].

Blood testis barrieris more than specialized junctional complexes, it is an interaction of the three components that comprise the fully functional BTB: anatomical/physical, physiological and immunological. The anatomical/physical barrier is composed, as referred, by the SCsbasolateral membrane, apical membrane and the tight junctional complex between these cells [17]. Each of these elements contributes for the selective passage of water, electrolytes, ions, nutrients, amino acids, hormones, paracrine and autocrine factors from the outside of the BTB into the adluminal compartment. A good example is the follicle-stimulating hormone (FSH), whichis essential for the maintenance of adult spermatogenesis[18], that does not pass through BTB readily [19, 20]. While anatomical barrier restricts passage of molecules into the lumen, physiological barrier reflects the permeability of both basolateral and apical membranes in which several specific transporters are responsible for the movement of molecules in or out of the lumen[17, 21].

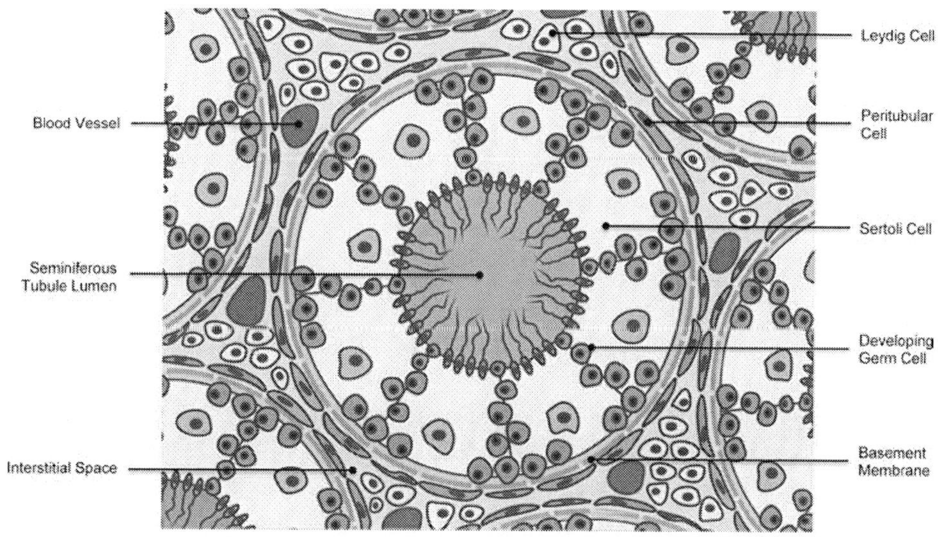

Figure 1. Schematic drawing illustration of the seminiferous tubule and the blood testis barrier (BTB).The BTB is a physical barrier between the blood vessels and the seminiferous tubule lumen and is formed by tight connections between Sertoli cells (SCs). Outside the BTB is the basal compartment where spermatogonial renewal occurs and inside the BTB is the apical compartment where meiosis, spermiogenesis and spermiation take place. At the interstitial space are located the blood vessels and the Leydig cells, which produce testosterone in the presence of luteinizing hormone (LH). The cytoplasmic extensions that enwrap the developing germ cells are responsible for the structural support through a microtubular filament present in the cytoplasm of SCs. This architecture is not static and depends on the stage of the seminiferous tubules. Additionally,the SCs regulate the internal microenvironment of the seminiferous tubule. External to the basement membrane are several layers of modified myofibroblastic cells, termed peritubular cells, responsible for the irregular contractions of the seminiferous tubules, which propel fluid secreted by the SCs. As a result of such particular organization, the establishment of anefficient and regulated BTB is essential to create a special environment for the normal development of a fully efficient sperm.

The flow of molecules is not done in only one direction, but by two, from the outside to inside and from the inside to outside of BTB. Brinster and colleagues [22] injected spermatogonia into the lumen of a single seminiferous tubule either directly or via the rete testis and they observed that injected germ cells can take up a position adjacent to the peritubular tissue and repopulate that area of the tubules with developing germ cells crossing the BTB in opposite direction of the "normal germ cellsroute". These observations supportthe idea that the movement across BTB occurs in both directions. Moreover, the physiological barrier is highly dynamic and the SCsare the responsible for this feature, since they present a dynamic changing in its structure during the spermatogenic events[17]. The last component is the immunological barrier. It has been reported that testes are an immunological privileged site which avoid auto-immune response[17, 23]. The immunological barrier created by BTB isolates the developing germ cells from circulating leukocytes and antibodies [21]. Spermatogonia and preleptotenespermatocytes, which are found in basal compartment, may be a target of host immune cells.Blood testis barrierceases at rete testis where immune cells and antibodies are able to enter[24]. However, Head and Billingham[25] placed allografts in the interstitial space, a location outside of the BTB, and found an extended survival without immune rejection. Indeed, there are other mechanisms responsible for the maintenance of the testicular immune privilege [23]. The SCs play a key role functioning as "sentinel cells" since they present some particular characteristics of immune cells [26], which are able to trigger an immunosuppressive response [23].

Together, the interaction of the three components of BTB, and all cellular events essentials for itsrestructuring, play a crucial role in the establishment of a fully functional BTB, which culminates in a special environment within seminiferous tubules required for normal development of spermatogenic process and consequent production of fully functional sperm.

Seminiferous Epithelium

Seminiferous tubules,the functional unities of testes,are highly convoluted structures where spermatogenesis takes place through close association of germ cells with epithelial somatic cells, the SCs[27]. Both ends of each seminiferous tubule open into *tubulirecti* and then connect in the *rete testis*[28]. Seminiferous tubules are lined by a complex stratified epithelium that is composed by two cell types: the SCs and several germ cells types [29].Sertoli cells occupy a volume of approximately 17–19% in the seminiferous epithelium of adult rats [30] and exhibit numerous cytoplasmic prolongations which allow them to sustain a vast number of developing germ cells [31]. Each SC supports up to 30–50 germ cells at different stages of development [13].Germ cells form intimate associations with SCsand multiple developing germ cells in different stages may be in contact with one SC (Figure 2). The various generations of germ cells are distributed in defined cellular associations called stages [32]. These stages represent all changes that germinal cells experiment throughout the spermatogenic event and the sequence of these cellular associations constitute the cycle of seminiferous epithelium [33]. In the rat, there are 14 stages to be considered whereas in the mouse there are only 12 stages and in man there are only 6 stages [32].

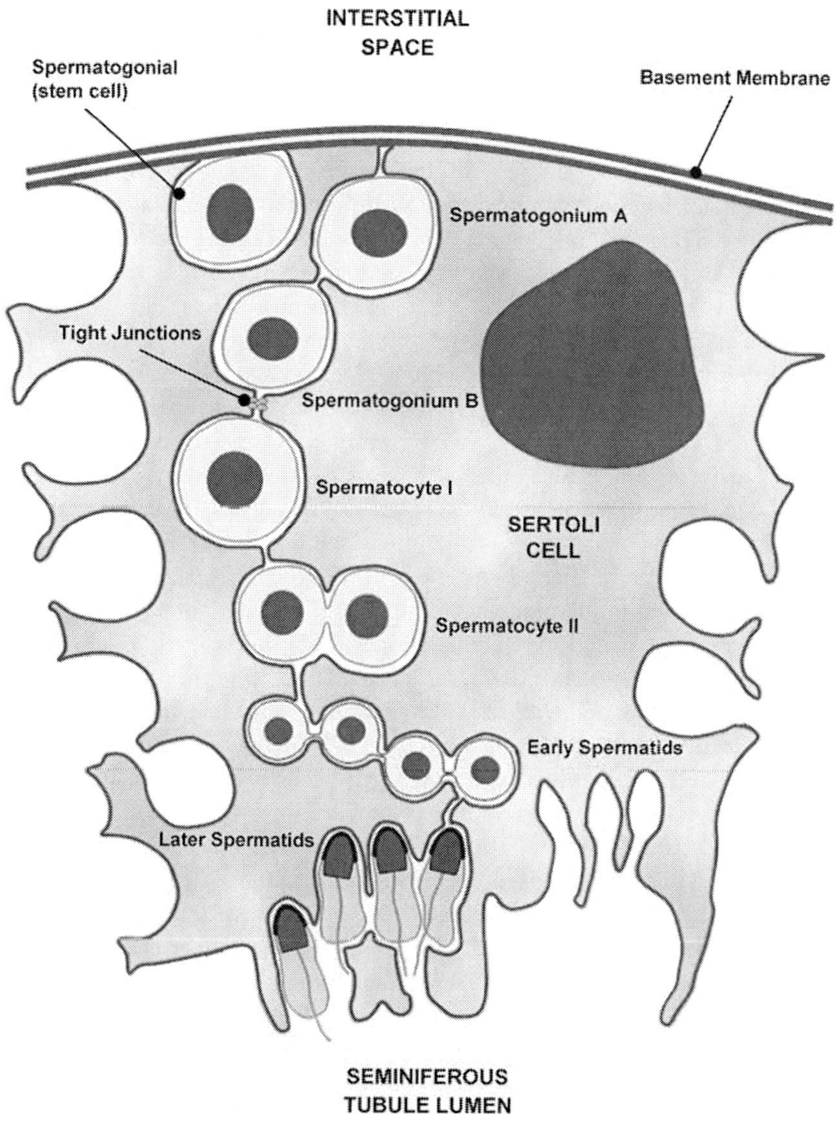

Figure 2. Schematic drawing illustration of spermatogenesis that occurs within the seminiferous epithelium.The seminiferous epithelium is composed by Sertoli cells (SCs) and germ cells at different stages of their development. The process by which male gametes (spermatozoa) are formed in the seminiferous tubules of the testis is called spermatogenesis and is a complex process of cellular transformation that produces male haploid germ cells from diploid spermatogonial stem cells. This process is usually described by 'stages' or 'phases', which progress through precisely timed and highly organized cycles. These cycles of spermatogenesis are essential for continuous sperm production, which is dependent upon several intrinsic (Sertoli and germ cells), extrinsic (hormonal, among others), as well as species-specific factors. The supporting SCs adhere to the basement membrane where spermatogonia are also adherent. Then, spermatogonia type A divide and develop into spermatogonia type B, which enter meiotic prophase and differentiate into primary spermatocytes that separate the homologous pairs of chromosomes in meiosis I(reduction division) to form the haploid secondary spermatocytes. The meiosis II yields four equal-sized spermatids that migrate toward the lumen where fully formed spermatozoa are finally released.

Sertoli cells are the main structural element of the seminiferous epithelium residing on basement membrane, acquiring a columnar shape that extend towards the lumen of the tubules [23, 32]. These branched cells were originally described by EnricoSertoli in 1865 [34, 35]. More than one century thereafter, Griswold [36]highlighted theevidences establishing the crucial role of SCs in spermatogenesis: 1) It has never been observed testes containing germ cells without SCs; 2) function and efficiency of SCs might be a limitative factor for germ cells; 3) in higher vertebrates, reproductive hormones such as FSH and testosterone, that are essential for normal spermatogenesis, act on SCs and not on germ cells

Often called as "nurse cells", SCs play distinct functions. In fetal life, SCs play a crucial role in the development of a functional testes and thus determining male phenotype. SCs are the first cells to differentiate in the indifferent fetal gonad and this differentiation results in seminiferous cord formation, prevention of germ cell entry into meiosis [37] by sequestering the germ cells (gonocytes) inside thenewly formed seminiferous tubules [36], and differentiation and function of the other somatic cells of the testes [37]. SCs ensure the regression of Müllerian ducts via secretion of anti-Müllerian hormone (AMH), influencing the testis formation in the embryo[36, 38]. During fetal or neonatal life, SCs show a highly proliferative capacity. There are only two periods of life when SCs proliferatein all species: in fetal or neonatal life, and in the peripubertal period [37]. The proliferation of SCs is likely controlled by numerous factors, including intratesticular factors and pituitary hormones, namely FSH that is a mitogenic factor for neonatal SCs[39].The immature SCs differs extensively from the mature cell with respect to both morphological and biochemical activity [38]. For instance, during proliferative activity of these cells there is a notable production of estrogens, leading to the suggestion that estrogens are involved in this process [32]. The proliferative activity of SCs is extremely important, because the number of SCs in adult testis will determine both testis size and sperm output [36, 37]. As puberty approaches, SCs experiment a radical change both in morphology and functions[37]. The formation of BTB initiates when SCs switches to a mature state [37]. Proliferative activity starts to decline at this time, the SCs become elongated and the nucleus becomes larger and irregularly shaped with a tripartite nucleolus. Abundant smooth and rough endoplasmic reticulum is also present[40]. After BTB is formed, the SCs begin to produce seminiferous tubular fluid (STF) [38].

Fully differentiated SCs must ensure a physical and nutritional support for germ cells, and the structural features of seminiferous epithelium. In regard to the physical support, SCs deposit extracellular matrix components (e.g. collagen and laminin), form specialized junctions and exhibit a well-organized cytoskeleton important for maintenance of seminiferous epithelium [41]. The SCs cytoskeleton consists of actin filaments, intermediate filaments and microtubules [41, 42].

Sertoli cells also secrete specific products that are necessary for germ cell survival. Amongst these products, are included severalglycoproteins that can be placed in differentcategories based on their biochemical properties [36]. These secreted glycoproteins correspond to 15% of total proteins produced by SCs[41]. Griswold and McLean [27] have summarized the recent knowledge regarding secreted proteins and secreted binding proteins by SCs. The first category of glycoproteins includes the transport or bioprotective proteins that are secreted in relative high abundance and metal ion transport proteins such as transferrin and ceruloplasmin. The second category of secretedproteins includes proteases (e.g. different types of cathepsins, plasminogen activator) and proteases inhibitors (e.g.

cystatin C, α_2-macroglobulin), essentials for tissue remodeling processes that occur during movement of preleptotenespermatocytes through BTB and spermiation. The third category, referred above, includes the glycoproteins that form the basement membrane between the SCs and the peritubular cells. Finally, SCs secrete a class of regulatory glycoproteins that can be made in very low abundance and still carry out their biochemical roles. These glycoproteins function as growth factors,paracrine or endocrine factors and include products such as AMH, *c kit* ligand and inhibin[27, 36]. In addition, the SCs secrete bioactive peptides such as prodynorphin and nutrients or metabolic intermediates [27, 36]. The transfer of nutritive products, which includes amino acids, carbohydrates, lipids, vitamins, and metal ions [41] from the SCs to the germ cells is possible due to the closest relationship between these two cell types. It is imperative that the germ cells receive an adequate level of energy substrates. Studies on the SCsmetabolism have shown that lactate play a major role in germ cells fate [43-45], an issue that will be discussed later.

Paradoxally to this protecting and nourishment facet of these "nurse cells", the SCs can induce apoptosis ofgerm cells[46], phagocytose apoptotic spermatogenic cells, and are also responsible for the processes of endocytose and degradation of the residual bodies[47].This could be probably because the number of spermatogenic cells that SCs can support for maturation is limited and when the number of cells reaches the threshold they should be removed to provide space in the seminiferous epithelium for the development of healthy spermatogenic cells [47]. Failure of functional SCs will result in a massive incapacity to support the development of various germ cells, and therefore inunsuccessful spermatogenic process.

Seminiferous Epithelium Environment

Seminiferous tubular fluid serves as a mean of transport for spermatozoa, and also allows keeping an appropriate microenvironment within tubules required for the normal occurrence of spermatogenesis.

Seminiferous tubular fluid starts to be produced by SCs when they switch to the maturational stage [37, 38, 48]. Its composition is controlled by SCs, as well as, the physico-chemical *milieu* where spermatogenesis occurs. It has been shown that the composition of the fluid within the seminiferous tubules is very stable due to the existence of the BTB [49] supporting the fundamental relevance of the STF composition. This luminal *milieu* is markedly distinct from the plasma and the testicular interstitial fluid and is critical to the occurrence of spermatogenesis [50]since meiosis can only be completed after the fluid secretion has been established [51]. Several reports have attempted to elucidate about the formation and composition of STF but unfortunately some of them lack in information concerning the origin and composition of STF. Recently, the knowledge on the mechanisms involved in the secretion, composition and regulation of SFT has been reviewed [48]. First reports from Tuck and collaborators [52] postulated that SCs were responsible for fluid secretion in the seminiferous tubules. These authors, using a variation of the micropuncture technique, suggested that the luminal fluid is composed by a K^+-rich solution [52]. Later Jenkins and collaborators [53] had observed the same results in their study. Clulow and Jones [54]determined a most suitable approach for defining the composition of the secretions of the seminiferous epithelium anddescribed a fluid rich in Na^+ and Cl^- content, with a K^+ concentration of at least twice that of blood concentration and indicated that this fluid is the main source of the luminal solutes in the extratesticular ducts.

Another important feature in seminiferous *milieu* is the control of its pH. Recent reports have elucidated the role of membrane transporters, both in controlling the STF pH, as in the control of the intracellular pH of SCs[55, 56]. This parameter is kept by intracellular buffers and the balancing between the elimination and production of protons [57]. Sertoli cellsexpress various types of ion membrane transporters directly involved on the movement of basic and acidic particles across the membrane [58]. The involvement of such transporters in the establishment of the STF is not yet completely disclosed, so it is essential further knowledge on their functioning, regulation and in the mechanisms responsible for determining the osmolarity and pH of STF [48].Furthermore, in the formation of STF, it should be taken into account the other functions of thisfluid: nourishment of germ cells, transport of secretion products and transport of the newly formed sperm towards the epididymis[59].

Metabolic Needs of Germ Cells

The adluminal compartment and intraluminal fluid contains both glucose and lactate, since SCs express specific membrane transporters that mediate the passage of these nutrients [43, 60]. Late developmental germ cells cannot metabolize glucose [61], thus they are dependent on SCs, but surprisingly mature spermatozoa show a high glycolytic activity and mainly metabolize fructose but also glucose existing in luminal fluid [62].

During spermatogenesis, germ cells experiment a series of changes that transform them from spermatogonia into fully functional spermatozoa. The spermatogonial cells, which lie at the basement membrane, proliferate by mitotic divisions giving rise a new population of stem cells (spermatogonia A), which is the guarantee of a germ cell line[63]. Other type of spermatogonia (spermatogonia B), are committed to differentiate and start moving down through seminiferous epithelium. Spermatogonia B also experiment mitosis producing primary spermatocytes, which will cross BTB to adluminal compartment[12]. After morphologic transformations, these cells will complete the first meiotic division given rise to secondary spermatocytes. Then they will experiment second meiosis producing round spermatids. After spermatids formation, cell division stops and starts the spermiogenesis, which originates elongated spermatids. Spermiation is the end step of spermiogenesis in which spermatozoa are released in the tubular lumen [33].

It is very interesting, as discussed below, but even more enigmatic the changing of the nutritional requirement of the germ cells during spermatogenesis. Spermatogonia, which lies in basal compartment of BTB, use glucose as a fuel for ATP production whereas spermatids cannot metabolize glucose, albeit they express all enzymes of the glycolytic pathway [61] and prefer lactate for maintaining their ATP content [61, 62]. Spermatocytes which are intermediate developing germ cells may depend on glycolysis, since the glycolytic activities are higher in spermatocytes than in spermatids [62], however it must not be overlooked the utilization of lactate by these cells at these stages of development, especially those that lies at adluminal compartment [64, 65].

Sertoli Cells Metabolism

As lactate is the preferred energy substrate for spermatocytes and spermatids [64, 66] SCs must ensure its supply to these developing germ cells. For this reason, the mechanisms that regulate SCs metabolism are very relevant in maintaining spermatogenesis and male fertility.

The carbohydrate metabolism in SCs presents some unique characteristics since they actively metabolize glucose but the majority of it is converted in lactate and is not oxidized in the Krebs cycle. Robinson and Fritz [67] were the first to describe glucose metabolism in SCs. These authors described that when in culture, the SCs convert the vast majority of glucose to lactate that is then secreted. Only a small fraction of the pyruvate produced from glucose is used in the Krebs cycle being instead converted to lactate. Later, Grootegoedand collaborators[68] continued those studies and concluded that only 25% of the pyruvate originated from glucose metabolism is oxidized via the Krebs cycle (Figure 3). Those authors described that, in SCs under in vitro conditions, the pentose phosphate pathway is not operating at its maximum rate, and the rate of the oxidative activity of the pentose phosphate pathway is determined by the rate of NADPH oxidation [67, 68]. They also found that exogenous pyruvate is oxidized at very low concentrations during incubation in the presence of glucose [68].

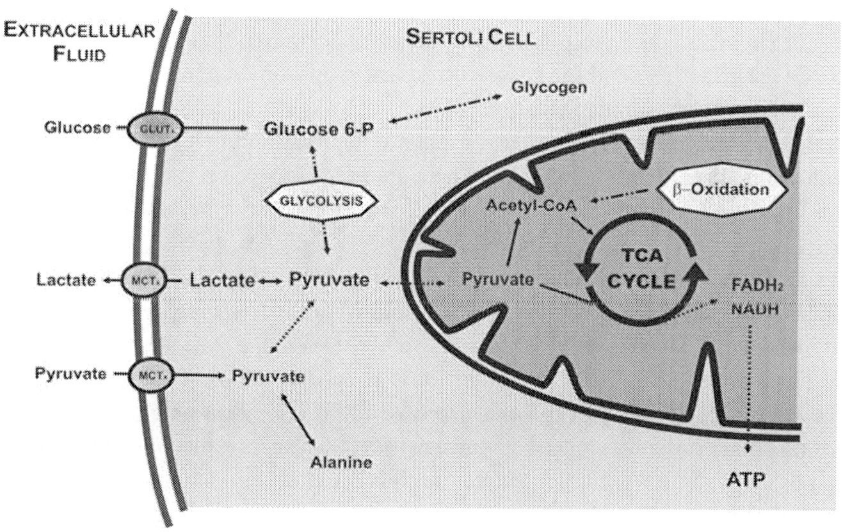

Figure 3.Schematic drawing illustration of Sertoli cell (SC) metabolism.TheSCs are capable of consuming a variety fuels including glucose, lactate, fatty acids and aminoacids. Nevertheless, the SCsactively metabolize glucose being the majority of it converted in lactate and not oxidized in the TCA cycle. The extracellular lactate and pyruvate are transported via the family of proton-linked plasma membrane transporters that carry molecules having one carboxylate group, the monocarboxylate transporters (MCTs), while glucose is imported via the family of membrane proteins called glucose transporters (GLUTs). Once glucose enters the glycolytic pathway, it is decomposed to pyruvate which can a) be converted into lactate via lactate dehydrogenase b) be converted into alaninevia alaninetransaminase or c)be transported to the mitochondrial matrix, oxidized and decarboxylated by the pyruvatedehydrogenase forming the two carbon intermediate Acetyl-CoAwhich can enter the TCA cycle to form $FADH_2$ and NADH. The oxidation of these substrates is coupled with ADP phosphorylation via the electron transport chain to form ATP. Abbreviations: TCA, tricarboxylic acid; GLUT, glucose transporter; MCT, monocarboxylate transporter.

Several mechanisms are known to be responsible for the modulation of SCs lactate production. Usually the rate-limiting step for glucose metabolism is the glucose transport of a family of structurally related glycoproteins designated by glucose transporters (GLUTs). Sertoli cells can take up glucose from the external medium [69] because hexose transporters are present on their plasma membranes [70]. Non-metabolizable analogues of glucose, such as 3H-3-O-methyl-D-glucose, can cross the BTB which indicates that glucose can actually be moved into the adluminal compartment, across both pro and antiluminal portion of the SCs plasma membrane [71]. Four glucose transporter isoforms (GLUT1, GLUT2, GLUT3 and GLUT8) have been so far identified in SCs[72-75]. However, GLUT8 is not expected to be involved in glucose transport from the extracellular *milieu* since it has not been identified in the plasma membrane of tissues [76, 77].

Also, lactate dehydrogenase (LDH) plays a crucial role in providing the substrate and later in the interconversion of pyruvate and lactate to developing germ cells. The lactate export and release from SCs by specific monocarboxylate transporters (MCTs) are also responsible for the improvement of lactate supply to germ cells.

The lactate production, in rat SCs, is under the control of FSH [66], epidermal growth factor (EGF) [78], insulin and insulin growth factor-I (IGF-I) [79], paracrine factor P-Mod-S [80], tri-iodothyronine[81], basic fibroblast growth factor (bFGF) [82], cytokines [44] and arachidonic acid [83]. Carnitine has also been described as a metabolic modulator of SCs metabolism. "In vitro" supplementation with carnitine increased the production of lactate and pyruvate, the LDH activity and the hexose transport, supporting that carbohydrate metabolism is increased by carnitine supplementation [84].

Sertoli cells can adapt to conditions of glucose deprivation to ensure the adequate lactate concentration in the microenvironment where germ cell develop and even when glucose levels are low or in the complete absence of glucose, the SCs still produce lactate [85]. Sertoli cells are known to change their metabolism by activating specific signal transduction pathways being the AMP-activated protein kinase (AMPK) a key mediator in cellular energy homeostasis [86]. The AMPK is present inSCs, and its activation increases lactate production with an increase of glucose uptake, increase of GLUT-1 and MCT4expression [87]. The decrease of glucose levels in SCs medium leads to an activation of AMPK, an increase in glucose uptake, increase in GLUT1 and decrease in GLUT3 expression [85]. Adenosine is described as a sign, possibly emitted by germ cells to SCs, that results in activation of AMPK that can simultaneously promote lactate production and the maintenance of junctional integrity thus allowing the optimal microenvironment for a successful spermatogenesis [88]. In the absence of glucose, production of lactate has been suggested to be a consequence of aminoacids or glycogen metabolism, since it has been described the presence of glycogen and glycogen phosphorylase activity in SCs[89, 90].

Grootegoedand collaborators[68] described that besides glucose, the oxidation of glutamine and leucine can yield much of the required energy by SCs. Others have also described an important role for the oxidation of other aminoacids such as alanine, leucine, and valine[91]. Nevertheless, the SCs have insignificant capacity to oxidize glycine[91] although the creatine synthesized in SCs from glycine and arginine, is crucial for providing energy to the spermatogenic cells [92]. The presence of glucose can modify the metabolism of these aminoacids. Kaiser and collaborators[91] suggested that low amounts of acetyl-CoA arising from glucose can modulate alanine and valine oxidation to CO_2 by competing with the acetyl-CoA, which results from these aminoacids metabolism. Also, the observed stimulation of the conversation of valine to lipids due to glucose metabolism was suggested to be a result of stimulation of the pentose cycle which is described as an important destination for glucose in SCs[67]. The presence of glutamine inhibited the oxidation of leucine, valine, and alanine but did not alter the conversion of these amino acids to lipids [91] although the incorporation of alanine to proteins was decreased [91]. Sertoli cells also convert leucine via transamination into 4-methyl-2-oxovalerate, spermatocytes and spermatids reduce exogenous 4-methyl- 2-oxovalerate to 2-hydroxy-4-methylvalerate, which is then released by the spermatogenic cells [93].

Sertoli cells are also important for the conversion of essential fatty acids in testis. When labeled fatty acids were injected in testis, they were firstly taken up and metabolized by SCs which proved that these cells have an important role in lipid metabolism [94]. The lipidic remodeling of rat spermatozoa has been shown to occur during their journey through the epididymis[95, 96]. Nevertheless, fatty acid metabolic activity in human testicular cells are low when compared with that found in rats [97], suggesting that human cells have a lower capacity for metabolizing fatty acid substrates. This may be related to the fact that adult rattestis are 0.7% of the body weight compared with approximately 0.04% of the body weight in adult men, and the daily sperm production per gram of testis is 5-6 times greater in rats than in men [98].

Lactate as Fuel for Germ Cell Development

Spermatogenic cells that undergo differentiation-related processes go through changes in their carbohydrate metabolism [62]. It has been suggested that there are specific metabolic signals required to the differentiation process of germ cells[99-101].

Germ cells undergoing meiosis exhibit a change in the type of substrate required for their energy metabolism and the equilibrium between glycolysis and the Krebs cycle shifts in favor of oxidative metabolism in post-meiotic cells [62]. The ratio between glycolysis and Krebs cycle is then greater in spermatocytes than in spermatids although in spermatozoa the equilibrium shifts again favoringglycolysis. Also, respiratory inhibitors reduce pyruvate/lactate consumption by spermatocytes while the spermatids completely degenerate in their presence [102].

Round spermatid energy metabolism is closely dependent on the presence of lactate in the extracellular medium and this lactate is supplied by SCs, as discussed above. That is why the STF is rich in lactate but has low glucose and pyruvate concentration[62]. Spermatids and spermatocytes are located inside the BTBwhilespermatogonia are outside the barrier. This may explain whyspermatogonia might utilize the glucose in the blood as energetic source, but as a large number of germ cells are developing into sperm inside seminiferous tubules, they must have others energy sources.

ATP levels in spermatids decrease in response to glucose metabolism [103]. The ATP is rapidly dephosphorylated when round spermatids are exposed to glucose in the absence of exogenous lactate [104]. Post-meiotic germ cells are unable to use glucose and rather prefer to use lactate as an energy source for their metabolism [105-107]. It has been reported that intratesticular infusion of lactate into the adult cryptorchid rat testis improves spermatogenesis [108]. Lactate is also responsible for RNA and protein synthesis stimulation in spermatids[106].

The spermatids exclusively use lactate probably due to their lower glycolytic potential whereas spermatocytes have greater glycolytic activity and have a limited dependence on lactate thus being able to use lactate, pyruvate and glucose. There is a metabolic recycling of lactate to glucose-6-P in round spermatids showing that both glycolytic and gluconeogenic pathways are functional in these cells [101]. Spermatozoa have the greatest glycolytic potential and the lowest citric acid cycle (TCA cycle) activity using only glucose or fructose for their energy metabolism [62]. It has been proposed that glyceraldehyde 3-phosphate dehydrogenase (GAPDH) regulates glycolysis in postmeiotic germ cells by being inactive in spermatids and active in spermatozoa [61]. However, the mechanism by which the enzyme becomes activated in sperm still remains unknown, although ATP and other adenine nucleotides might be responsible for that process [61].

The pentose phosphate pathway is activated in germ cells, as indicated by the glucose-6-phosphate activity. This pathway is required for the biosynthesis of nucleotides for RNA and for production of NADPH and ribose 5-phosphate. Also, during their journey through the epididymis, rat spermatozoa undergo considerable lipid remodeling and pachytenespermatocytes are more active in metabolizing the fatty acids than round spermatids[109]. Whether the germ cells are able to synthesize fatty acids from linoleic and α-linolenic acid or whether they must be provided with these fatty acids from theSCs is unknown, but several authors proposed that there is a transport of long polyenes from SCs to the germ cells although it has not been fully proved yet [110-112]. That transport may require a cell-to-cell contact as in vivo the germ cells are in close contact with SCs throughout all steps of development [112].

Conclusion

Formation of competent spermatozoa is a complex multistep process that is initiated in the seminiferous epithelium. One critical feature in this process is the establishment of the BTB that physically and physiologically supports the proliferation and differentiation of germ cells into mature spermatids.

Germ cells have special metabolic needs that render them dependent on the nurturing provided by SCs, the structural element of the seminiferous epithelium. Developing germ cells are unable to use glucose for their energy metabolism and preferentially use lactate. Thus, carbohydrate metabolism in SCs presents some distinctive characteristics as they actively metabolize glucose, but the majority of it is converted tolactate. Moreover, these cells must be capable of modulating their metabolism in order to ensure the adequate lactate concentration in the microenvironment where germ cell develop, even in conditions of glucose limitation. Production of lactate has been suggested to be derived also from aminoacids or glycogen metabolism, although these processes are not entirely understood.

Further knowledge on the functioning and regulation of these biochemical mechanisms is essential for the enlightenment of a process that is central to the propagation of life, spermatogenesis. This will be an important step in understanding infertility linked to BTB dysfunction and the role of cellular metabolism in some pathological conditions.

Aknowledgements

PF Oliveira was financed by Foundation for Science and Technology (FCT) under the program Ciência 2008. L Rato was financed by FCT(SFRH/BD/72733/2010).

References

[1] Kormano M. Dye permeability and alkaline phosphatase activity of testicular capillaries in the postnatal rat. *Histochemie*. 1967;9(4):327-38.

[2] Setchell B, Davies R, Gladwell R, Hinton B, Main S, Pilsworth L, et al. The movement of fluid in the seminiferous tubules and rete testis. *Annales de biologie animale, biochimie, biophysique*. 1978;18(2B):623-32.

[3] Setchell BP, Hinton BT, Jacks F, Davies RV. The restricted penetration of iodinated rat FSH and LH into the seminiferous tubules of the rat testis. *Molecular and CellularEndocrinology*. 1976 Nov;6(1):59-69.

[4] Voglmayr JK, Waites GM, Setchell BP. Studies on spermatozoa and fluid collected directly from the testis of the conscious ram. *Nature*. 1966;210(5038):861-63.

[5] Waites G, Gladwell R. Physiological significance of fluid secretion in the testis and blood-testis barrier. *Physiol. Rev*. 1982;62(2):624-71.

[6] Su L, Mruk DD, Cheng CY. Drug transporters, the blood-testis barrier and spermatogenesis. *Journal of Endocrinology*. 2011;208(3):207-23.

[7] Setchell BP. The Functional-Significance of the Blood-Testis Barrier. *Journal of andrology*. 1980;1(1):3-10.

[8] Wong CH, Cheng CY. The blood-testis barrier: its biology, regulation, and physiological role in spermatogenesis. *Curr. Top Dev. Biol*. 2005;71:263-96.

[9] Toyama Y, Mackawa M, Yuasa S. Ectoplasmic specializations in the Sertoli cell: new vistas based on genetic defects and testicular toxicology. *Anatomical Science International*. 2003 Mar;78(1):1-16.

[10] Mazaud-Guittot S, Meugnier E, Pesenti S, Wu X, Vidal H, Gow A, et al. Claudin 11 deficiency in mice results in loss of the Sertoli cell epithelial phenotype in the testis. *Biology of Reproduction*. 2010 Jan;82(1):202-13.

[11] Lui WY, Cheng CY. Regulation of cell junction dynamics by cytokines in the testis: a molecular and biochemical perspective. *Cytokine Growth Factor Review*. 2007 Jun-Aug;18(3-4):299-311.

[12] Cheng CY, Mruk DD. An intracellular trafficking pathway in the seminiferous epithelium regulating spermatogenesis: a biochemical and molecular perspective. *Critical Reviews in Biochemistry and Molecular Biology*. 2009 Sep-Oct;44(5):245-63.

[13] Cheng CY, Wong EW, Yan HH, Mruk DD. Regulation of spermatogenesis in the microenvironment of the seminiferous epithelium: new insights and advances. *Molecular and Cellular Endocrinology*. 2010 Feb 5;315(1-2):49-56.

[14] Siu MKY, Cheng CY. Extracellular matrix and its role in spermatogenesis. *Molecular Mechanisms in Spermatogenesis*. 2009:74-91.

[15] Russell LD. The blood-testis barrier and its formation relative to spermatocyte maturation in the adult rat: a lanthanum tracer study. *Anatomical Record.* 1978 Jan;190(1):99-111.
[16] Yao PL, Lin YC, Richburg JH. Mono-(2-ethylhexyl) phthalate-induced disruption of junctional complexes in the seminiferous epithelium of the rodent testis is mediated by MMP2. *Biology of Reproduction.* 2010;82(3):516-27.
[17] Mital P, Hinton BT, Dufour JM. The Blood-Testis and Blood-Epididymis Barriers Are More Than Just Their Tight Junctions. *Biology of Reproduction.* 2011 Jan 5;84(5):851-58.
[18] Sharpe R. Follicle-stimulating hormone and spermatogenesis in the adult male. *Journal of Endocrinology.* 1989;121(3):405.
[19] Means AR, Huckins C. Coupled events in the early biochemical actions of FSH on the Sertoli cells of the testis. *Current topics in molecular endocrinology.* 1974;1:145-65.
[20] Means AR, Dedman JR, Tash JS, Tindall DJ, van Sickle M, Welsh MJ. Regulation of the testis sertoli cell by follicle stimulating hormone. *Annual Reviews of Physiology.* 1980;42:59-70.
[21] Setchell BP. Blood-testis barrier, junctional and transport proteins and spermatogenesis. *Adv. Exp. Med. Biol.* 2008;636:212-33.
[22] Brinster RL, Zimmermann JW. Spermatogenesis following male germ-cell transplantation. *Proceeding of the National Academy ofScience.* 1994 Nov 22;91(24):11298-302.
[23] Wang T, Feng X, Han D. Mechanisms of testicular immune privilege. *Frontiers inBiology.* 2011;6(1):19-30.
[24] Dym M, Romrell LJ. Intraepithelial lymphocytes in the male reproductive tract of rats and rhesus monkeys. *Journal of Reproduction and Fertility.* 1975 Jan;42(1):1-7.
[25] Head JR, Billingham RE. Immune privilege in the testis. II. Evaluation of potential local factors. *Transplantation.* 1985 Sep;40(3):269-75.
[26] Nagaosa K, Nakashima C, Kishimoto A, Nakanishi Y. Immune response to bacteria in seminiferous epithelium. *Reproduction.* 2009 May;137(5):879-88.
[27] Griswold M, McLean D. The Sertoli cell. In: Neill J, editor. Knobil and Neill's physiology of reproduction. San Diego: Elsevier; 2006. p. 949-75.
[28] Dym M. The fine structure of monkey Sertoli cells in the transitional zone at the junction of the seminiferous tubules with the tubuli recti. *American Journal Anatomy.* 1974 May;140(1):1-25.
[29] Russell L, Ettlin R, Sinha Hikim A, Clegg E. Histological and histopathological evaluation of the testis. Clearwater: Cache River Press; 1990.
[30] Russell LD, Ren HP, Hikim IS, Schulze W, Hikim APS. A comparative study in twelve mammalian species of volume densities, volumes, and numerical densities of selected testis components, emphasizing those related to the Sertoli cell. *American Journal ofAnatomy.* 1990;188(1):21-30.
[31] Weber JE, Russell LD, Wong V, Peterson RN. Three-dimensional reconstruction of a rat stage V Sertoli cell: II. Morphometry of Sertoli--Sertoli and Sertoli--germ-cell relationships. *American Journal Anatomy.* 1983 Jun;167(2):163-79.
[32] O'Donnell L, Robertson K, Jones M, Simpson E. Estrogen and spermatogenesis. *Endocr. Rev.* 2001;22(3):289-318.

[33] Hess R, de Franca L. Spermatogenesis and cycle of the seminiferous epithelium. Molecular Mechanisms in Spermatogenesis Austin, TX: Landes Bioscience/Springer *Science*. 2008:1–15.
[34] Sertoli E. Sulla struttura dei canalicoli seminiferi dei testicoli *Arch Sc Med.* 1878;2:107-19.
[35] Sertoli E. Dell'esistenza di particolari cellule ramifi cate nei canalicoli seminiferi del testicolo umano. *Morgagni*. 1865;7:31-3.
[36] Griswold M, editor. The central role of Sertoli cells in spermatogenesis1998: London: Academic Press.
[37] Sharpe R, McKinnell C, Kivlin C, Fisher J. Proliferation and functional maturation of Sertoli cells, and their relevance to disorders of testis function in adulthood. *Reproduction*. 2003;125(6):769-84.
[38] Petersen C, Soder O. The Sertoli cell-a hormonal target and'super'nurse for germ cells that determines testicular size. *Hormone Research*. 2006;66(4):153-61.
[39] Gondos B, WE B. Postnatal and pubertal development In: Russell LD GM, editor. The Sertoli Cell. Clearwater, FL: Cache River Press; 1994. p. 116-53.
[40] Nistal M, Abaurrea MA, Paniagua R. Morphological and histometric study on the human Sertoli cell from birth to the onset of puberty. *Journal of Anatomy*. 1982;134:351-63.
[41] Mruk DD, Cheng CY. Sertoli-Sertoli and Sertoli-germ cell interactions and their significance in germ cell movement in the seminiferous epithelium during spermatogenesis. *Endocr. Rev*. 2004 Oct;25(5):747-806.
[42] Vogl AW, Vaid KS, Guttman JA. The Sertoli Cell Cytoskeleton. In: Cheng CY, editor. Molecular Mechanisms in Spermatogenesis: Landes Bioscience/ Springer Science + *Business Media*; 2008. p. 186-211.
[43] Riera M, Meroni S, Schteingart H, Pellizzari E, Cigorraga S. Regulation of lactate production and glucose transport as well as of glucose transporter 1 and lactate dehydrogenase A mRNA levels by basic fibroblast growth factor in rat Sertoli cells. *Journal of Endocrinology*. 2002;173(2):335-43.
[44] Riera MF, Meroni SB, Gomez GE, Schteingart HF, Pellizzari EH, Cigorraga SB. Regulation of lactate production by FSH, iL1beta, and TNFalpha in rat Sertoli cells. *General and Comparative Endocrinology*. 2001 Apr;122(1):88-97.
[45] Erkkila K, Aito H, Aalto K, Pentikainen V, Dunkel L. Lactate inhibits germ cell apoptosis in the human testis. *Molecular human reproduction*. 2002;8(2):109-17.
[46] Lee J, Richburg JH, Younkin SC, Boekelheide K. The Fas system is a key regulator of germ cell apoptosis in the testis. *Endocrinology*. 1997 May;138(5):2081-8.
[47] Xiong WP, Wang HK, Wu H, Chen YM, Han DS. Apoptotic spermatogenic cells can be energy sources for Sertoli cells. *Reproduction*. 2009 Mar;137(3):469-79.
[48] Rato L, Socorro S, Cavaco JE, Oliveira PF. Tubular fluid secretion in the seminiferous epithelium: ion transporters and aquaporins in Sertoli cells. *Journal of Membrane Biology*. 2010 Jul;236(2):215-24.
[49] Koskimies A, Kormano M. The proteins in fluids from the seminiferous tubules and rete testis of the rat. *Reproduction*. 1973;34(3):433-34.
[50] Fisher D. New light shed on fluid formation in the seminiferous tubules of the rat. *Journal of Physiology*. 2002 July 15, 2002;542(2):445-52.

[51] Setchell B. The secretion of fluid by the testes of rats, rams and goats with some observations on the effect of age, cryptorchidism and hypophysectomy. *Reproduction.* 1970;23(1):79-85.

[52] Tuck RR, Setchell BP, Waites GM, Young JA. The composition of fluid collected by micropuncture and catheterization from the seminiferous tubules and rete testis of rats. *Pflugers Arch.* 1970;318(3):225-43.

[53] Jenkins A, Lechene C, Howards S. Concentrations of seven elements in the intraluminal fluids of the rat seminiferous tubules, rete testis, and epididymis. *Biology of Reproduction.* 1980;23(5):981-87.

[54] Clulow J, Jones RC. Composition of luminal fluid secreted by the seminiferous tubules and after reabsorption by the extratesticular ducts of the Japanese quail, Coturnix coturnix japonica. *Biology of Reproduction.* 2004 Nov;71(5):1508-16.

[55] Oliveira P, Sousa M, Barros A, Moura T, Rebelo da Costa A. Intracellular pH regulation in human Sertoli cells: role of membrane transporters. *Reproduction.* 2009;137(2):353.

[56] Oliveira PF, Sousa M, Barros A, Moura T, Rebelo da Costa A. Membrane transporters and cytoplasmatic pH regulation on bovine Sertoli cells. *Journal of Membrane Biology.* 2009a Jan;227(1):49-55.

[57] Roos A, Boron W. Intracellular pH. *Am. Physiological. Soc*; 1981. p. 296-434.

[58] Boron W. Regulation of intracellular pH. American Physiological Society. 2004;28(4):160-79.

[59] Jegou B, Le Gac F, de Kretser D. Seminiferous tubule fluid and interstitial fluid production. I. Effects of age and hormonal regulation in immature rats. *Biology ofReproduction.* 1982;27(3):590-95.

[60] Brauchi S, Rauch MC, Alfaro IE, Cea C, Concha, II, Benos DJ, et al. Kinetics, molecular basis, and differentiation of L-lactate transport in spermatogenic cells. *American Journal Physiology Cell Physiology.* 2005 Mar;288(3):523-34.

[61] Boussouar F, Benahmed M. Lactate and energy metabolism in male germ cells. *TRENDS in Endocrinology and Metabolism.* 2004;15(7):345-50.

[62] Bajpai M, Gupta G, Setty B. Changes in carbohydrate metabolism of testicular germ cells during meiosis in the rat. *European journal ofendocrinology.* 1998;138(3):322-27.

[63] Tokas J, Mukhopadhyay C, and, Verma A. Self Renewal of Spermatogonial Stem Cells: the Most Promising Multipotent Cells - A Review. *Veterinary World.* 2011;4:234-40.

[64] Jutte N, Grootegoed J, Rommerts F, Van der Molen H. Exogenous lactate is essential for metabolic activities in isolated rat spermatocytes and spermatids. *Reproduction.* 1981;62(2):399-405.

[65] Jutte N, Jansen R, Grootegoed J, Rommerts F, Clausen O, Van der Molen H. Regulation of survival of rat pachytene spermatocytes by lactate supply from Sertoli cells. *Reproduction.* 1982;65(2):431-38.

[66] Mita M, Price JM, Hall PF. Stimulation by follicle-stimulating hormone of synthesis of lactate by Sertoli cells from rat testis. *Endocrinology.* 1982 May;110(5):1535-41.

[67] Robinson R, Fritz IB. Metabolism of glucose by Sertoli cells in culture. *Biology of Reproduction.* 1981;24(5):1032.

[68] Grootegoed J, Oonk R, Jansen R, Van der Molen H. Metabolism of radiolabelled energy-yielding substrates by rat Sertoli cells. *Reproduction.* 1986;77(1):109.

[69] Hall PF, Mita M. Influence of follicle-stimulating hormone on glucose transport by cultured Sertoli cells. *Biol. Reprod.* 1984 Dec;31(5):863-9.

[70] Angulo C, Rauch MC, Droppelmann A, Reyes AM, Slebe JC, Delgado-Lopez F, et al. Hexose transporter expression and function in mammalian spermatozoa: cellular localization and transport of hexoses and vitamin C. *J. Cell Biochem.* 1998 Nov 1;71(2):189-203.

[71] Turner TT, D'Addario DA, Howards SS. The transepithelial movement of 3H-3-O-methyl-D-glucose in the hamster seminiferous and cauda epididymidal tubules. *Fertil. Steril.* 1983 Oct;40(4):530-5.

[72] Carosa E, Radico C, Giansante N, Rossi S, D'Adamo F, Di Stasi SM, et al. Ontogenetic profile and thyroid hormone regulation of type-1 and type-8 glucose transporters in rat Sertoli cells. *Int. J. Androl.* 2005 Apr;28(2):99-106.

[73] Galardo MN, Riera MF, Pellizzari EH, Chemes HE, Venara MC, Cigorraga SB, et al. Regulation of expression of Sertoli cell glucose transporters 1 and 3 by FSH, IL1 beta, and bFGF at two different time-points in pubertal development. *Cell Tissue Res.* 2008 Nov;334(2):295-304.

[74] Ulisse S, Jannini EA, Pepe M, De Matteis S, D'Armiento M. Thyroid hormone stimulates glucose transport and GLUT1 mRNA in rat Sertoli cells. *Mol. CellEndocrinol.* 1992 Sep;87(1-3):131-7.

[75] Kokk K, Verajankorva E, Wu XK, Tapfer H, Poldoja E, Pollanen P. Immunohistochemical detection of glucose transporters class I subfamily in the mouse, rat and human testis. *Medicina* (Kaunas, Lithuania). 2004;40(2):156.

[76] Piroli GG, Grillo CA, Hoskin EK, Znamensky V, Katz EB, Milner TA, et al. Peripheral glucose administration stimulates the translocation of GLUT8 glucose transporter to the endoplasmic reticulum in the rat hippocampus. *J. Comp. Neurol.* 2002 Oct 14;452(2):103-14.

[77] Reagan LP, Gorovits N, Hoskin EK, Alves SE, Katz EB, Grillo CA, et al. Localization and regulation of GLUTx1 glucose transporter in the hippocampus of streptozotocin diabetic rats. *Proc. Natl. Acad. Sci. U S A.* 2001 Feb 27;98(5):2820-5.

[78] Mallea LE, Machado AJ, Navaroli F, Rommerts FF. Epidermal growth factor stimulates lactate production and inhibits aromatization in cultured Sertoli cells from immature rats. *Int. J. Androl.* 1986 Jun;9(3):201-8.

[79] Oonk RB, Jansen R, Grootegoed JA. Differential effects of follicle-stimulating hormone, insulin, and insulin-like growth factor I on hexose uptake and lactate production by rat Sertoli cells. *J. Cell Physiol.* 1989 Apr;139(1):210-8.

[80] Mullaney BP, Rosselli M, Skinner MK. Developmental regulation of Sertoli cell lactate production by hormones and the testicular paracrine factor, PModS. *Mol. Cell Endocrinol.* 1994 Aug;104(1):67-73.

[81] Palmero S, Prati M, Bolla F, Fugassa E. Tri-iodothyronine directly affects rat Sertoli cell proliferation and differentiation. *J. Endocrinol.* 1995 May;145(2):355-62.

[82] Schteingart HF, Meroni SB, Canepa DF, Pellizzari EH, Cigorraga SB. Effects of basic fibroblast growth factor and nerve growth factor on lactate production, gamma-glutamyl transpeptidase and aromatase activities in cultured Sertoli cells. *Eur. J. Endocrinol.* 1999 Nov;141(5):539-45.

[83] Meroni SB, Riera MF, Pellizzari EH, Schteingart HF, Cigorraga SB. Possible role of arachidonic acid in the regulation of lactate production in rat Sertoli cells. *Int. J. Androl.* 2003 Oct;26(5):310-7.

[84] Palmero S, Bottazzi C, Costa M, Leone M, Fugassa E. Metabolic effects of L-carnitine on prepubertal rat Sertoli cells. *Horm. Metab. Res.* 2000 Mar;32(3):87-90.

[85] Riera MF, Galardo MN, Pellizzari EH, Meroni SB, Cigorraga SB. Molecular mechanisms involved in Sertoli cell adaptation to glucose deprivation. *Am. J. Physiol. Endocrinol. Metab.* 2009 Oct;297(4):E907-14.

[86] Hardie DG. Minireview: the AMP-activated protein kinase cascade: the key sensor of cellular energy status. *Endocrinology.* 2003 Dec;144(12):5179-83.

[87] Galardo MN, Riera MF, Pellizzari EH, Cigorraga SB, Meroni SB. The AMP-activated protein kinase activator, 5-aminoimidazole-4-carboxamide-1-b-D-ribonucleoside, regulates lactate production in rat Sertoli cells. *J. Mol. Endocrinol.* 2007 Oct;39(4):279-88.

[88] Galardo MN, Riera MF, Pellizzari EH, Sobarzo C, Scarcelli R, Denduchis B, et al. Adenosine regulates Sertoli cell function by activating AMPK. *Mol. Cell Endocrinol.* 2010 Dec 15;330(1-2):49-58.

[89] Leiderman B, Mancini RE. Glycogen content in the rat testis from postnatal to adult ages. *Endocrinology.* 1969 Sep;85(3):607-9.

[90] Slaughter GR, Means AR. Follicle-stimulating hormone activation of glycogen phosphorylase in the Sertoli cell-enriched rat testis. *Endocrinology.* 1983 Oct;113(4):1476-85.

[91] Kaiser GR, Monteiro SC, Gelain DP, Souza LF, Perry ML, Bernard EA. Metabolism of amino acids by cultured rat Sertoli cells. *Metabolism.* 2005 Apr;54(4):515-21.

[92] Moore NP, Gray TJ, Timbrell JA. Creatine metabolism in the seminiferous epithelium of rats. I. Creatine synthesis by isolated and cultured cells. *J. Reprod. Fertil.* 1998 Mar;112(2):325-30.

[93] Grootegoed JA, Jansen R, van der Molen HJ. Intercellular pathway of leucine catabolism in rat spermatogenic epithelium. *Biochem. J.* 1985 Mar 15;226(3):889-92.

[94] Beckman JK, Coniglio JG. The metabolism of polyunsaturated fatty acids in rat Sertoli and germinal cells. *Lipids.* 1980 Jun;15(6):389-94.

[95] Hall JC, Hadley J, Doman T. Correlation between changes in rat sperm membrane lipids, protein, and the membrane physical state during epididymal maturation. *J. Androl.* 1991 Jan-Feb;12(1):76-87.

[96] Aveldano MI, Rotstein NP, Vermouth NT. Lipid remodelling during epididymal maturation of rat spermatozoa. Enrichment in plasmenylcholines containing long-chain polyenoic fatty acids of the n-9 series. *Biochem. J.* 1992 Apr 1;283 (Pt 1):235-41.

[97] Retterstol K, Haugen TB, Woldseth B, Christophersen BO. A comparative study of the metabolism of n-9, n-6 and n-3 fatty acids in testicular cells from immature rat. *Biochim. Biophys. Acta.* 1998 May 20;1392(1):59-72.

[98] Sharpe R. Regulation of spermatogenesis. In: Knobil E, Neill J, editors. The physiology of reproduction. New York: Raven Press Ltd, ; 1994. p. 1363-434.

[99] Herrera E, Salas K, Lagos N, Benos DJ, Reyes JG. Energy metabolism and its linkage to intracellular Ca2+ and pH regulation in rat spermatogenic cells. *Biol. Cell.* 2000 Sep;92(6):429-40.

[100] Reyes JG, Herrera E, Lobos L, Salas K, Lagos N, Jorquera RA, et al. Dynamics of intracellular calcium induced by lactate and glucose in rat pachytene spermatocytes and round spermatids. *Reproduction*. 2002 May;123(5):701-10.

[101] Yanez AJ, Bustamante X, Bertinat R, Werner E, Rauch MC, Concha, II, et al. Expression of key substrate cycle enzymes in rat spermatogenic cells: fructose 1,6 bisphosphatase and 6 phosphofructose 1-kinase. *J. Cell Physiol.* 2007 Sep;212(3):807-16.

[102] Grootegoed JA, Jansen R, Van der Molen HJ. The role of glucose, pyruvate and lactate in ATP production by rat spermatocytes and spermatids. *Biochim. Biophys. Acta.* 1984 Nov 26;767(2):248-56.

[103] Nakamura M, Fujiwara A, Yasumasu I, Okinaga S, Arai K. Regulation of glucose metabolism by adenine nucleotides in round spermatids from rat testes. *J. Biol. Chem.* 1982 Dec 10;257(23):13945-50.

[104] Grootegoed JA, Jansen R, van der Molen HJ. Effect of glucose on ATP dephosphorylation in rat spermatids. *J. Reprod. Fertil.* 1986 May;77(1):99-107.

[105] Mita M, Hall PF. Metabolism of round spermatids from rats: lactate as the preferred substrate. *Biology of Reproduction*. 1982;26(3):445.

[106] Jutte NH, Grootegoed JA, Rommerts FF, van der Molen HJ. Exogenous lactate is essential for metabolic activities in isolated rat spermatocytes and spermatids. *J. Reprod. Fertil.* 1981 Jul;62(2):399-405.

[107] Nakamura M, Okinaga S, Arai K. Metabolism of round spermatids: evidence that lactate is preferred substrate. *Am. J. Physiol.* 1984 Aug;247(2 Pt 1):E234-42.

[108] Courtens JL, Ploen L. Improvement of spermatogenesis in adult cryptorchid rat testis by intratesticular infusion of lactate. Biol Reprod. 1999 Jul;61(1):154-61.

[109] Retterstol K, Haugen TB, Tran TN, Christophersen BO. Studies on the metabolism of essential fatty acids in isolated human testicular cells. *Reproduction*. 2001 Jun;121(6):881-7.

[110] Lynch KM, Jr., Scott WW. Lipid distribution in the Sertoli cell and Leydig cell of the rat testis as related to experimental alterations of the pituitary-gonad system. *Endocrinology*. 1951 Jul;49(1):8-14.

[111] Beckman JK, Coniglio JG. A comparative study of the lipid composition of isolated rat Sertoli and germinal cells. *Lipids*. 1979 Mar;14(3):262-7.

[112] Retterstol K, Tran TN, Haugen TB, Christophersen BO. Metabolism of very long chain polyunsaturated fatty acids in isolated rat germ cells. *Lipids*. 2001 Jun;36(6):601-6.

In: Endothelium and Epithelium
Editors: J. Carrasco and M. Mota, pp. 157-173
ISBN 978-1-61470-874-2

Chapter VII

Interleukin-6 (IL6) Signalling and Its Role in the Development of Cardiovascular Disease

Jamie J. L. Williams and Timothy M. Palmer*
Institute of Cardiovascular and Medical Sciences, College of Medical, Veterinary and Life Sciences, University of Glasgow, Glasgow G12 8QQ, Scotland, U.K.

Abstract

The pleiotropic cytokine interleukin-6 (IL6) is elaborated by many cells involved in the development of atherosclerosis, including smooth muscle and endothelial cells as well as monocytes and T cells. In turn, IL6 controls several aspects of vascular and haematopoietic cell function that become dysregulated during the development of the chronic inflammatory phenotype that drives atherogenesis. However many studies have linked IL6 to a range of often contradictory effects on cardiovascular function. In this chapter, we will examine how the unique position of IL6 at the junction of the adaptive and innate immune responses, and its ability to signal via two biochemically distinct IL6-IL6 receptor complexes, may explain the effects ascribed to it in various models of chronic inflammatory disease. We will also assess how an increased understanding of IL6 signalling and its inhibitory regulation, particularly by suppressor of cytokine signalling (SOCS) proteins, might be useful in suggesting new therapeutic strategies for treatment of specific cardiovascular diseases such as atherosclerosis.

Introduction

IL6 is a multifunctional mediator of a myriad of complex biological responses, including hepatic induction of acute phase proteins, differentiation of B cells, as well as T cell

*E-mail: Tim.Palmer@glasgow.ac.uk

activation, growth, and differentiation. In a pro-inflammatory role IL6 initiates inflammation *via* the activation of endothelial cells by up regulating expression of the adhesion molecules ICAM-1, E-selectin and VCAM-1, chemokine MCP-1/CCL2 and also by induction of smooth muscle cell proliferation and migration [1-3]. Anti-inflammatory effects include negative regulation of IL1 and TNFα synthesis as well as quenching of the pro-inflammatory effects of IL1 by up-regulation of IL1R antagonist IL1ra [4]. IL6 mediates its effects by targeting the largest family of cytokine receptors, the haematopoietin/type 1 receptors, of which the membrane-bound receptor gp130 (IL6Rβ, CD130) is the most common signalling unit. Gp130 acts as a signal transducer for the other members of the four α-helical IL6 family of cytokines, including IL11, leukaemia inhibitory factor (LIF), oncostatin M (OSM), cardiotrophin-1 (CT-1), ciliary neurotrophic factor (CNTF), and cardiotrophin-like cytokine (CLC). IL6 itself is a 22-28 kDa protein, depending on its glycosylation state, secreted by leukocytes, fibroblasts, endothelial cells, keratinocytes, hepatocytes, and bone marrow cells. Five isoforms exist in human, IL-6, IL6alt, IL6Δ2, IL6Δ2,4, and IL6Δ4 due to alternative splicing of gene transcripts [5,6]. Increased signalling diversity is achieved from their interaction with membrane-bound, soluble and associated isoforms of both IL6R and gp130 subunits.

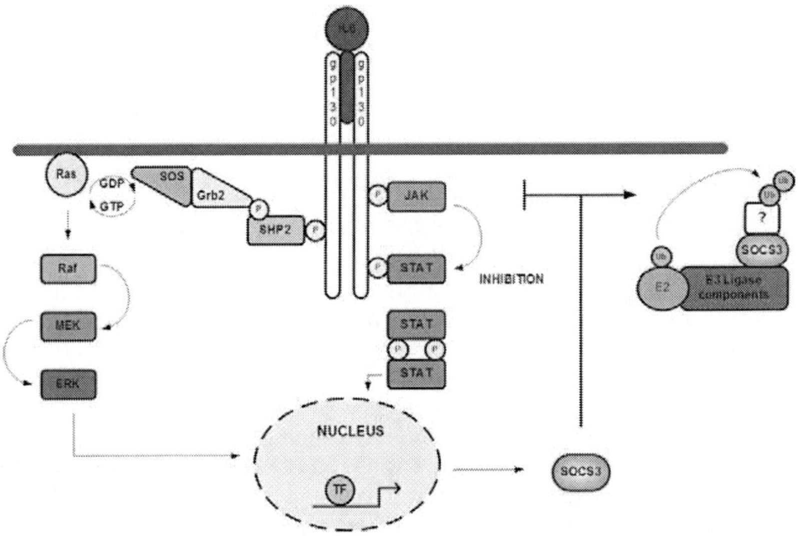

Figure 1. IL6 family cytokine signalling. IL6 family cytokines activate JAK/STAT and ERK pathways. IL6 binds either soluble or membrane-associated IL6 receptor alpha chains followed by binding a gp130 receptor which itself lacks kinase activity but is constitutively associated with JAKs. Cytokine-receptor ligation forms receptor dimers which brings JAKs into close proximity. Following trans-phosphorylation and activation of JAKs, the receptor subunits are phosphorylated producing docking sites for SH2-domain containing proteins, STATs and SHP2. Activation of STATs by JAKs results in receptor dissociation, dimerisation and translocation to the nucleus where they drive cytokine-induced gene transcription. STAT-induced SOCS3 negatively regulates cytokine signalling either by direct inhibition of JAKs or via the degradation of signalling intermediates in its role as a specificity factor for the ECS family of E3 ubiquitin ligase. In a similar fashion to STATs, bound, phosphorylated SHP2 induces gene transcription by activation of the MAPK pathway *via* the recruitment of the growth factor receptor bound protein 2-Son of Sevenless (Grb2-SOS) complex Grb2 is an adaptor protein required for SHP2 association and SOS is a ras guanine nucleotide exchange factor (GEF). SHP2 is a phosphatase so might act on gp130 and associated factors to negatively regulate signalling.

Most IL6 family members transduce signals *via* gp130 either as a homodimer, as in the case of IL6, or as a heterodimer of gp130 and a second signalling receptor protein. Although each ligand-receptor partnership imparts a specific response, in common is the activation of the janus kinase/signal transducers and activator of transcription (JAK/STAT) and extracellular signal-regulated kinase (ERK) pathways. As well as binding and interacting with membrane-localised IL6R to trigger signalling *via* gp130 (this is termed "classical" IL-6 signalling), IL6 is unique in that it can also function *via* a so-called trans-signalling mechanism. This is achieved because gp130 does not recognise IL6 directly. Instead, IL6 must first form a low affinity complex with the 80kDa non-signalling α-receptor, IL6Rα (CD126), which then forms a high affinity tetramer with a homodimer of gp130. Gp130 is constitutively and ubiquitously expressed whereas membrane bound/non-signalling IL6R is restricted to hepatocytes, leukocytes, and lymphocytes. However soluble IL6R (sIL6R) variants can also be generated either *via* alternate splicing of the IL6Rα primary transcript or by limited proteolysis of the membrane-localised IL6Rα protein by metalloproteases ADAM10 or ADAM17, resulting in soluble IL6Rα (sIL6Rα) being shed from hepatocytes or macrophages. This enables signalling in non-IL6R-expressing cells, thus increasing the repertoire of IL6-responsive cells during an inflammatory response [7].

The gp130 signalling receptor lacks intrinsic kinase activity but is constitutively associated with JAK family tyrosine kinases. Cytokine-receptor ligation brings gp130-JAK into close proximity allowing trans-phosphorylation and activation of JAKs which then tyrosine phosphorylate gp130 producing docking sites for the SH2-domain-containing signalling mediators STAT (signal transducers and activators of transcription) and SH2 domain-containing protein-tyrosine phosphatase-2 (SHP2). IL6-family cytokines predominantly signal from gp130 through JAK1/STAT3 and to a lesser extent JAK2/STAT1 [7]. JAK1-mediated phosphorylation of gp130 at Y^{767}, Y^{814}, Y^{905} and Y^{915} at a pYXXQ consensus sequence enables binding of STAT3 which competes with STAT1 binding at Y^{905} and Y^{915} at a more constrained pYXLQ consensus sequence [7,8]. This could account for the more potent activation and dimer formation of transcriptionally active STAT3 compared to STAT1. Docked STATs are tyrosine phosphorylated by JAKs within their SH2 domains (STAT1, Y^{701} and STAT3, Y^{705}) which enables receptor dissociation and formation of homo- or hetero- dimers, *via* their SH2 domain, or trimers with DNA-binding adapter proteins such as the IFN regulatory factors (IRF) before translocating to the nucleus where they drive cytokine-inducible gene transcription from palindromic DNA elements with a $TTN_{4-6}AA$ consensus sequence [9] as well as IFN-stimulated response elements (ISRE sites, $AGTTTN_3TTTC$) and IFNγ-activated sequence (GAS sites, $TTCN_3GAA$). The exact preference depends on STAT dimer composition, adapter protein interaction, IFN type as well as STAT-specific sequence variations[10]. Formation of dimers and trimers further increases the functional capacity of STATs to modulate gene expression patterns in a cytokine dependent fashion.

JAKs also activate SHP2 *via* tyrosine phosphorylation of Y^{542} and Y^{580} and at Y^{759} on gp130, all of which overcome the auto-inhibitory mechanism of the internal SH2 domains within SHP2. JAK-phosphorylated SHP2 then induces ERK-mediated gene transcription *via* the recruitment of the Growth factor Receptor-Bound 2-Son of Sevenless (Grb2-Sos) complex. Grb2 functions as an adaptor protein for Sos, a Ras guanine nucleotide exchange factor (GEF). Co-localisation of SOS with Ras at the plasma membrane enables activation of Ras and initiation of the ERK signalling pathway [7].

Regulating the JAK/STAT Pathway

Chronic activation of the JAK/STAT pathway can initiate and perpetuate chronic inflammatory diseases such as rheumatoid arthritis and Crohn's disease as well as certain forms of cancer, and so many inhibitory mechanisms operate at several levels to ensure that signalling responses are transient in nature.

Negative regulation occurs through multiple routes within both extracellular and intracellular domains *via* inhibitory and degradative mechanisms. For example, soluble gp130 (sgp130) can trap circulating sIL6Rα/IL6 complexes and thus quench trans-signalling. Furthermore, active phagocytic cells, such as neutrophils at sites of inflammation, secrete proteases that degrade cytokines including IL6 [7,11].

Intracellular signalling intermediates can be made functionally mute by protein tyrosine phosphatases such as SHP2, and also by direct inhibition by protein inhibitors of activated STAT (PIAS) [12,13] and suppressor of cytokine signalling (SOCS) proteins, the latter of which can also drive ubiquitin-directed proteasomal degradation.

Protein Tyrosine Phosphatases (PTPs)

While SHP2 can activate the ERK pathway, it can also negatively regulate the JAK/STAT pathway *via* its Tyr phosphatase activity against gp130, JAK, and STAT proteins. SHP2 is ubiquitously expressed and consists of two N-terminal SH2 domains and a C-terminal catalytic phosphatase domain. In its inactive state, the SH2 domains bind to the phosphatase domain to block substrate interaction. This inhibitory action is removed upon tyrosine phosphorylation by cytokine-activated JAKs (see above). Other phosphatases might also have roles in JAK/STAT regulation such as PTPεC, PTP1B, CD45, SHP1 and nuclear-localised phosphatase TCPTP [7].

Protein Inhibitors of Activated Stats (PIAS)

Protein inhibitors of activated STAT (PIAS), as their name suggests, block the function of active/phosphorylated STATs but they also interact with a wide range of non-STAT proteins, most of which are transcription factors such as NFκB, and so have wider actions as transcriptional regulators[12]. Five PIAS family members (PIAS1, PIAS3, PIASxα, PIASxβ, and PIASy) have been defined.

All PIAS bind to active/tyrosine phosphorylated, nuclear-localised STAT proteins although the inhibitory mechanism appears to be PIAS-specific. IFNγ-induced STAT1 and IL6-induced STAT-dependent gene transcription are suppressed by PIAS1 and PIAS3 respectively *via* inhibition of transcription factor-DNA binding. In contrast PIASx and PIASy inhibit gene transcription by acting as transcriptional co-repressors.

All PIAS proteins also have E3 SUMO ligase activity enabling the reversible covalent attachment of "small ubiquitin-related modifier" (SUMO) to the ε-amino groups of lysine residues residing within a Ψ-K-x-D/E consensus sequence for SUMOylation on target proteins (Ψ=hydrophobic residue). This post-translational modification modulates cellular localisation, function, and protein-protein interactions, thus altering the composition and activity of transcription factor complexes. Regulation by PIAS proteins is complex with certain regulatory roles being SUMOylation independent and others relying on specific PIAS domains [13].

Suppressors of Cytokine Signalling (SOCS)

SOCS proteins negatively regulate multiple pathways involved in processes such as immunity, cell growth and metabolism amongst others. Eight SOCS family members have been defined (SOCS1-7 and cytokine-inducible SH2 domain-containing protein (CIS)) of which CIS and SOCS1-3 are best characterised [14]. SOCS proteins were discovered as part of a search for novel genes rapidly induced in hematopoietic cells in response to cytokines [15].

In general, SOCS proteins have low basal expression levels that are rapidly elevated by several inflammatory mediators. In a classical negative feedback loop, STAT-inducible SOCS proteins ablate intracellular pro-inflammatory signalling by either direct inhibition of signalling components or by ubiquitin-mediated proteasomal degradation of intermediates *via* its action as the specificity factor for E3 ubiquitin ligases [7,16]. The list of inducers of SOCS proteins is growing and the variety of SOCS-regulated pathways becoming more diverse.

Moreover, differing expression kinetics and the potential for cross-regulation within the SOCS family adds a further layer of complexity. At present, inducers of SOCS include, but are not limited to, cytokines (IL1, IL6, LIF, OSM, IFNγ, TNF, EPO, prolactin), chemo-attractants (IL8, N-formyl-Met-Leu-Phe), bacterial components (LPS, unmethylated CpG DNA) and the intracellular second-messenger cyclic AMP. Ligand-activated receptor tyrosine kinases (RTKs) such as EGFR and PDGFR might also induce and be regulated by SOCS[17]. SOCS proteins bind tyrosine-phosphorylated receptors that have Immunoreceptor Tyrosine-based Inhibitory Motif (ITIM)-like sites such as gp130, EpoR, ObR and others such as CD33 of the SIGLEC glycoprotein family [18]. Due to the requirement for tyrosine phosphorylation, it is possible that in certain circumstances inhibition might be indirect i.e. cross-talk between related signalling pathways [17].

Structurally, each SOCS family member consists of a conserved central SH2 domain, an extended SH2 sub-domain (ESS), a variable-length non-conserved N-terminal domain, and a 40 amino acid C-terminal domain SOCS-box (Figure 2). The SOCS-box is considered to be a multi-functional region, as it can regulate stability and the cross-regulation of SOCS proteins as well as SH2-dependent SOCS/substrate recognition and regulation.

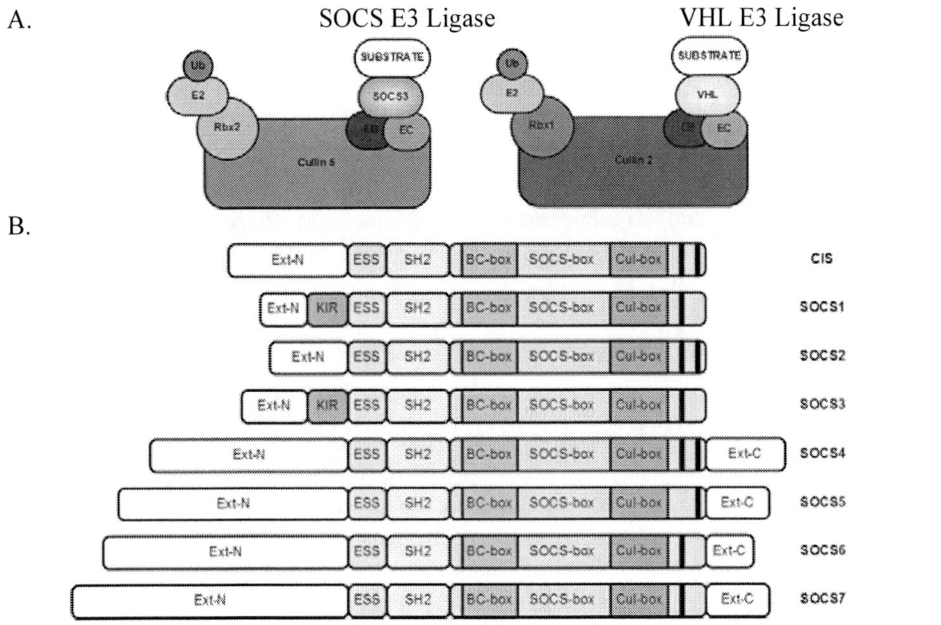

Figure 2. SOCS: Elongin-Cullin-SOCS-box (ECS) family of E3 ubiquitin ligases. A. Basic E3 ubiquitin ligase architecture. VHL and SOCS proteins act as specificity units for the ECS family of E3 ubiquitin ligases. SOCS proteins bind specific phosphorylated substrates or phospodegrons *via* an SH2 domain. The SOCS-box motif enables interaction with components an E3 ubiquitin ligase. SOCS binds to the N-terminal of the Cullin5 scaffold protein *via* the adaptor proteins Elongin B, Elongin C. Cullin5, *via* its C-terminus, also binds the RING finger-containing protein Rbx2 which enables interaction with the E2 conjugation protein. In complex with the E3 ligase scaffold, the substrate and E2-bound ubiquitin are in close proximity thus facilitating ubiquitin conjugation to bound substrate. It is now known that VHL lacks part of the SOCS-box sequence and is currently referred to as a VHL-box. This difference alters the specificity for Cullin family members where SOCS proteins interact with Cullin5 and Rbx2 whereas VHL interacts with Cullin2 and Rbx1. Such a variation adds further target specificity to the complete E3 ubiquitin ligase [19]. Furthermore, variation between SOCS1 and SOCS3 Cul-boxes result in a 100 or 10 fold reduction in affinity for Cullin5 respectively compared to other SOCS proteins[20]. B. Organisation of domains within the SOCS protein family, see text for description.

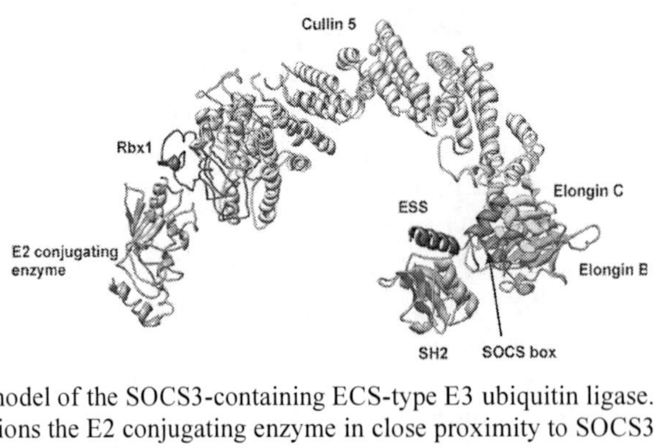

Figure 3. Structural model of the SOCS3-containing ECS-type E3 ubiquitin ligase. The central Cullin 5 scaffold protein positions the E2 conjugating enzyme in close proximity to SOCS3 which binds a target substrate (not shown) via its SH2 domain. SOCS3 is attached to Cullin 5 via a complex of ElonginB and Elongin C. The triple α-helix structure of the C-terminal domain SOCS-box forms a 4-helix bundle with a single helix, helix 4, of Elongin C. Reproduced from reference 16 with permission.

```
                       BC-box          Cul5-box
CIS    VQPFVRRSSARSLQHLCRLVINRLVA--DVDCLPLPRRMADYLRQYPFQL
SOCS1  LGAPLRQRRVRPLQELCRQRIVAAVGRENLARIPLNPVLRDYLSSFPFQI
SOCS2  TKPLY--TSAPSLQHLCRLTINKCTG--AIWGLPLPTRLKDYLEEYKFQV
SOCS3  VLSRPLSSNVATLQHLCRKTVNGHLDS-YEKVTQLPGPIREFLDQYDAPL
SOCS4  LSTPLIRTFPFSLQHICRTVICNCTTYDGIDALPIPSSMKLYLKEYHYKS
SOCS5  LTISLNRTFPFSLQYICRAVICRCTTYDGIDGLPLPSMLQDFLKEYHYKQ
SOCS6  TNPVSRFMQVRSLQYLCRFVIRQYTRIDLIQKLPLPNKMKDYLQEKHY--
SOCS7  LYPVSRFSNVKSLQHLCRFRIRQLVRIDHIPDLPLPKPLISYIRKFYYYD
```

Figure 4. Alignment of BC and Cul5 boxes within the SOCS family. Sequence alignment of the SOCS-box for all SOCS family members. The alignment was performed using the ClustalW2 multiple sequence alignment tool freely available at http://www.ebi.ac.uk/Tools/msa/clustalw2/. Highlighted sections indicate the BC-box and Cul5-box.

In its role as an E3 ligase, the SOCS-box domain is required for interaction with the ubiquitin machinery (Cullin5, Elongin B, and Elongin C) via Cullin-binding and B/C-box subdomains, while tyrosine-phosphorylated substrates are specifically targeted *via* the SH2 domain. SOCS1 and SOCS3 are unique in that they also contain an N-terminal kinase inhibitory region (KIR) that shares homology with the pseudosubstrate domain of JAKs. SOCS4-7 also have extended C-terminal domains which have no known functional motifs. Crystal structures of Elongin-bound SOCS2 and SOCS4 suggest that extended N- and C-terminal regions might function to stabilise the tertiary structure. SOCS2 has a conserved C-terminus that is predicted to stabilise the SH2-SOCS-box interaction allowing exposure of the N-terminal ESS and KIR. The SOCS4 SH2-SOCS-box interaction is stabilised by the extended N-terminal domain, the extended C-terminus then interacts with the N-terminal domain to further stabilise the structure [16]. These structural characteristics are thought to apply to similar SOCS family members i.e. SOCS1, SOCS3, and CIS and SOCS5-7 respectively. Over the full length of the protein, SOCS family members display significant sequence homology with SOCS4-5 being most similar (92%) followed by SOCS6-7 (57%), CIS and SOCS2 (45%), and finally SOCS1 and SOCS3 (35%), although certain domains are more conserved than others [21] (Figure 2). The Elongin binding site is the most conserved site on all SOCS proteins[16,21] which points to its importance. Consistent with this observation, mutation of a single leucine within the B/C-box has been demonstrated to completely block Elongin interaction, substrate ubiquitination and degradation [22]. As an inhibitory molecule, SOCS proteins bind receptors and receptor-bound substrates to suppress signalling. However, sequence deviations between family members bring about different mechanisms of action, and combined with differential expression kinetics, suggest that SOCS family members might each provide non-redundant functionality. Direct inhibition of substrates requires the SH2 domain, ESS, and KIR, but uniquely, CIS also requires the SOCS-box and pY^{253} [16]. Furthermore, interaction of the CIS SH2 domain with pY^{253} is hypothesised to regulate function in a fashion analogous to JAKs, keeping CIS in an inactive state until tyrosine-phosphorylated substrates are available. SOCS1 and SOCS3 are the only family members that can inhibit cytokine signalling via a SOCS-box-independent route, inhibiting JAK catalytic activity by acting as pseudosubstrates and thus blocking STAT1 and STAT3 activation. SOCS1 binds to the tyrosine-phosphorylated activation loop of JAK2 via its KIR, SH2, and ESS. SOCS3 similarly binds JAK1 but its interaction is weaker and requires further interaction via its SH2 domain with tyrosine phosphorylated receptor residues such as pY^{759} on gp130. Consequently, mutation to phenylalanine of Y^{759} (Y^{757} on murine gp130) reduces the inhibitory effect of SOCS3 but not SOCS1, potentially explaining why SOCS1 is a more potent inhibitor of JAK/STAT signalling [17]. IL6 and IFNγ both induce SOCS1 and SOCS3 in a STAT-dependent fashion and overexpression of either protein inhibits JAK/STAT signalling. In contrast, in vivo studies indicate that SOCS1 specifically targets IFNγ-dependent JAK2/STAT1 signalling whereas SOCS3 targets IL6-dependent JAK1/STAT3 signalling. Interestingly, in cells from conditional SOCS3 knock-out mice, IL6 has similar response and gene expression profile as IFNγ due to prolonged STAT1 signalling suggesting that SOCS3 might be required to limit a specific IFNγ response [17].

The SOCS3 SH2 domain shares 39% homology with that of SHP2 [18]. Also SOCS3 binds at an overlapping sequence and as such can compete with SHP2 to ablate SHP2-induced ERK activation by gp130. SHP2 itself might also contribute to signal suppression via the tyrosine dephosphorylation of either receptors or receptor-bound substrates such as JAKs. SOCS1 has been shown to direct proteasomal degradation of active/tyrosine phosphorylated JAK2 [23]. JAK2 is mono-ubiquitinated in unstimulated cells but becomes polyubiquitinated following tyrosine phosphorylation (Y^{1007}) and subsequent association with SOCS1. It is possible that SOCS3 similarly regulates JAK1 [14] but this has yet to be demonstrated. SOCS3 can also prolong ERK signalling by binding, and thus inhibiting, the ras inhibitor p120 RasGAP. Upon SOCS3 induction by IL-2, IL-6, Epo, insulin, EGF and PDGF, SOCS3 is tyrosine-phosphorylated within the SOCS-box domain enabling binding to p120 RasGAP, thereby blocking its ability to accelerate conversation of active Ras-GTP to inactive Ras-GDP. This function does not impact on SOCS3-dependent inhibitory effects such as its ability of block STAT5 activation. Unregulated ras-mediated ERK activation is an initiator of tumourigenesis and ras is found mutated in over 15% of all cancers [24]. Since IL6 activates ERK signalling but also maintains ERK activation via SOCS3, sustained JAK/STAT signalling due to chronic inflammation might also contribute to tumourigenesis.

CIS is thought to compete with STAT5 for phosphotyrosine residues on active erythropoietin (Epo) and growth hormone (GH) receptors [15,17]. However, this is disputed in the case of GHR due to non-overlapping binding sites of STAT5 and CIS [16]. Although CIS might equally disrupt STAT5 binding via steric hindrance [25]. A degradation-dependent regulatory mechanism is now preferred as CIS can only bind its substrates when in complex with ElonginB/C. SOCS2, being most similar to CIS with which it shares sharing 35% identity, has been shown to function by inhibiting GH-induced STAT5 activation in a SOCS-box dependent fashion. Furthermore, $SOCS2^{-/-}$ mice have an overgrown phenotype [16,26] which indicates its importance in GH receptor regulation. Inhibitory mechanisms and functional relevance of SOCS4-7 are as yet undefined. However, both SOCS4 and SOCS5 have been reported to increase EGF-dependent EGFR degradation [16] while SOCS5, being specifically expressed in Th1 but not Th2 cells, might affect T-helper cell equilibrium by repressing Th2 cell differentiation [17]. This is relevant in the study of allergic diseases such as asthma, where an increase in the Th2 compartment due to atopic SOCS protein expression is thought to be an initiating factor in disease progression [27]. Furthermore, SOCS5 might also regulate IL4 signalling by inhibiting IL4R-bound JAK1[25]. SOCS6 and SOCS7 are most similar in relation to SH2 and SOCS-box domains and both bind tyrosine-phosphorylated insulin receptors to suppress signalling following insulin treatment, suggesting similar or redundant functions. Suggestions of E3 ligase functionality resulted from analysis of SOCS3 which identified a conserved C-terminal domain, the SOCS-box, also present in a known E3 ligase Von-Hippel-Lindau (VHL), the VHL-box and others, MUF-1, and Elongin-A [17]. SOCS-box domains have been identified within over 70 human proteins [16,28] conjugated to a variety of protein-protein interaction domains[16,28] including SH2 domains, WD-40 repeats, SPRY domains, leucine rich repeats and ankyrin repeats that are thought to add target specificity and diversity. E3 ligase functionality has been demonstrated for SOCS1 and SOCS3 but has yet to be confirmed for other SOCS family members [20] although studies on mice that have been genetically manipulated to remove the SOCS3 SOCS-box i.e. SOCS3 SOCS-box knockout mice, have immunological defects suggesting and important proteasome-dependent regulatory roles (see below). Most SOCS members form ECS type E3 ligases and as such specifically bind E3 ligase components ElonginB/C, Cullin5 and the RING finger-containing protein Rbx2 via the SOCS-box. SOCS binds the Cullin5 scaffold protein via the ElonginB/C dimer that binds the N-terminal region of Cullin5. Cullin5 also binds Rbx2 via its C-terminus and enables interaction with the E2 conjugation protein. SOCS1 has also been found in complex with Cullin5/Rbx1 [17]. Both Rbx1 and Rbx2 bind Cullin2 and Cullin5 when overexpressed but under physiological conditions preferentially binds Cullin2 and Cullin5 respectively [29]. Unbound SOCS-box domains exist in a disordered state and only become stable/ordered upon binding ElonginB/C, requiring that this interaction precede the binding of Cullin5 and E3 ligase complex formation [22]. This was initially hypothesised from observations of crystal structures of VHL, SOCS2 and SOCS4 that were found to interact with Elongins through mainly (+70%) hydrophobic interactions. Uncoupling of these proteins vastly increases exposure of non-polar residues and, as such, was expected to contribute to unfolding and also increase the difficulty in producing unbound crystal structures [16]. Two subclasses of SOCS proteins can be defined based on their differing affinity for Cullin5 [20]. SOCS1 and SOCS3 binds ElonginB/C-Cullin5-Rbx2 weakly (K_d=10^{-6} M (SOCS1), 10^{-7} M (SOCS3)) compared to all other family members (K_d=10^{-8} M) due to slower on rates, and they also have shorter half lives. As such, it is thought that SOCS2, SOCS4-7 and CIS regulate signalling exclusively via ubiquitin-dependent pathways.

Table 1. Ubiquitin-regulated SOCS substrates

Substrate	Reference	SOCS Family Member
CD33	[30]	SOCS3
Siglec-7	[18]	
IRS-1/2	[31]	
JAK2	[23]	SOCS1
Tel-JAK2	[32]	
Vav	[33]	
NFκB subunit p65/Rel A	[34]	
Mal	[35]	
HPV, E7	[36]	
FAK	[37]	SOCS1, SOCS3

The SOCS proteins act as a specificity unit within the ElonginBC-Cullin5-Rbx2 E3 ligase complex. Binding to the E3 ligase scaffold is via the SOCS box domain. SOCS proteins associate with tyrosine-phosphorylated target substrates via a conserved central SH2 domain. It is currently regarded as a multifunctional domain implicated in SOCS stability, adaptor protein interaction, proteasomal degradation, receptor binding, and regulation of the ERK pathway. The full spectrum of ubiquitin-regulated SOCS substrates is unknown, but some of those indentified thus far are shown in Table 1.

The importance of the SOCS-box component of individual SOCS proteins has been demonstrated using several animal models. Zhang et al [38] showed SOCS1-SOCS-box-deleted (SOCS1$^{\Delta SB/\Delta SB}$) knock-in mice developed a less severe phenotype than SOCS1 knock-out mice suggesting an essential role for the SOCS-box domain [38]. The severity of phenotype has been linked to hyper-responsiveness to IFNγ resulting from a more sustained activation of STAT1 signalling in SOCS1$^{\Delta SB/\Delta SB}$, but intermediate to that observed for the WT and conditional SOCS1$^{-/-}$ mice. Similarly, SOCS3$^{\Delta SB/\Delta SB}$ mice, although viable, have prolonged STAT3 signalling intermediate to that of conditional SOCS3$^{-/-}$ and WT mice. Hyper-responsiveness to G-CSF also resulted in enhanced granulopoiesis, tissue infiltration, and arthritis, although the phenotype was reduced compared to the conditional SOCS3$^{-/-}$ mice (39). The increased half-life of active pJAK1 found in SOCS3$^{\Delta SB/\Delta SB}$ embryonic stem cells (ES) suggests that SOCS3 can regulate JAK1 *via* E3 ubiquitin ligase activity [14]. Furthermore, comparison of WT, SOCS3$^{-/-}$ and SOCS3$^{\Delta SB/\Delta SB}$ ES cells has shown that the SOCS box is specifically required to target LIF-induced pJAK for proteasomal degradation. These results indicated that in ES cells, pJAK dissociates from gp130 and is cleared *via* proteasomal degradation by SOCS3 and not simply inhibited by the KIR domain [14].

SOCS3 negatively regulates IL6-induced JAK/STAT signalling *in vitro* and *in vivo* [40]. This is supported from the finding that IL6-activated STAT1 and STAT3 signalling is prolonged in conditional SOCS3$^{-/-}$ mice. *In vitro* studies have found that over-expressed SOCS1, SOCS3 or CIS can inhibit JAK/STAT signalling although *in vivo* function is more specific. If over-expressed, SOCS1 and SOCS3 inhibits IL6 and IFNγ signalling although in conditional SOCS3$^{-/-}$ mice IL6 signalling is prolonged whereas IFNγ is unaffected. The opposite occurs in SOCS1$^{-/-}$ mice i.e. IFNγ signalling is prolonged whereas IL6 signalling is unaffected, suggesting that SOCS1 and SOCS3 have distinct non-overlapping roles. Interestingly, IL6-induced gene products from SOCS3$^{-/-}$ mice overlapped strongly with those induced by IFNγ: this observation might be explained by the extended IL-6-mediated activation of STAT1, since it is potently activated by IFNγ but minimally activated by IL6 in the presence of SOCS3. Suppression of the IL6-activated JAK/STAT pathway by SOCS3 might therefore prevent an IFNγ-type response by IL6 [40].

Regulation/Cross-Regulation of SOCS Protein Function

Across all SOCS proteins a high proportion of tyrosine residues within the SOCS-box (70%) are conserved and might have important regulatory roles including substrate binding, SOCS protein stability, ERK pathway regulation or auto-regulation in complex with the ESS or SH2 domains[16]. The interaction of the CIS SH2 domain with pY^{253} has already been mentioned with regard to the potential for auto-regulation. Furthermore, SOCS3 SOCS-box-dependent maintenance of ERK signalling has also been discussed. Other phosphorylation events are thought to regulate interaction of SOCS with the ubiquitination machinery and also modulate SOCS stability. In these cases, phosphorylation is thought to affect auto-ubiquitination or interaction with other E3 ligase complexes including those constructed from other SOCS family members. A complete list of protein kinases responsible for SOCS phosphorylation has yet to be documented although cytokine-induced kinases such as the Pim family of serine/thronine-directed kinases are thought to stabilise SOCS1 [41]. In support of this, all Pim family members (Pim1-Pim3) interact and phosphorylate SOCS1, while reduced SOCS1 levels were detected in Pim1/2 knockout mice. It should be noted that not all phosphorylation events are sufficient to affect stability [14], since hyperphosphorylation of SOCS3 by a constitutively active V617F JAK2 mutant is insufficient to cause destabilisation [14,42].

Ubiquitin-directed proteasomal degradation of SOCS proteins, which is intimately linked to phosphorylation, might occur in an auto-regulatory fashion or *via* another SOCS protein or E3 ligase. Ubiquitination typically occurs at specific lysine residues on target proteins but unlike SUMOylation, there appears to be no consensus sequences for which ubiquitination can be predicted with any reliability. SOCS3 stability can be increased following an N-terminal domain truncation that resulted in the loss of Lys6, which appears to be a major site for ubiquitination [17]. Furthermore SOCS2, 6 and 7 can bind all other SOCS proteins and have thus been implicated in targeting them for ubiquitination and subsequent proteasomal degradation [16].

However the mechanisms underlying ubiquitin-dependent regulation of SOCS proteins have yet to be resolved, with current literature painting a complex picture of both auto- and cross-regulation, mainly from the finding that formation of SOCS E3 ligases, specifically the SOCS-Elongin interaction, can either stabilise or destabilise SOCS proteins. Furthermore, RING E3 ligases often perform auto-ubiquitination but are also found in complex with other E3 ligases [43] and so both mechanisms of regulation are equally possible. For example, concomitant degradation of CIS and SOCS3 with their target proteins, GHR and SIGLEC 7 respectively [16], as well as the observation that disruption of the Elongin B/C interaction leads to stabilisation, suggests that CIS/SOCS3-E3 ligases perform auto-ubiquitination. Conversely, a similar disruption to SOCS1 has been reported to destabilise the protein [44] suggesting that the Elongin interaction is protective in this instance. In addition, SOCS3 is phosphorylated within the SOCS-box at Y^{204} and Y^{221} in response to IL2, IL3, Epo, EGF, PDGF and IL6. Such phosphorylation events fail to prevent binding to gp130 or JAK inhibition, instead blocking binding to ElonginB/C and thereby reducing protein stability. In MEFs, IL6-inducible SOCS3 is phosphorylated independently from IL6 stimulation, gp130 recruitment and JAKs. Instead Src family tyrosine kinases were shown to be involved, although residual SOCS3 phosphorylation following their inhibition suggested the involvement of additional kinases. It might therefore be speculated that SOCS3 could be regulated *via* crosstalk from other signalling pathways [7]. Furthermore, SOCS3 was found to be stabilised in $SOCS3^{\Delta SB/\Delta SB}$ cells suggesting that the SOCS-box, and possibly SOCS-box phosphorylation, is necessary for its degradation [39].

Finally, the observed differences in the kinetics of expression of SOCS proteins suggests the potential for additional regulation. SOCS1,3 and CIS are rapidly and transiently induced, whereas SOCS2, 6 and 7 are expressed later. SOCS2, expression in also typically much more sustained compared to the other SOCS family members [16]. For example, Tannahill et al [45] showed that while SOCS3 is induced after 30 minutes in IL2-treated peripheral blood mononuclear cells (PBMC) SOCS2 was only detectable after 4 hours. Furthermore, whereas SOCS3 expression diminished to undetectable levels after 4 hours, SOCS2 could still be detected after 24 hours. All SOCS proteins interact with the E3 ubiquitin ligase machinery, although SOCS1 and SOCS3 do so more weakly [20,20].Thus a situation is possible where stronger interacting members could potentially out-complete weaker members for limiting levels of the ubiquitin machinery, thus preventing their stabilisation and triggering their proteasomal degradation. For example, SOCS2 is expressed later than SOCS1/3, and overexpression was found to have the unexpected effect of restoring JAK/STAT signalling following GH stimulation [17]. This would suggest a role for SOCS2 in the negative regulation of SOCS1/3, and although new data disputes such a mechanism [46], SOCS2 has been demonstrated to regulate CIS in this fashion [16]. Furthermore, different expression kinetics might allow temporal regulation of signalling. For example, delayed expression of SOCS proteins would prolong the action of target signalling intermediates by delaying their inhibition/degradation. SOCS2-mediated regulation is further complicated by the observation that low levels of SOCS2 can suppress GH signalling, indicating a dual function for SOCS2 in GH receptor function [45]. As well as ubiquitination, specific regions such as PEST (proline, glutamate, serine, threonine-rich) motifs are also associated with the regulation of stability via proteasomal degradation. An unstructured PEST motif is found within the SH2 domain of SOCS3. Mutation of the PEST motif does not impact SH2 binding of phosphorylated tyrosine residues and complete removal of the motif does not prevent STAT3 inhibition. However, it does seem to impact stability since its removal increases the half-life of SOCS3, suggesting a destabilising role [22].

Therapeutic Intervention

Dysfunctional JAK/STAT signalling contributes to diseases such as rheumatoid arthritis, numerous lung inflammatory diseases, kidney failure, growth retardation, Crohn's disease and cancer. JAK/STAT signalling has been therapeutically targeted using anti-IL6 mAbs such as elsilimomab, anti-IL6R mAbs such as tocilizumab, as well as inhibitors of JAK and STAT [47]. Therefore, the multi-suppressive qualities and specificity of SOCS proteins might offer an important new approach for treating a wide spectrum of disorders if targeted effectively. How to harness these qualities therapeutically remains challenging, but there are several disease in which such therapeutics might prove useful, as outlined below.

Rheumatoid Arthritis

IL6 is a mediator of rheumatoid arthritis (RA) and elevated STAT3 mRNA is found in synovial tissue of RA patients. Adenoviral delivery of the SOCS3 gene to the joints of arthritic mice was demonstrated to reduce synovial fibroblast proliferation, IL6 production and inflammation via inhibition of STAT3 signalling [48]. Recently, liposome delivery of SOCS3 has also been used successfully as a treatment for endotoxic shock in mice [49]. Thus in this particular scenario, overexpression or selective activation of SOCS3 might increase survival rate by suppressing LPS signalling from TLR4 and also from the inhibition of autocrine signalling at cytokine receptors which occurs as a consequence of pro-inflammatory cytokine secretion following TLR4 activation.

Idiopathic Pulmonary Fibrosis

Lung tissue from idiopathic pulmonary fibrosis (IPF) patients have decreased levels of SOCS1 suggesting its role in pathogenesis. In support of this hypothesis, adenoviral delivery of SOCS1 to lung tissue was shown to reduce lymphocytic inflammation, pulmonary fibrosis and mortality in mouse models of IPF (SOCS1-haplodeficient mice treated with bleomycin) [50]. A reduction in lung-localised cytokine expression was noted but a link to STAT signalling was not pursued. However, due to the transient response of trans-gene expression following adenoviral gene delivery, regulated induction of SOCS *via* cAMP or a structurally similar molecule might also be another potential therapeutic approach [51].

Atherosclerosis

Numerous studies have shown that activation of the transcription factor NF-κB in the endothelium is an early event in the pathogenesis of cardiovascular diseases (CVDs) [52,53]. However, IL6 is one of several important class I cytokines that utilise alternative signalling pathways to control endothelial cell (EC) function. Interestingly, prolonged elevation of circulating IL6 levels are as strongly associated with increased risk of CVD as elevated cholesterol levels and hypertension [54]. Additionally, since plasma concentrations of IL-6 have been shown to positively correlate with body fat mass [55] consistent with visceral adipose tissue constituting an important source of IL6 in vivo [56], the status of IL-6 trans-signalling in vascular cells may constitute an important aspect of the link between obesity and pre-disposition to atherosclerosis and CVD.

However, as described earlier in this article, the biology of IL6 is complex, with the cytokine able to initiate pro- and anti-inflammatory effects due to both systemic (e.g. induction and release of acute phase proteins) and localised effects (e.g. on vascular cells in atherogenic regions [57]). As discussed earlier, in B cells, T cells monocytes and hepatocytes, IL-6 exerts its action mainly by binding cell membrane-localised IL-6Rα which then engages with the gp130 transmembrane signalling complex.

However, because IL-6 can also bind to sIL-6Rα to form the agonistic trans-signalling complex, the ubiquitous expression of gp130 means that the presence of sIL-6Rα renders cells that do not express membrane-localised IL-6Rα, such as vascular smooth muscle and endothelial cells, responsive to IL-6. Moreover, a growing body of evidence suggests that the detrimental effects of IL-6 in chronic inflammatory conditions such as rheumatoid arthritis, Crohn's disease and even various types of cancer are due mainly to trans-signalling-initiated events [58]. STAT3 in particular initiates important NF-κB-independent changes in transcription of target genes in vascular cells, including the chemokine MCP-1/CCL2 [59,60]. Genetic disruption of MCP-1 function, either by deletion of the genes encoding the chemokine itself or its cognate receptor CCR2 [61,62], reduces the development of atherosclerosis in mice, indicating the importance of signalling pathways that mediate MCP-1 induction as targets for generating anti-atherogenic therapeutics.

Several recent studies have demonstrated the importance of SOCS3 in regulating IL6 trans-signalling in vascular cells [60,63]. Moreover, immunohistochemical analysis of atherosclerotic plaques from both human carotid artery and the aortic arch of atherosclerosis-prone $ApoE^{-/-}$ mice have demonstrated that both SOCS1 and SOCS3 proteins accumulate in both the intima and media of pro-inflammatory shoulder regions, co-localising with smooth muscle cells and invading macrophages [63]. These studies argue that suppression of IL6 trans-signalling in vascular cells would limit adhesion molecule and chemokine induction, hence reducing inflammatory cell infiltration, remodelling and plaque development.

However since both SOCS1 and SOCS3 can target multiple intracellular pathways and regulate target protein function by several mechanisms (blocking SH2 domain binding to phosphorylated Tyr residues, JAK kinase inhibition, ubiquitination and proteasomal degradation), the key targets responsible for their protective effects remain to be elucidated. Approaches aimed at identifying the SOCS1 and SOCS3 interactomes following their induction by defined stimuli will help resolve these questions, as will the generation of new mouse models in which individual SOCS genes are specifically deleted in vascular cells.

Conclusion

The significance of enhanced IL6 signalling, particularly trans-signalling, in a variety of pathologies linked to the development and progression of a chronic pro-inflammatory state, including atherosclerosis, is now well established and several IL6-targeted antibody therapeutics are now available for treatment of rheumatoid arthritis. Exploiting the various inhibitory mechanisms invoked to limit IL6 signalling therapeutically with the aim of generating small molecules capable of either arresting or reversing disease progression is now an important goal. Understanding the molecular mechanisms underlying the regulation and function of key inhibitory regulators involved in disease, such as SOCS3, will undoubtedly help inform these approaches.

Acknowledgments

Work in TMP's laboratory is supported by the British Heart Foundation and Chest, Heart and Stroke Scotland. JJLW is supported by a PhD studentship from the UK Biotechnology and Biological Sciences Research Council.

References

[1] Lau, D. C., Dhillon, B., Yan, H., Szmitko, P. E., and Verma, S. (2005) *Am. J. Physiol. Heart Circ. Physiol* 288, H2031-H2041.

[2] Zhang, J., Alcaide, P., Liu, L., Sun, J., He, A., Luscinskas, F. W., and Shi, G. P. (2011) *PLoS. One.* 6, e14525.

[3] Wung, B. S., Ni, C. W., and Wang, D. L. (2005) *J. Biomed. Sci.* 12, 91-101.

[4] Borish, L. C. and Steinke, J. W. (2003) *J. Allergy Clin. Immunol.* 111, S460-S475.

[5] Bihl, M. P., Heinimann, K., Rudiger, J. J., Eickelberg, O., Perruchoud, A. P., Tamm, M., and Roth, M. (2002) *Am. J. Respir. Cell Mol. Biol.* 27, 48-56.

[6] Yatsenko, O. P., Filipenko, M. L., Khrapov, E. A., Voronina, E. N., Kozlov, V. A., and Sennikov, S. V. (2004) *Cytokine* 28, 190-196.

[7] Heinrich, P. C., Behrmann, I., Haan, S., Hermanns, H. M., Muller-Newen, G., and Schaper, F. (2003) *Biochem. J.* 374, 1-20.

[8] Sommer, U., Schmid, C., Sobota, R. M., Lehmann, U., Stevenson, N. J., Johnston, J. A., Schaper, F., Heinrich, P. C., and Haan, S. (2005) *J. Biol. Chem.* 280, 31478-31488.

[9] Brierley, M. M. and Fish, E. N. (2005) *J. Interferon Cytokine Res.* 25, 733-744.

[10] Ihle, J. N. (1996) *Cell* 84, 331-334

[11] Rose-John, S., Waetzig, G.H., Scheller, J., Grötzinger, J., and Seegert, D. (2007) *Expert Opin. Ther. Targets* 11, 613-624.

[12] Rytinki, M. M., Kaikkonen, S., Pehkonen, P., Jaaskelainen, T., and Palvimo, J. J. (2009) *Cell Mol. Life Sci.* 66, 3029-3041.

[13] Palvimo, J. J. (2007) *Biochem. Soc. Trans.* 35, 1405-1408.

[14] Boyle, K., Zhang, J. G., Nicholson, S. E., Trounson, E., Babon, J. J., McManus, E. J., Nicola, N. A., and Robb, L. (2009) *Cell Signal.* 21, 394-404.

[15] Yoshimura, A., Ohkubo, T., Kiguchi, T., Jenkins, N. A., Gilbert, D. J., Copeland, N. G., Hara, T., and Miyajima, A. (1995) *EMBO J.* 14, 2816-2826.

[16] Piessevaux, J., Lavens, D., Peelman, F., and Tavernier, J. (2008) *Cytokine Growth Factor Rev* 19, 371-381.

[17] Johnston, J. A. (2004) *J. Leukoc. Biol* 75, 743-748.

[18] Orr, S. J., Morgan, N. M., Buick, R. J., Boyd, C. R., Elliott, J., Burrows, J. F., Jefferies, C. A., Crocker, P. R., and Johnston, J. A. (2007) *J. Biol. Chem* 282, 3418-3422.

[19] Kamura, T., Maenaka, K., Kotoshiba, S., Matsumoto, M., Kohda, D., Conaway, R. C., Conaway, J. W., and Nakayama, K. I. (2004) *Genes Dev.* 18, 3055-3065.

[20] Babon, J. J., Sabo, J. K., Zhang, J. G., Nicola, N. A., and Norton, R. S. (2009) *J. Mol. Biol.* 387, 162-174.

[21] Alexander, W. S. (2002) *Nat. Rev Immunol.* 2, 410-416.

[22] Babon, J. J., McManus, E. J., Yao, S., DeSouza, D. P., Mielke, L. A., Sprigg, N. S., Willson, T. A., Hilton, D. J., Nicola, N. A., Baca, M., Nicholson, S. E., and Norton, R. S. (2006) *Mol. Cell* 22, 205-216.

[23] Ungureanu, D., Saharinen, P., Junttila, I., Hilton, D. J., and Silvennoinen, O. (2002) *Mol. Cell Biol.* 22, 3316-3326.

[24] Bos, J. L. (1998) *EMBO J.* 17, 6776-6782.

[25] Kubo, M., Hanada, T., and Yoshimura, A. (2003) *Nat. Immunol.* 4, 1169-1176.

[26] Elliott, J. and Johnston, J. A. (2004) *Trends Immunol.* 25, 434-440.

[27] Seki, Y., Inoue, H., Nagata, N., Hayashi, K., Fukuyama, S., Matsumoto, K., Komine, O., Hamano, S., Himeno, K., Inagaki-Ohara, K., Cacalano, N., O'Garra, A., Oshida, T., Saito, H., Johnston, J. A., Yoshimura, A., and Kubo, M. (2003) *Nat. Med.* 9, 1047-1054.

[28] Hilton, D. J., Richardson, R. T., Alexander, W. S., Viney, E. M., Willson, T. A., Sprigg, N. S., Starr, R., Nicholson, S. E., Metcalf, D., and Nicola, N. A. (1998) *Proc. Natl. Acad. Sci. U. S. A* 95, 114-119.

[29] Wei, D. and Sun, Y. (2010) *Genes Cancer* 1, 700-707.

[30] Orr, S. J., Morgan, N. M., Elliott, J., Burrows, J. F., Scott, C. J., McVicar, D. W., and Johnston, J. A. (2007) *Blood* 109, 1061-1068.

[31] Kawaguchi, T., Yoshida, T., Harada, M., Hisamoto, T., Nagao, Y., Ide, T., Taniguchi, E., Kumemura, H., Hanada, S., Maeyama, M., Baba, S., Koga, H., Kumashiro, R., Ueno, T., Ogata, H., Yoshimura, A., and Sata, M. (2004) *Am. J. Pathol* 165, 1499-1508.

[32] Kamizono, S., Hanada, T., Yasukawa, H., Minoguchi, S., Kato, R., Minoguchi, M., Hattori, K., Hatakeyama, S., Yada, M., Morita, S., Kitamura, T., Kato, H., Nakayama, K., and Yoshimura, A. (2001) *J. Biol. Chem.* 276, 12530-12538.

[33] De, S. P., Ilangumaran, S., and Rottapel, R. (2000) *J. Biol. Chem.* 275, 14005-14008.

[34] Ryo, A., Suizu, F., Yoshida, Y., Perrem, K., Liou, Y. C., Wulf, G., Rottapel, R., Yamaoka, S., and Lu, K. P. (2003) *Mol. Cell* 12, 1413-1426.

[35] Mansell, A., Smith, R., Doyle, S. L., Gray, P., Fenner, J. E., Crack, P. J., Nicholson, S. E., Hilton, D. J., O'Neill, L. A., and Hertzog, P. J. (2006) *Nat. Immunol.* 7, 148-155.

[36] Kamio, M., Yoshida, T., Ogata, H., Douchi, T., Nagata, Y., Inoue, M., Hasegawa, M., Yonemitsu, Y., and Yoshimura, A. (2004) *Oncogene* 23, 3107-3115.

[37] Liu, E., Cote, J. F., and Vuori, K. (2003) *EMBO J.* 22, 5036-5046.

[38] Zhang, J. G., Metcalf, D., Rakar, S., Asimakis, M., Greenhalgh, C. J., Willson, T. A., Starr, R., Nicholson, S. E., Carter, W., Alexander, W. S., Hilton, D. J., and Nicola, N. A. (2001) *Proc. Natl. Acad. Sci. U. S. A* 98, 13261-13265.

[39] Boyle, K., Egan, P., Rakar, S., Willson, T. A., Wicks, I. P., Metcalf, D., Hilton, D. J., Nicola, N. A., Alexander, W. S., Roberts, A. W., and Robb, L. (2007) *Blood* 110, 1466-1474.

[40] Croker, B. A., Krebs, D. L., Zhang, J. G., Wormald, S., Willson, T. A., Stanley, E. G., Robb, L., Greenhalgh, C. J., Forster, I., Clausen, B. E., Nicola, N. A., Metcalf, D., Hilton, D. J., Roberts, A. W., and Alexander, W. S. (2003) *Nat. Immunol.* 4, 540-545.

[41] Chen, X. P., Losman, J. A., Cowan, S., Donahue, E., Fay, S., Vuong, B. Q., Nawijn, M. C., Capece, D., Cohan, V. L., and Rothman, P. (2002) *Proc. Natl. Acad. Sci. U. S. A* 99, 2175-2180.

[42] Irandoust, M. I., Aarts, L. H., Roovers, O., Gits, J., Erkeland, S. J., and Touw, I. P. (2007) *EMBO J.* 26, 1782-1793.

[43] Woelk, T., Sigismund, S., Penengo, L., and Polo, S. (2007) *Cell Div* 2, 11.

[44] Kamura, T., Sato, S., Haque, D., Liu, L., Kaelin, W. G., Jr., Conaway, R. C., and Conaway, J. W. (1998) *Genes Dev.* 12, 3872-3881.

[45] Tannahill, G.M., Elliott, J., Barry, A.C., Hibbert, L., Cacalano, N.A., and Johnston, J.A. (2005) *Mol. Cell Biol.* 25, 9115-9126.

[46] Kiu, H., Greenhalgh, C. J., Thaus, A., Hilton, D. J., Nicola, N. A., Alexander, W. S., and Roberts, A. W. (2009) *Growth Factors* 27, 384-393.

[47] Barton, B.E. (2005) *Expert Opin. Ther. Targets* 9, 737-752.

[48] Shouda, T., Yoshida, T., Hanada, T., Wakioka, T., Oishi, M., Miyoshi, K., Komiya, S., Kosai, K., Hanakawa, Y., Hashimoto, K., Nagata, K., and Yoshimura, A. (2001) *J. Clin. Invest* 108, 1781-1788

[49] Fang, M., Dai, H., Yu, G., and Gong, F. (2005) *Cell Mol. Immunol.* 2, 373-377.

[50] Nakashima, T., Yokoyama, A., Onari, Y., Shoda, H., Haruta, Y., Hattori, N., Naka, T., and Kohno, N. (2008) *J. Allergy Clin. Immunol.* 121, 1269-1276.

[51] Yarwood, S. J., Borland, G., Sands, W. A., and Palmer, T. M. (2008) *J. Biol. Chem* 283, 6843-6853.

[52] Collins, T. and Cybulsky, M.I. (2001) *J. Clin. Invest.* 107, 255-264.

[53] de Winther, M.P., Kanters, E., Kraal, G., and Hofker, M.H. (2005) *Arterioscler Thromb Vasc. Biol.* 25, 904-914.

[54] Danesh, J., Kaptoge, S., Mann, A.G., Sarwar, N., Wood, A., Angleman, S.B. et al. (2008) *PLoS Med.* 5, e78.

[55] Kern, P.A., Ranganathan, S., Li, C., Wood, L., and Ranganathan, G. (2001) *Am. J. Physiol. Endocrinol. Metab.* 280, E745-751.

[56] Galic, S., Oakhill, J.S., and Steinberg, G.R. (2010) *Mol. Cell Endocrinol* 316, 129-139.

[57] Schuett, H., Luchtefeld, M., Grothusen, C., Grote, K., and Schieffer, B. (2009) *Thromb Haemost* 102, 215-222.

[58] Scheller, J., Chalaris, A., Schmidt-Arras, D., and Rose-John, S. (2011) *Biochim. Biophys. Acta* 1813, 878-888.

[59] Romano, M., Sironi, M., Toniatti, C., Polentarutti, N., Fruscella, P., Ghezzi, P. et al. (1997) *Immunity* 6, 315-325.

[60] Sands, W.A., Woolson, H.D., Milne, G.R., Rutherford, C., and Palmer, T.M. (2006) *Mol. Cell Biol.* 26:6333-6346.

[61] Gosling, J., Slaymaker, S., Gu, L., Tseng, S., Zlot, C.H., Young, S.G., Rollins, B.J., and Charo, I.F. (1999) *J. Clin. Invest.* 103, 773-778.

[62] Dawson, T.C., Kuziel, W.A., Osahar, T.A., and Maeda, N. (1999) *Atherosclerosis* 143, 205 211.

[63] Ortiz-Muñoz, G., Martin-Ventura, J.L., Hernandez-Vargas, P., Mallavia, B., Lopez-Parra, V., Lopez-Franco, O., Muñoz-Garcia, B., Fernandez-Vizarra, P., Ortega, L., Egido, J., and Gomez-Guerrero, C. (2009) *Arterioscler Thromb Vasc Biol* 29, 525-531.

In: Endothelium and Epithelium
Editors: J. Carrasco and M. Mota, pp. 175-191

ISBN 978-1-61470-874-2

Chapter VIII

Endothelium-Dependent Effects of Heat Shock on Vasomotor Tone: Role of eNOS and HSP70

Salvatore Salomone[1,2,*] ***and Jean-Christophe Plumier***[1,3]

[1]Stroke and neurovascular regulation laboratory, Massachusetts General Hospital, Charlestown, MA, USA

[2]Department ofClinical and Molecular Biomedicine, Catania University, Catania, Italy

[3]Laboratory for Animal Physiology, Université de Liège, Liège, Belgium

Abstract

Heat shock-mediated delayed protection against myocardial and cerebral ischemia has been attributed to induction of heat shock proteins (HSP), especially HSP70, in myocytesand neurons. However, heat shock preconditioning only induces HSP70 in endothelium. The purpose of this study was to examine the effect of heat shock on endothelium-dependent vasomotor tone.

Isolated mouse aortas were heated to 43.5°C for 15 min and returned to 37°C for 18 h.They were then mounted in a wire myograph to assess vasomotor responses to vasoconstrictors and vasodilatators. Intact heat shocked aortas showed reduction in phenylephrine- or 5-hydroxytryptamine-induced vasoconstriction,while no difference in vascular reactivity was detected in heat shocked endothelium-denuded aortas. Reduced vasoconstriction in heat shocked aortas was restored to control level by incubation with L-NAME, a nitric oxide synthase (NOS) inhibitor, but not by geldanamycin, a HSP90 inhibitor.Furthermore, the effect of heat shock was not observed in aortas fromeNOSnull mice, suggesting that vasomotor responsiveness reduced by heat shock was related to increased basal eNOS activity. Immunoblot of lysates from heat shocked aortas did not show upregulation of eNOS or other NOS isoforms, while showingrobust expression of HSP70; immunoprecipitation using eNOS polyclonal or monoclonal antibodies demonstrated HSP70 interaction witheNOSafter heat shock. In enzymatic assay

[*]Tel 39-095-7384085, Fax 39-095-7384228, E-mail: salomone@unict.it

(conversion of L-arginine to L-citrulline), HSP70 increased enzymatic activity of recombinanteNOS.

We conclude that heat shock decreases vasomotor responses to vasoconstrictors by enhancing eNOS basal activity, through interaction with HSP70, that is robustly induced. Such an increase of eNOS might be related to protective effect of heat shock against ischemia reported *in vivo*. Finally, interaction between HSP70 and eNOS opens the intriguing possibility that HSP70 could have vasoregulatory properties in addition to its cytoprotective functions.

Keywords: Preconditioning, heat shock, endothelium, eNOS, HSP70

Introduction

Heat shock preconditioning enhances myocardial resistance against ischemia by reducing incidence of arrhythmias and infarct volume, as well as by improving post-ischemic contractile recovery [1, 2]. This protective effect closely correlates with the induction of a stress response characterized by *de novo* expression of the inducible 70-kD heat shock protein, HSP70 [3]. In addition to its protein chaperone activity, HSP70 has also shown clear cytoprotective properties in cultured myocytes, neurons and other cell types wereit confers resistance to caspase-mediated (apoptotic) cell death [4, 5]and necrotic cell death induced by oxidative stress or simulated ischemia [6]. The protective potential of HSP70 against myocardial and cerebral ischemia has also been established *in vivo* using gene transfer [7] and transgenic mice [8-10]. Constitutive expression of either human or rat hsp70 gene reduces myocardial and cerebral ischemic injury and improves functional recovery[8-10]. From this evidence, it has been suggested that heat shock preconditioning *in vivo* induces HSP70 accumulation in myocytes or neurons, thereby increasing cellular resistance to ischemic injury. This simple explanation has been challenged by reports demonstrating that heat shock preconditioning *in vivo* does not induce a stress response (measured by HSP70 expression) in myocytes[11]or neurons [12] but rather in endothelial cells [11]. These observations therefore suggest that additional protective mechanism(s) of heat shock preconditioning should be explored within the endothelium and endothelium-related functions.

Besides heat shock preconditioning, induction of a stress response in blood vessels also occurs during balloon angioplasty [13]and in several pathophysiological conditions, such as stroke [14]or atherosclerosis [15, 16]. However, the precise impact of HSP70 and other stress-induced proteins to endothelial patho-physiology is still poorly understood.

Nitric oxide production has a central role in ischemic preconditioning [17-19].Furthermore, mice deficient in endothelial nitric oxide synthase (eNOS or NOS type III) [20]or rats treated with eNOS inhibitors [21] have larger cerebral infarctions when submitted to permanent occlusion of middle cerebral artery [20]. In contrast, increased endothelial NO production by upregulation of eNOS (by administration of statins, inhibitors of 3-hydroxy-3-methylglutaryl-coenzyme A reductase, HMG-CoA)[22],or by enhanced eNOS activity (by infusion of eNOS substrate L-arginine [23, 24], reduces the cerebral area at risk after middle cerebral artery occlusion. A key step in eNOS activation relies on eNOS subcellular localization and interaction with scaffold/chaperone proteins, such as caveolins.Caveolins localize in membrane invaginations, called caveolae, forming cellular compartments enriched with proteins involved in signal transduction. In endothelial cells, caveolin-1 binds to eNOS, preventing its interaction with calmodulin and its subsequent activation [25, 26]. Upon stimulation (either following acetylcholine treatment or shear stress), caveolin-1 inhibitory binding with eNOS is displaced by the 90-kDa heat shock protein (HSP90) which favors interaction with calmodulin and increases eNOS activity [27].

Using isolated mouse aortas, here we examined the effect of induction of a stress response on endothelial function. More specifically, we investigated the effect of heat shock and heat shock proteins on NO-mediated vasomotor tone. We observed that heat shock preconditioning decreases the response to vasoconstrictor challenges, increases basal eNOS activity in a HSP90-independent manner and promotes eNOS-HSP70 interaction.

Methods

Animals

The animal protocols have been reviewed and approved by the Massachusetts General Hospital Committee on research, Subcommittee on research Animal Care (SRAC).All mice (6-8 week-old) used in this study had C57Bl/6J genetic background, including eNOS$^{-/-}$ mice (Charles River Laboratories, MA).Mice were fed regular chow, and water was available ad libitum.

Reagents

Antibodies against heat shock proteins (HSP70 and HSP90) and recombinant human HSP70 were obtained from Stressgen Biotechnologies (Victoria, BC, Canada).Anti-NOSIII (anti-eNOS) and anti-caveolin polyclonal antibodies were obtained from Transduction Laboratories (Lexington, NY, USA), anti-iNOS (anti-NOS II; sc-650) antibody from Santa Cruz Biotechnology (Santa Cruz, CA, USA) and anti-nNOS (anti-NOS I; 61-700) and anti-eNOS (anti-NOS III monoclonal antibody; 33-4600) antibodies from Zymed (Oxnard, CA, USA). All additional reagents including phenylephrine, acetylcholine, N(G)-nitro-L-arginine methyl ester, sodium nitroprusside and geldanamycin were purchased from Sigma (St Louis, MO, USA).

In Vitro Preparation and Heat Shock Treatment

Mice were killed by decapitation. Aortas were removed and dissected under a binocular microscope into 1.5-2 mm long rings. Aortic rings were placed in 100μl phosphate buffered saline (PBS; pH 7.4) and incubated at room temperature, 42°C or 43.5°C for 15 min, using a PCR thermal cycler (PTC-200, MJ Research, Waltham, MA, USA). After treatment, rings were transferred to 5 ml DMEM supplemented with 10% fetal bovine serum and antibiotics and maintained 18 h in a humidified atmosphere of 5% CO_2 at 37°C.After incubation, rings were rinsed in physiological solution before mounting in a wire myograph.

Measurement of Contractile Tension in Isolated Aortas

Aortic rings were immersed in physiological solution (composition, mM: NaCl, 118; KCl, 4.6; $NaHCO_3$, 25; $MgSO_4$, 1.2; KH_2PO_4, 1.2; $CaCl_2$, 1.25; glucose, 10; EDTA, 0.025; pH 7.4 at 37°C), threaded onto stainless steel wires (40 μm diameter) and mounted in an isometric myograph (610M, Danish Myo Technology, Aarhus, Denmark) in physiological solution, maintained at 37°C and aerated with a gas mixture of 95% O_2- 5% CO_2.After equilibration, the normalized passive resting force and the corresponding diameter were determined for each preparation from its own length-pressure curve, according to Mulvany and Halpern [28]. In this aim, the preparation was distended stepwise while recording micrometer reading and force. The stepwise distension was stopped when the effective transmural pressure exceeded 100 mm Hg (13.3 kPa). An exponential curve was then fitted to the internal circumference pressure data and the point of the curve corresponding to 100 mm Hg was determined (IC_{100}). The preparation was then relaxed to a circumference equal to 0.9 IC_{100} and held at this degree of stretch for the remainder of the experiment. Contractile responses were recorded into a computer, by using a data acquisition and recording software (Myodaq and Myodata, Danish Myo Technology).

After normalization and 30-min equilibration in physiological solution, the preparations were stimulated with a 100 mMKCl-depolarizing solution (composition, mM: NaCl, 22.6; KCl, 98.8; $NaHCO_3$, 25; $MgSO_4$, 1.2; KH_2PO_4, 1.2; $CaCl_2$, 1.25; glucose, 10; EDTA, 0.025, pH 7.4 at 37 °C). After washout and 30-min recovery, they were exposed to cumulative concentrations of phenylephrine (PE, 1 nM - 1 μM) followed by cumulative concentrations of acetylcholine (ACh, 1 nM-10 μM). After washout and recovery, rings were exposed to N(G)-nitro-L-arginine methyl ester (L-NAME; 100 μM) and 30 min later to cumulative concentration of PE followed by cumulative concentration of sodium nitroprusside (SNP, 0.1 nM - 1 μM). In some instances, rings were collected, opened and immersed in SDS-loading buffer for Western blot analysis.

Removal of Endothelium

To remove endothelium, anesthetized mice were perfused intracardially with 3 ml phosphate buffered saline (PBS; pH 7.4) followed by 1.5 ml of 0.03% Triton-X100 in water.Aortas were then removed and rings prepared. This treatment did not affect the maximal contractile response to PE, but abolished the relaxing response to 3 μM ACh.

HUVEC Cell Cultures

Human umbilical vein endothelial cells (HUVEC) were purchased from ATCC (Manassas, VA, USA) and cultured in M199 supplemented with 20% fetal bovine serum, 0.1 mg/ml heparin, 0.03 mg/ml endothelial cell growth supplement (ECGS) on collagen-coated dishes (BD Biosciences, San Jose, Ca, USA). Heat shock was performed on exponentially growing cells by placing the culture dish in a 43.5°C water bath for 15 min.After heat shock, cultured cells returned to their incubator for an additional 18 h before being harvested in

RIPA buffer (10 mM HEPES, pH 7.4, 0.1% CHAPS, 2 mM EDTA, 1 mM EGTA, 5 mM DTT, 1 mM PMSF, 10 μg/ml proteinase inhibitor cocktail).

Western Blot Analysis

Western blot analysis was performed using whole aortas lysed in RIPA buffer or single aortic rings collected in denaturing buffer. Lysates were centrifuged at 11,000 x g for 15 min at 4ºC. Protein concentration in supernatant was determined using Bio-Rad protein assay reagent (Bio-Rad Laboratories, Hercules, CA, USA). Proteins were denatured in SDS-denaturing buffer (5M Tris, pH 6.8), 20% SDS, 20% glycerol, 1M β-mercaptoethanol), boiled for 2 min, separated using SDS-PAGE on 4-20% Tris-glycine gel (Invitrogen, Carlsbad, CA, USA) and transferred onto Immobilon^{TM}-P membranes (Millipore Corporation, Bedford, MA, USA). After blocking with 5% skim milk in Tris-buffered saline (TBS; 50 mMTris·HCl, 150 mMNaCl) with 0.05% Tween 20 (Sigma), the membranes were incubated with primary antibody (see reagents). After incubation, membranes were washed in TBS-Tween and incubated with secondary antibodies (1:1,000; Amersham) for 1 h at room temperature. Detection was achieved using a chemiluminescence system (ECL system; Amersham, Buckinghamshire, UK) and exposing to Hyperfilm (Amersham).Equal loading of protein was confirmed by re-probing membranes using an anti-β-Actin antibody (1:5,000; Amersham).

Immunoprecipitation

Immunoprecipitation was performed according to Garcia-Gardena et al. [29]. The samples were placed in immunoprecipitation (IP) buffer (100mM Tris-HCl, pH 7.4, 1% v/v nonidet P-40; 10μl/ml proteinase inhibitor cocktail), homogenized on ice and centrifuged at 12,000 × g for 10 min at 4ºC. Equal amounts of proteins were pre-cleaned by incubation with protein G-agarose and normal rabbit IgG for 2 h at 4°C, and then incubated with 2μg of anti-eNOS polyclonal antibody (N30030; Transduction Laboratories) overnight. Protein complexes were then precipitated with protein G-agarose for 2 h at 4°C. The immunoprecipitants were washed 5 times in IP buffer, boiled 2 min in denaturing buffer, separated by 4-20% Tris-glycine SDS-PAGE and transferred to Immobilon^{TM}-P membrane. Membranes were then incubated with primary antibodies and processed as described above for Western blot analysis.

Statistical Analysis

Data are presented as mean ± standard error of mean (SEM). Concentration-response curves were analyzed using two-way ANOVA. Difference between maximal contractions was analyzed using unpaired Student *t*-test or one-way ANOVA, followed by Tukey test.Statistical significance was set at $P<0.05$.

Results

Induction of Stress Response in Isolated Mouse Aortas by Heat Shock

In order to provide a better control for individual variability, we first established in vitro incubation of isolated mouse aortas; this allowed us to compare untreated vascular segments (control) versus segments in which the stress response (after heat shock) was induced; in this protocol the stress response was defined as a condition where heat shock induced a robust HSP70 expression, which was monitored in each experiment by western blot. In control (not subjected to heat shock) aortic rings, contractile responses to KCl-induced depolarization, phenylephrine, or acetylcholine were not changed by 18 h-incubation in culture medium (data not shown). Since incubation did not change physiological responses of isolated aortic rings, we used this *in vitro* model to examine the effect of heat shock on vasomotor function. Our *in vitro* approach offered the additional advantage of excluding possible confounding effects, observed after heat shock *in vivo*, such as heat stress-mediated activation of neuronal and/or hormonal axis.In a first series of experiments, aortic rings were subjected for 15 min to either 42°C or 43.5°C. Eighteen hours later, the stress response induction and vasomotor function were examined (Figure 1). A 15-min exposure to 42°C barely affected HSP70 expression in aortic rings, as revealed by Western blot analysis; in contrast, a robust induction of HSP70 expression in isolated mouse aortas was obtained after treatment to 43.5°C for 15 min (Figure 1). Vasoconstriction induced by 1 μM phenylephrine (PE) was not different in 42°C-heat shocked aortic rings and controls.

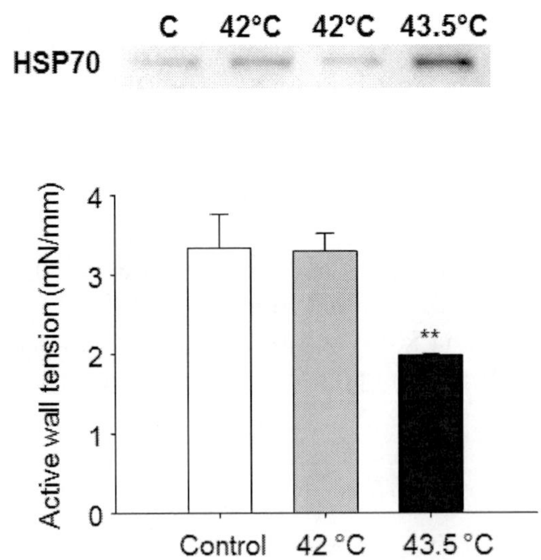

Figure 1. Induction of a stress response as detected by HSP70 expression was accompanied by changes in vasomotor functions.Western blot analysis of HSP70 expression in 4 aortic rings incubated at 37°C (Control, C), 42°C (2 samples in middle lanes) or 43.5°C. The bar graph shows comparison of 1μM phenylephrine (PE)-induced contractions between aortic rings incubated at 37°C, 42°C or 43.5°C for 15 min, and then maintained in DMEM (with 10% FBS) for 18 h before physiological analysis. Each column represents the mean (±SEM, vertical bars)from 6-8 different rings. Notice the lack of change in

PE-induced vasoconstriction following incubation at 42 °C, paralleled by lack of HSP70 induction as detected by western blot. **P<0.01 versus Control; one-way ANOVA and Tukey test.

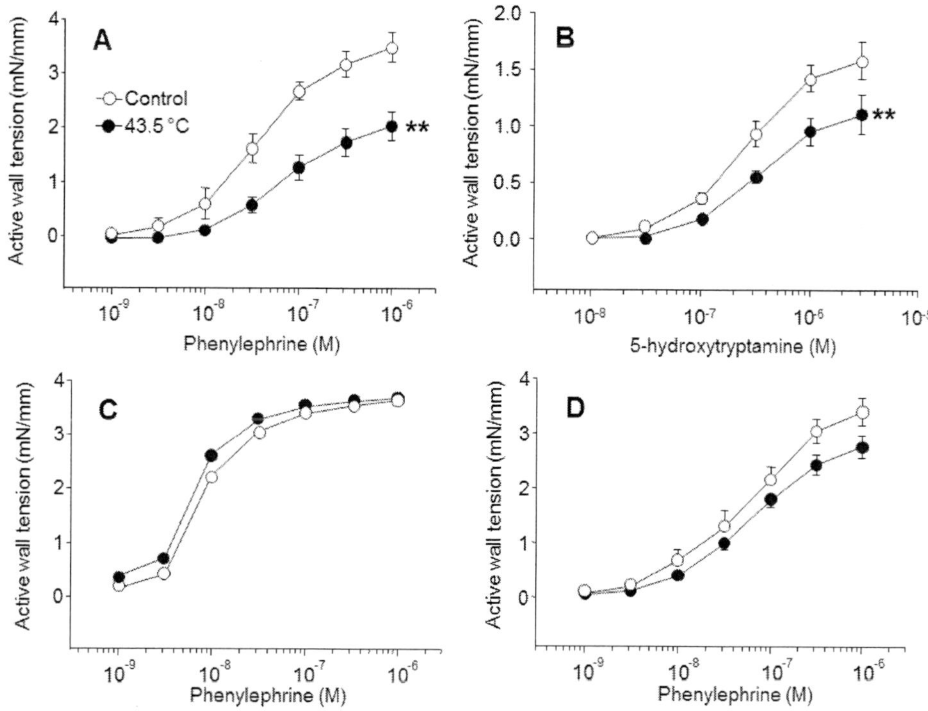

Figure 2. Heat shock effect on vascular tone was mediated by endothelial nitric oxide synthase (eNOS).Active arterial wall tension, induced *in vitro* by cumulative concentrations of phenylephrine (A) or 5-hydroxytryptamine(B), was reduced in heat-shocked (43.5 °C) mouse aortic rings compared to controls. In endothelium-denuded rings (C) and in rings from eNOS deficient mice (D) the effect of heat shock on phenylephrine-induced tone was not significant. Each curve represents the mean (± SEM, vertical bars)from 8-12 different rings. **P<0.01 versus Control; two-way ANOVA.

Conversely, exposure to 43.5°C, which induced a stress response as detected by HSP70 expression, was accompanied by significant decrease in vasoconstriction to either PE or 5-hydroxytryptamine (5-HT, Figure 1 and 2 A, B). Therefore, 43.5°C temperature was used as heat shock treatment throughout the experiments described. Contraction to 100 mMKCl was also reduced in heat shocked aortas compared to controls (2.54±0.12 mN/mm vs. 3.30±0.34 mN/mm, respectively; p<0.05).

Heat Shock Effect on Vascular Tone Was Mediated by Endothelial Nitric Oxide Synthase (eNOS)

Basal (unstimulated) release of NO from endothelium modulates the responses of isolated vessels to vasoconstrictors, such as PE and 5-HT [30]. To assess if the decrease of contraction to PE after heat shock involved endothelial mechanisms, we tested some endothelium-denuded isolated aortas (see also methods) in the same experimental paradigm.

Control and heat shocked aortic rings without endothelium showed similar contraction to cumulative concentrations of PE (Figure 2C), indicating that heat shock exerted its effect through endothelium. More detailed evidence was obtained using eNOS deficient mice,were heat shock did not change PE-induced contraction (Figure 2D). Furthermore, the treatment with L-NAME, a NOS inhibitor, abolished the difference in PE-induced contraction between heat shocked and control aortas (Figure 3A), suggesting that heat shock did not affect contractile function *per se* but rather increased basal NO production.In order to exclude the potential involvement of inducible NOS (iNOS or NOS II), we tested also aortas from iNOS null mice. These preparations responded to heat shock in a manner similar to aortas from wild type (Figure 3C), ruling out iNOS as potential mediator of increased basal NO production in this system.

Finally, because heat shocked and control aortas showed similar relaxation to the NO donor, sodium nitroprusside (Figure 3D), we could exclude an effect of heat shock on signaling downstream to NO and/or in the ability of smooth muscle cells to relax in response to NO. Thus, these results indicated that heat shock treatment increased NO productionin isolated aortas, through an increase of eNOS activity under basal conditions.

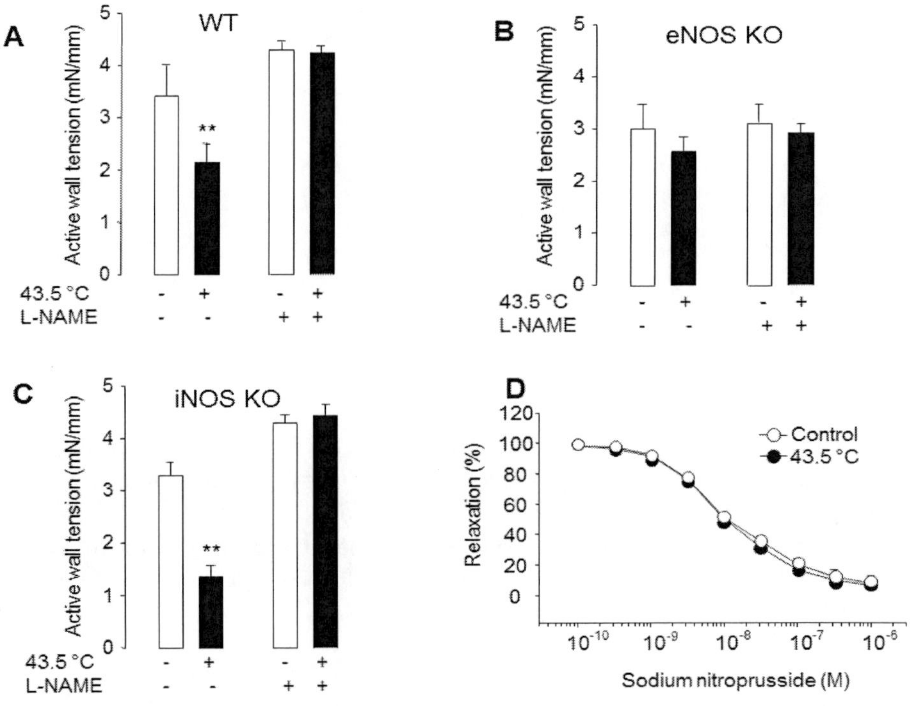

Figure 3. Heat shock effect on vascular tone increased basal NO (was abolished by L-NAME), without changing sensitivity to exogenously added NO.Active arterial wall tensionto 1μM phenylephrine, reduced in heat-shocked (43.5 °C)aortas from wild type mice (WT, A) or iNOS deficient mice (iNOS KO, C), was restored to control level by incubation with the nitric oxide synthase inhibitor N(G)-nitro-L-arginine methyl ester (L-NAME; 100 μM). Notice, in(B), the absence of effect of heat shock and/or L-NAME in rings from eNOS deficient mice. In (D), concentration-dilatation curves to cumulative concentrations of the NO-donor sodium nitroprusside, in control and heat-shocked rings from control wild type mice. Each column or curve represents the mean (± SEM, vertical bars)from 8-12 different rings. **P<0.01 versus Control; one-way ANOVAand Tukey test.

Stress Response Induction Increased the Expression of Heat Shock Proteins but Not of Nitric Oxide Synthase Isoforms

Both HSP90 and HSP70 were constitutively expressed in aortas as shown by Western blot analysis. After heat shock, HSP90 was slightly increased (Figure 4) while HSP70 expression was markedly increased (Figure 1). We also examined whether heat shock altered the expression of caveolin-1, a protein exhibiting an inhibitory interaction with endothelial NOS [25, 29]. Western blot analysis revealed that caveolin-1 was unchanged after heat shock in isolated aortas (not shown). The effect of heat shock on expression of NOS isoforms was also examined by Western blot analysis (Figure 4). Neuronal NOS (nNOS, NOS type I) or inducible NOS (iNOS, NOS type II) were barely detected in control aortas and did not increase in heat shocked aortas. Finally and more importantly, eNOSexpression was not increased after heat shock (Figure 4). These data therefore showed that increased basalNO production after heat shock could not be explained by an increased expression of any NOS isoform (including non-endothelial isoforms).

Inhibition of HSP90 Did Not Inhibitthe Effect of Heat Shock on Phenylephrine-Induced Vasoconstriction

Under stimulated conditions, i.e. when challenging with endothelium-dependent vasodilators, HSP90has an activator effect on eNOS activity, which can be blocked bygeldanamycin, an inhibitor of HSP90 [27]. To determine whether or not the increase in HSP90 observed after heat shock (Figure 4) was related to endothelium-dependent inhibiton of vasoconstriction to PE, we challenged some aortic rings with geldanamycin (20 μM).

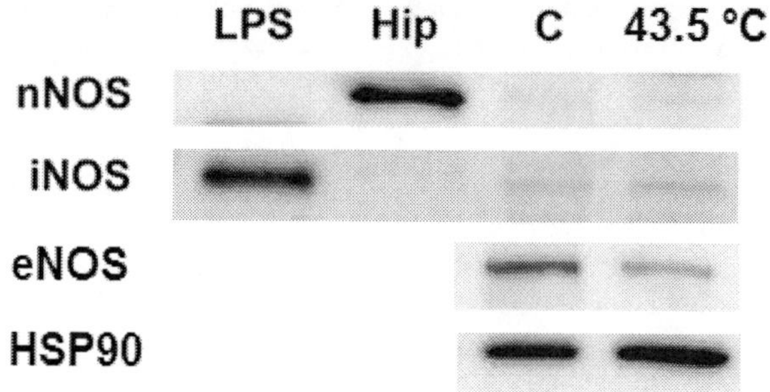

Figure 4. Induction of a stress response increased expression of heat shock proteins (HSP)without changing the expression of any NOS isoform.Western blot analysis for nitric oxide synthase isoforms was carried out in control (C) and heat-shocked (43.5 °C) aortic rings. Expression of neuronal NOS or NOS I (nNOS), inducible NOS or NOS II (iNOS) and endothelial NOS or NOS III (eNOS) were not increased by heat shock. BecausenNOS and iNOSwere barely detectable in mouse aorta, mouse hippocampus (Hip) and LPS-treated mouse microglial N11 cells (LPS) were run as positive controls for

nNOSand iNOS, respectively. Expression of 90 kD heat shock protein (HSP90) was increased by heat shock, as expected.

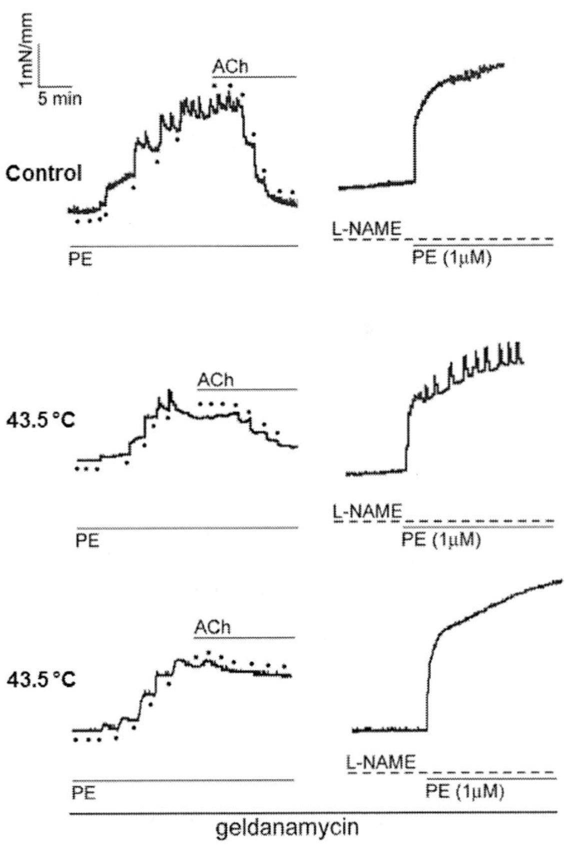

Figure 5. Inhibition of HSP90 did not inhibitthe effect of heat shock on phenylephrine-induced vasoconstriction. Typical experiment repeated three times with similar results. Cumulative concentrations (1 nM-1 μM) of phenylephrine (PE) followed by ACh, were added to the organ chamber by half log increase, as indicated by black dots on the tracing. After a first run, preparations were washed and allowed to recover for 30 min, incubated with 100 μM L-NAME, and challenged again with 1 μM PE. Upper trace, Control; middle trace heat-shock; lower trace heat-shock + 20 μM geldanamycin (HSP90 inhibitor). Notice that the contractile response evoked by PE in the absence of L-NAME is depressed by heat shock, regardless of geldanamycin-treatment (as further demonstrated by the increase in contraction with L-NAME, in both middle and lower traces). In contrast, relaxation to AChis abolished by geldanamycinbut is unaffected by heat shock.

As expected [27], geldanamycinblocked acetylcholine(ACh)-induced relaxation, indicating that inhibition of HSP90 and thereby of ACh-stimulated eNOS activity, effectively occurred in our system (Figure 5); however, geldanamycin failed to restore the contractile response to PEin heat shocked aortic rings, indicating that increased basal NO production was unrelated to HSP90,while L-NAME was able to restore PE-induced maximal contraction in heat shocked aortas, in a similar manner, with or without geldanamycin(Figure 5).

HSP70 Interacts with eNOS after Heat Shock

Protein-protein interaction between eNOS and the inducible HSP70 was examined in isolated aortas after heat shock using immunoprecipitation with anti-eNOS polyclonal antibody (Figure 6).HSP70 was detected in eNOS- immunoprecipitants from heat shocked aortas but not from controls, although there was constitutive HSP70 expression in isolated aortas without heat shock treatment (Figure 1). Similar results were also observed in eNOS-immunoprecipitants from heat shocked HUVEC using either monoclonal or polyclonal anti-eNOS antibodies (not shown). In addition, eNOS expression was similar in control and heat shocked aortas (Figure 6), while HSP90 was not co-immunoprecipitated with eNOS, either in control or heat shocked aortic rings (not shown).

HSP70-eNOS in vitro interaction was further examined by measuring eNOS enzymatic activity as conversion ofL-arginine to L-citrulline. As shown in Figure 7, addition of recombinant human HSP70 to recombinant bovine eNOSsignificantly increased eNOS enzymatic activity, which was strongly inhibited when L-NAME had been included in the incubation medium.

Discussion

Induction of heat shock proteins, especially HSP70, in blood vessels is a relatively common phenomenon observed in various experimental models and in some pathophysiological conditions[31, 32]. Induction of HSP70 in vascular endothelium occurs following heat shock preconditioning, suggesting a central role of blood vessels in heat shock-mediated protection against myocardial ischemia [11]. In this report, we show that heat shock preconditioning modified blood vessel physiology and endothelial function. More specifically, heat shock reduced blood vessel vasoconstriction tovarious stimuli (high K^+-induced depolarization, 5-hydroxytryptamine, phenylephrine) by increasing eNOS activity under basal conditions, via a mechanism independent of HSP90. The other potential candidate as modulator of eNOS activity was HSP70, because we found that it was strongly induced by heat shock in our system. In this respect, we showed, by immunoprecipitation, that eNOS and HSP70 interacted in heat shocked aortas or HUVEC, and confirmed, in a cell-free assay,that HSP70 can increase eNOS enzymatic activity. Theseobservations indicate that HSP70 interacts with eNOS and increases its enzymatic activity, opening new therapeutic targets which could be relevant in pathological conditions such as myocardial infarction and stroke.

The *in vitro*induction of the stress response in aortic segments, as defined by HSP70 expression, occurred after heat shock at 43.5 °C in our system, but not after heat shock at 42 °C. This is consistent with induction of a stress response in cultured endothelial cells where HSP70 expression and cell resistance to injury have been observed after exposure to 43°C for 1.5 h [33]. In contrast, *in vivo*, a single hyperthermic episode at 42°C for 15 min may be sufficient to induce an endothelial stress response measured by HSP70 expression [11, 34], suggesting that additional factors *in vivo*contribute to lower the threshold for induction of a stress response.

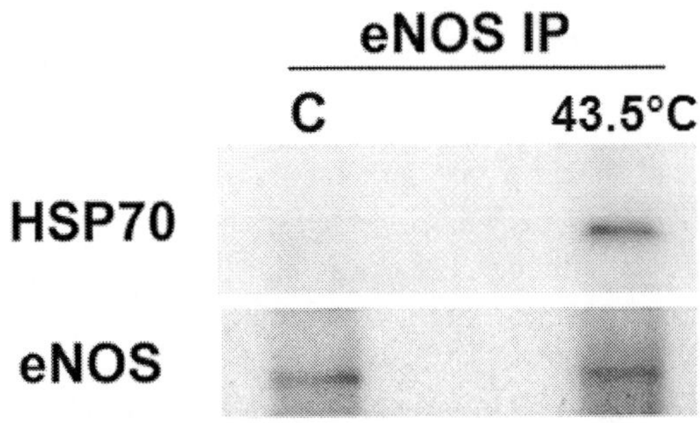

Figure 6. Interaction between HSP70 and eNOS in vascular endothelium. Typical experiment repeated three times with similar results. Protein lysates(120 µg) from control (C) and heat shocked (43.5°C) mouse aorta were immunoprecipitated using eNOS polyclonal antibody and further analyzed by SDS-PAGE and Western blot, using HSP70 antibody. The same membrane was rinsed and probed again using a specific eNOS monoclonal antibody to show equal amount of eNOS in both lysates. Similar results were obtained by immunoprecipitation of lysates from heat shocked HUVEC.

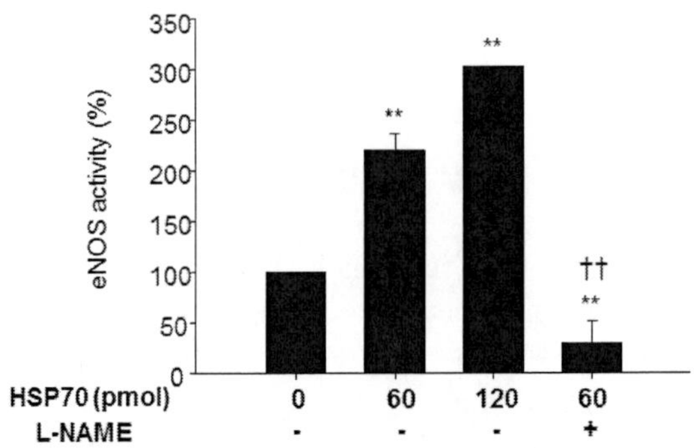

Figure 7. Interaction between HSP70 and eNOS in a cell free system. Recombinant bovine eNOS was incubated in Tris buffer with various amount of recombinant human HSP70 in absence or presence of L-NAME. After 15 min incubation, NOS activity was determined by measuring conversion of L-arginine to L-citrulline. Each column represent the mean (± SEM, vertical bars)from two different experiments, each performed in triplicate. **P<0.01 versus HSP70 0, ††P<0.01 versus HSP70 60 pmol; one-way ANOVAandTukey test.

Our data show that the reduction of contractile responses induced byheat shock was caused by increased NO production, but was not due to higher expression of NOS isoforms, in particular of eNOS, as detected by western blotting. Therefore, heat shock preconditioning differsfrom other treatments, such as endotoxic shock and LPS-treatment,(that induce iNOS gene expression)[35], or statin administration (that stabilizeseNOS mRNA)[36], because it increases NO production in blood vessels without changing the abundance of eNOSor iNOS. The fact that heat shock-induced vasomotor changes required the presence of endothelium and were absent in eNOS deficient mice, indicates that heat shock acted through a modification of eNOS enzymatic activity.

Interestingly, in our system, heat shock increasedbasal eNOS activity without changing eNOS activity stimulated byACh, at variance with the effect ofthe activator protein HSP90, which has been shown to interact with eNOS following agonist-stimulated endothelium-dependent vasodilation [27]. Although HSP90 was slightly increased by heat shock in our system, three arguments indicate that heat shock-increased basal eNOS activity did not involve HSP90. Firstly, agonist (ACh)-mediated relaxation in heat shocked aortas was not increased but, rather, it was reduced (not shown). This effect is opposite to that reported in HSP90-transfected endothelial cells [27]. Secondly, inhibition of HSP90 (using a specific HSP90 inhibitor, geldanamycin) did not restore contractility in heat shocked aortas while stillblocking ACh-mediated relaxation, confirming that HSP90 regulates eNOS upon agonist stimulation but not basal eNOS activity [27]. Finally, HSP90 only barely co-immunoprecipitated with eNOS in control or heat shocked aortas (not shown), confirming other reports where HSP90-eNOS interaction, drastically increased after agonist stimulation, was barely detectable under basal conditions[27, 37]. From these results, it appeared that heat shock enhanced eNOS activity via a mechanism independent of HSP90.

Another mechanism reported to increase basal eNOS activity involves downregulation of caveolin[38]. However, when probing for caveolin-1, we found that heat shock did not modify the total amount of caveolin-1 in isolated aortas (not shown). Since heat shock has been reported to acutely modify caveolin localization in cultured mouse NIH3T3 fibroblasts and to induce internalization of caveolin followed by a rapid (within 4 h) return to plasma membrane [39], it is also possible that long-term effects could take place after heat shock and account for increased eNOS basal activity. Further investigations are needed to clarify the role of caveolin and caveolin-eNOS interaction after heat shock preconditioning.

We showed that HSP70 co-immunoprecipitated with eNOS in mouse aortas and HUVEC. Although reverse immunoprecipitation experiment (detection of eNOS in immunoprecipitated proteins using HSP70 antibody) was not feasible due to high amount of proteins competing for HSP70 binding after heat shock preconditioning, we observed HSP70-eNOS interaction in various paradigms (cell culture or cell-free assay), in different species (mouse and human) and using monoclonal or polyclonal anti-eNOS antibodies, which strengthens the observation in isolated blood vessels. Furthermore, addition of recombinant HSP70 could increaseeNOSactivity in a cell-free assay, indicating that HSP70-eNOS interaction may have a functional outcome.

Obviously HSP70 is known to chaperone denatured proteins after treatment with elevated temperatures, which helps in guiding their refolding and/or triage to the proteasome[40].In our system, however, eNOSdid not appear to be denatured, as indicated by a maintained ACh-mediated relaxation in mouse aortas immediately or 5 h after heat shock (not shown). Moreover, one would not expect a recovered activity to be significantly increased compared to initial conditions. It seems more likely that HSP70 interacts with native eNOS. We propose that after heat shock, HSP70-binding domains are exposed due to post-translational modifications of eNOS or removal of masking protein(s). Indeed, HSP70 is able to interact with co-chaperones, such as CHIP (carboxyl terminus of Hsp70-interactingprotein), which regulates eNOS activity and trafficking between subcellular compartments [41]. Further experiments are needed to determine the precise molecular mechanism through which HSP70modulateseNOS activity following heat shock preconditioning.

Our data cannot rule out that additional mechanisms regulating eNOS activity, such as fatty acylation or phosphorylation of eNOS or substrate/co-factor availability, could also contribute to heat shock-mediated effects. Analyses of eNOS phosphorylation state, eNOS subcellular localization and in vitro NOS activity will help in clarifying these issues. Preliminary experiments revealed that heat shock did not increase the amount of Akt or phophorylated isoforms of Akt (not shown), a kinases that phosphorylates eNOS and increases its activity [42].

Our findings are important for several reasons.Firstly, we provide a mechanistic basis to explain heat shock (and HSP70) protective effect *in vivo* against myocardial and cerebral ischemia. Indeed, increase in basal endothelial NO production enhances blood supply to ischemic tissue as shown after eNOSupregulation, and results in decreased infarct size [24]. Similarly, NO increase could also provide explanation for heat shock-mediated prevention of leukocyte adherence after ischemia/reperfusion injury [43].

Secondly, we showed that HSP70 interacts with eNOS under stressful conditions, but it is likely that similar interactions also occurs in pathological conditions known to induce HSP70 synthesis in endothelial cells and blood vessels. For example, endothelial cells overexpress HSP70 in response to exposure to oxidized LDL [16]and HSP70 has been observed in atherosclerotic human arteries [44] or aortas after physiological stress [45].Our data provide evidence that HSP70 interacts with eNOS and could thereby modify eNOS function, giving a new perspective to previous observations.

Thirdly, regulation of NO production by endothelium is of paramount importance in pathological conditions such as stroke, atherosclerosis, hypertension to mention a few.For example, stroke prevention favors treatments with statins that increase basal eNOS activity. Here we observed that heat shock preconditioning also increase basal eNOS activity without eNOSupregulation, suggesting that alternative (and maybe synergistic) mechanisms may be explored and provide new therapeutic targets in the treatment of vascular diseases.

Acknowledgments

J.-C. P. was supported by the American Heart Association (New England affiliate).
S.S. was supported by a grant N.A.T.O./Consiglio Nazionale delle Ricerche

References

[1] Currie, R. W., Karmazyn, M., Kloc, M., and Mailer, K. (1988) Heat-shock response is associated with enhanced postischemic ventricular recovery. *Circ. Res.* 63, 543-549.

[2] Iliodromitis, E. K., Karavolias, G. K., Bofilis, E., Yellon, D. M., and Kremastinos, D. T. (1999) Enhanced protection of heat shock in myocardial infarction: inhibition of detrimental effect of systemic hyperthermia. *Cardiovasc. Drugs Ther.* 13, 223-231.

[3] Hutter, M. M., Sievers, R. E., Barbosa, V., and Wolfe, C. L. (1994) Heat-shock protein induction in rat hearts. A direct correlation between the amount of heat-shock protein induced and the degree of myocardial protection. *Circulation,* 89, 355-360.

[4] Beere, H. M. (2001) Stressed to death: regulation of apoptotic signaling pathways by the heat shock proteins. *Sci STKE*2001, re1.

[5] Mosser, D. D., Caron, A. W., Bourget, L., Meriin, A. B., Sherman, M. Y., Morimoto, R. I., and Massie, B. (2000) The chaperone function of hsp70 is required for protection against stress-induced apoptosis. *Mol. Cell Biol.* 20, 7146-7159.

[6] Yellon, D. M., and Marber, M. S. (1994) Hsp70 in myocardial ischaemia. *Experientia,* 50, 1075-1084.

[7] Yenari, M. A., Fink, S. L., Sun, G. H., Chang, L. K., Patel, M. K., Kunis, D. M., Onley, D., Ho, D. Y., Sapolsky, R. M., and Steinberg, G. K. (1998) Gene therapy with HSP72 is neuroprotective in rat models of stroke and epilepsy. *Ann. Neurol.* 44, 584-591.

[8] Plumier, J. C., Ross, B. M., Currie, R. W., Angelidis, C. E., Kazlaris, H., Kollias, G., and Pagoulatos, G. N. (1995) Transgenic mice expressing the human heat shock protein 70 have improved post-ischemic myocardial recovery. *J. Clin. Invest.* 95, 1854-1860.

[9] Marber, M. S., Mestril, R., Chi, S. H., Sayen, M. R., Yellon, D. M., and Dillmann, W. H. (1995) Overexpression of the rat inducible 70-kD heat stress protein in a transgenic mouse increases the resistance of the heart to ischemic injury. *J. Clin. Invest,* 95, 1446-1456.

[10] Kamii, H., Kinouchi, H., Sharp, F. R., Koistinaho, J., Epstein, C. J., and Chan, P. H. (1994) Prolonged expression of hsp70 mRNA following transient focal cerebral ischemia in transgenic mice overexpressing CuZn-superoxide dismutase. *J. Cereb. Blood Flow Metab.* 14, 478-486.

[11] Leger, J. P., Smith, F. M., and Currie, R. W. (2000) Confocal microscopic localization of constitutive and heat shock-induced proteins HSP70 and HSP27 in the rat heart. *Circulation,* 102, 1703-1709.

[12] Krueger, A. M., Armstrong, J. N., Plumier, J., Robertson, H. A., and Currie, R. W. (1999) Cell specific expression of Hsp70 in neurons and glia of the rat hippocampus after hyperthermia and kainic acid-induced seizure activity. *Brain Res. Mol. Brain Res.* 71, 265-278.

[13] Kirby, L. B., Mondy, J. S., and Brophy, C. M. (1999) Balloon angioplasty induces heat shock protein 70 in human blood vessels. *Ann. Vasc. Surg.* 13, 475-479.

[14] Masada, T., Hua, Y., Xi, G., Ennis, S. R., and Keep, R. F. (2001) Attenuation of ischemic brain edema and cerebrovascular injury after ischemic preconditioning in the rat. *J. Cereb. Blood Flow Metab.* 21, 22-33.

[15] Roma, P., and Catapano, A. L. (1996) Stress proteins and atherosclerosis. *Atherosclerosis,* 127, 147-154.

[16] Zhu, W., Roma, P., Pellegatta, F., and Catapano, A. L. (1994) Oxidized-LDL induce the expression of heat shock protein 70 in human endothelial cells. *Biochem. Biophys. Res. Commun.* 200, 389-394.

[17] Guo, Y., Jones, W. K., Xuan, Y. T., Tang, X. L., Bao, W., Wu, W. J., Han, H., Laubach, V. E., Ping, P., Yang, Z., Qiu, Y., and Bolli, R. (1999) The late phase of ischemic preconditioning is abrogated by targeted disruption of the inducible NO synthase gene. *Proc. Natl. Acad. Sci. U S A,* 96, 11507-11512.

[18] Gidday, J. M., Shah, A. R., Maceren, R. G., Wang, Q., Pelligrino, D. A., Holtzman, D. M., and Park, T. S. (1999) Nitric oxide mediates cerebral ischemic tolerance in a neonatal rat model of hypoxic preconditioning. *J. Cereb. Blood Flow Metab.* 19, 331-340.

[19] Xuan, Y. T., Tang, X. L., Qiu, Y., Banerjee, S., Takano, H., Han, H., and Bolli, R. (2000) Biphasic response of cardiac NO synthase isoforms to ischemic preconditioning in conscious rabbits. *Am. J. Physiol. Heart Circ. Physiol.* 279, H2360-2371.

[20] Huang, Z., Huang, P. L., Ma, J., Meng, W., Ayata, C., Fishman, M. C., and Moskowitz, M. A. (1996) Enlarged infarcts in endothelial nitric oxide synthase knockout mice are attenuated by nitro-L-arginine. *J. Cereb. Blood Flow Metab.* 16, 981-987.

[21] Yamamoto, S., Golanov, E. V., Berger, S. B., and Reis, D. J. (1992) Inhibition of nitric oxide synthesis increases focal ischemic infarction in rat. *J. Cereb. Blood Flow Metab.* 12, 717-726.
[22] Endres, M., Laufs, U., Huang, Z., Nakamura, T., Huang, P., Moskowitz, M. A., and Liao, J. K. (1998) Stroke protection by 3-hydroxy-3-methylglutaryl (HMG)-CoA reductase inhibitors mediated by endothelial nitric oxide synthase. *Proc. Natl. Acad. Sci. U S A,* 95, 8880-8885.
[23] Morikawa, E., Rosenblatt, S., and Moskowitz, M. A. (1992) L-arginine dilates rat pial arterioles by nitric oxide-dependent mechanisms and increases blood flow during focal cerebral ischaemia. *Br. J. Pharmacol.* 107, 905-907.
[24] Morikawa, E., Moskowitz, M. A., Huang, Z., Yoshida, T., Irikura, K., and Dalkara, T. (1994) L-arginine infusion promotes nitric oxide-dependent vasodilation, increases regional cerebral blood flow, and reduces infarction volume in the rat. *Stroke,* 25, 429-435.
[25] Garcia-Cardena, G., Martasek, P., Masters, B. S., Skidd, P. M., Couet, J., Li, S., Lisanti, M. P., and Sessa, W. C. (1997) Dissecting the interaction between nitric oxide synthase (NOS) and caveolin. Functional significance of the nos caveolin binding domain in vivo. *J. Biol. Chem.* 272, 25437-25440.
[26] Michel, J. B., Feron, O., Sacks, D., and Michel, T. (1997) Reciprocal regulation of endothelial nitric-oxide synthase by Ca2+-calmodulin and caveolin. *J. Biol. Chem.* 272, 15583-15586.
[27] Garcia-Cardena, G., Fan, R., Shah, V., Sorrentino, R., Cirino, G., Papapetropoulos, A., and Sessa, W. C. (1998) Dynamic activation of endothelial nitric oxide synthase by Hsp90. *Nature,* 392, 821-824.
[28] Mulvany, M. J., and Halpern, W. (1977) Contractile properties of small arterial resistance vessels in spontaneously hypertensive and normotensive rats. *Circ. Res.* 41, 19-26.
[29] Garcia-Cardena, G., Fan, R., Stern, D. F., Liu, J., and Sessa, W. C. (1996) Endothelial nitric oxide synthase is regulated by tyrosine phosphorylation and interacts with caveolin-1. *J. Biol. Chem.* 271, 27237-27240.
[30] Martin, W., Furchgott, R. F., Villani, G. M., and Jothianandan, D. (1986) Depression of contractile responses in rat aorta by spontaneously released endothelium-derived relaxing factor. *J. Pharmacol. Exp. Ther.* 237, 529-538.
[31] Lu, X., and Kakkar, V. (2010) The role of heat shock protein (HSP) in atherosclerosis: Pathophysiology and clinical opportunities. *Curr. Med. Chem.* 17, 957-973.
[32] Bielecka-Dabrowa, A., Barylski, M., Mikhailidis, D. P., Rysz, J., and Banach, M. (2009) HSP 70 and atherosclerosis--protector or activator? *Expert Opin. Ther. Targets,* 13, 307-317.
[33] Wong, H. R., Mannix, R. J., Rusnak, J. M., Boota, A., Zar, H., Watkins, S. C., Lazo, J. S., and Pitt, B. R. (1996) The heat-shock response attenuates lipopolysaccharide-mediated apoptosis in cultured sheep pulmonary artery endothelial cells. *Am. J. Respir. Cell Mol. Biol.* 15, 745-751.
[34] Amrani, M., Latif, N., Morrison, K., Gray, C. C., Jayakumar, J., Corbett, J., Goodwin, A. T., Dunn, M. J., and Yacoub, M. H. (1998) Relative induction of heat shock protein in coronary endothelial cells and cardiomyocytes: implications for myocardial protection. *J. Thorac. Cardiovasc. Surg.* 115, 200-209.

[35] Thiemermann, C. (1997) Nitric oxide and septic shock. *Gen. Pharmacol.* 29, 159-166.
[36] Laufs, U., La Fata, V., Plutzky, J., and Liao, J. K. (1998) Upregulation of endothelial nitric oxide synthase by HMG CoA reductase inhibitors. *Circulation,* 97, 1129-1135.
[37] Gratton, J. P., Fontana, J., O'Connor, D. S., Garcia-Cardena, G., McCabe, T. J., and Sessa, W. C. (2000) Reconstitution of an endothelial nitric-oxide synthase (eNOS), hsp90, and caveolin-1 complex in vitro. Evidence that hsp90 facilitates calmodulin stimulated displacement of eNOS from caveolin-1. *J. Biol. Chem.* 275, 22268-22272.
[38] Feron, O., Dessy, C., Moniotte, S., Desager, J. P., and Balligand, J. L. (1999) Hypercholesterolemia decreases nitric oxide production by promoting the interaction of caveolin and endothelial nitric oxide synthase. *J. Clin. Invest.* 103, 897-905.
[39] Kang, Y. S., Ko, Y. G., and Seo, J. S. (2000) Caveolin internalization by heat shock or hyperosmotic shock. *Exp. Cell Res.* 255, 221-228.
[40] Lanneau, D., Wettstein, G., Bonniaud, P., and Garrido, C. (2010) Heat shock proteins: cell protection through protein triage. *ScientificWorldJournal,* 10, 1543-1552.
[41] Jiang, J., Cyr, D., Babbitt, R. W., Sessa, W. C., and Patterson, C. (2003) Chaperone-dependent regulation of endothelial nitric-oxide synthase intracellular trafficking by the co-chaperone/ubiquitin ligase CHIP. *J. Biol. Chem.* 278, 49332-49341.
[42] Atochin, D. N., Wang, A., Liu, V. W., Critchlow, J. D., Dantas, A. P., Looft-Wilson, R., Murata, T., Salomone, S., Shin, H. K., Ayata, C., Moskowitz, M. A., Michel, T., Sessa, W. C., and Huang, P. L. (2007) The phosphorylation state of eNOS modulates vascular reactivity and outcome of cerebral ischemia in vivo. *J. Clin. Invest.* 117, 1961-1967.
[43] Chen, G., Kelly, C., Stokes, K., Wang, J. H., Leahy, A., and Bouchier-Hayes, D. (1997) Induction of heat shock protein 72kDa expression is associated with attenuation of ischaemia-reperfusion induced microvascular injury. *J. Surg. Res.* 69, 435-439.
[44] Berberian, P. A., Myers, W., Tytell, M., Challa, V., and Bond, M. G. (1990) Immunohistochemical localization of heat shock protein-70 in normal-appearing and atherosclerotic specimens of human arteries. *Am. J. Pathol.* 136, 71-80.
[45] Udelsman, R., Blake, M. J., Stagg, C. A., Li, D. G., Putney, D. J., and Holbrook, N. J. (1993) Vascular heat shock protein expression in response to stress. Endocrine and autonomic regulation of this age-dependent response. *J. Clin. Invest.* 91, 465-473.

In: Endothelium and Epithelium
Editors: J. Carrasco and M. Mota, pp. 193-211
ISBN 978-1-61470-874-2

Chapter IX

Gasdermin Superfamily: A Novel Gene Family Functioning in Epithelial Cells

Norihisa Saeki and Hiroki Sasaki
Division of Genetics, National Cancer Center Research Institute.,
5-1-1 Tsukiji, Tokyo, Japan

Abstract

The Gasdermin Superfamily is a novel gene group consisting of Gasdermin Family genes (*GSDMA*, *GSDMB*, *GSDMC* and *GSDMD* in humans) and Gasdermin-related genes (*DFNA5* and *DFNB59* in humans). The members of the Superfamily are expressed in the epithelial cells of a variable type of tissue and may have a differentiation-status specific role; however, their actual functions remain unknown. In this chapter, the current status of studies on the Superfamily genes is overviewed and their biological significance is discussed.

In the esophageal and gastric epithelia, each member of the Gasdermin Family shows a distinct expression pattern according to the differentiation status of the epithelial cells; *GSDMA* is expressed in differentiated cells in both the epithelia and, in the gastric mucosa, it induces apoptosis of the surface pit cells. On the contrary, *GSDMB* expression can be observed in the region harboring rapidly amplifying epithelial cells. According to this expression pattern and function in the normal epithelium, *GSDMA* is frequently suppressed in esophageal and gastric cancers but *GSDMB* is even over-expressed in them. The expression of Gasdermin-related genes is also observed in several organs including the nervous system. *DFNA5* is a causative gene of autosomal dominant deafness, whose expression was detected in the cochlea. *DFNB59* is responsible for autosomal recessive deafness, expressed in the inner ear and in neurons of the auditory pathway.

The Gasdermin Superfamily genes are expressed in many tissues in a manner of tissue-type and differentiation-status specificities, which implies their important function in cell differentiation. Therefore, phylogenetic conservation of the genes in the animals other than human and mouse is also expected; however, that is yet to be unveiled. Since Placozoa (a primitive form of the invertebrate) as well as amphioxus has at least one sequence coding hypothetical protein similar to human DFNB59, it is very likely that the superfamily genes diverged from a common ancestor gene and now harbor in the genome of other lower animals. A phylogenetic approach may disclose the biological significance of the superfamily.

Introduction

The epithelium is observed in a variety of tissue types in the human body. Each epithelium in each organ has a distinct function, which is specialized to that required for the function of the organs. The epidermal epithelium protects the body from environmental insults mechanically and physiologically. The pancreas harbors several types of epithelial cells, some secreting enzymes required for digesting foods and others secreting hormons for regulating blood glucose level. The epithelium in the kidney has essential roles in maintaining fluid and electrolyte balances. Differentiation of epithelial cells, from tissue stem cells to mature functional cells, progresses through epithelial-mesenchymal interaction; in other words, molecular interaction between epithelial cells and niches provided by mesenchymal cells, in the processes for maintaining the homeostasis of the organs in the adult body and for the organogenesis at an embryonal period. In the differentiation process of the epithelia, a number of the molecules, such as membrane-type receptors, protein kinases and transcription factors, are expressed depending on the type of tissues, which is regulated through epigenetic and other mechanisms, and execute their tasks in cell fate determination. Carcinogenesis, the malignant transformation of epithelial cells, is one of the forms of deviation of that cells take from their usual fate. To understanding this, it is important to identify the functions of each molecule that appears in the normal differentiation process, and the information of the molecules could be also applied to tissue engineering, which definitely needs a practical apparatus to control cell differentiation.

Table 1. Gasdermin Superfamily genes in human

Gene Name [other name]	Mouse ortholog	Reported expression site in human (in mice)	Reference
Gasdermin Family genes			
GSDMA [Gasdermin (*GSDM*), Gasdermin1 (*GSDM1*)]	*Gsdma1, Gsdma2, Gsdma3*	Esophagus, stomach, skin, mammary gland	[1,2,3]
GSDMB [PRO2521 (*PRO2521*), Gasdermin-like (*GSDML*)]	(not identified)	Esophagus, stomach, liver, colon	[3,6,7]
GSDMC [Melanoma-derived leucine Zipper, extranuclear factor (*MLZE*)]	*Gsdmc1, Gsdmc2 Gsdmc3, Gsdmc4*	Esophagus, stomach, trachea, spleen, skin	[3,8]
GSDMD [Gasdermin domain-containing 1 (*GSDMDC1*), Dfna5-like (*DFNA5L*)]	*Gsdmd*	Esophagus, stomach	[3,9]
Gasdermin-related genes			
DFNA5	*Dfna5*	Placenta, brain, heart, kidney (cochlea)	[10]
DFNB59 [pejvakin]	*Dfnb59*	(Brain, eye, inner ear, heart, lung, kidney, etc)	[11]

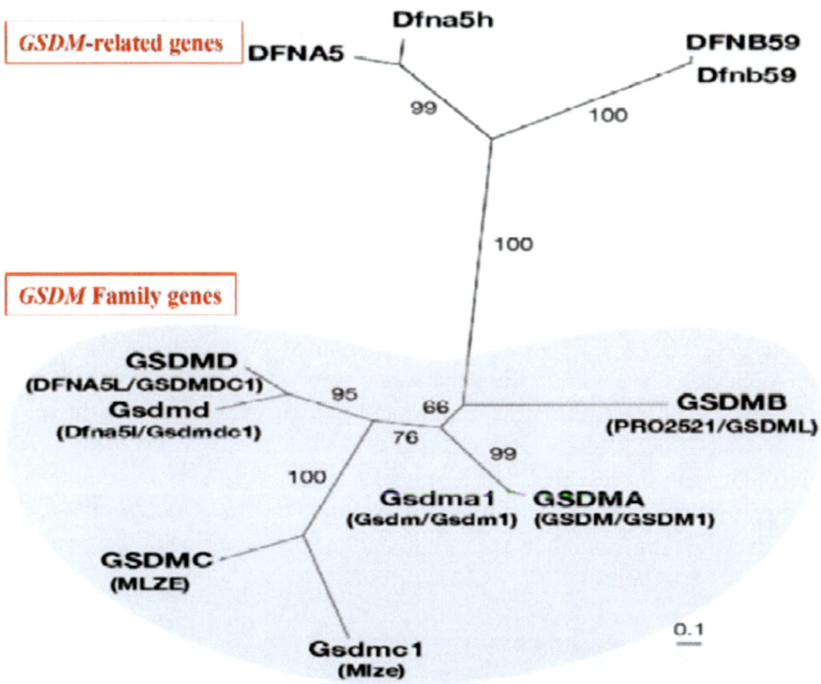

Figure 1. Phylogeny of Gasdermin (*GSDM*/*Gsdm*) Superfamily genes. Unrooted gene tree inferred from amino acid sequence data using the neighbor-joining method. The bootstrap values (percentage) obtained from 1000 replications are shown at the branches determining each clade. The scale bar shows the estimated evolutionary distance. (Adapted from Tamura M, et al. Members of a novel gene family, *Gsdm*, are expressed exclusively in the epithelium of the skin and gastrointestinal tract in a highly tissue-specific manner, Genomics 89 (2007), with permission from Elsevier).

For more than 10 years, the authors have been involved in studies on Gasdermin Family genes, each of which showed a distinct expression pattern in the upper gastrointestinal epithelium. The Family is now classified as Gasdermin Superfamily. Gasdermin Superfamily is a novel gene group consisting of Gasdermin Family genes and Gasdermin-related genes (Table 1, Figure 1, refs. 1-11, the definition of Gasdermin-related genes being originally introduced for this book and different from that used in ref. 4), whose discovery and investigation have been heavily contributed to by Japanese researchers. The members of the superfamily are expressed in the epithelium of a variable type of tissue and may have a differentiation-status specific role; however, their functions remain unknown. This chapter overviews the current status of studies on the superfamily, which have been performed by our and other research groups.

Discovery of Gasdermin Family Genes

In 2000, Saeki N and Shiroishi T at the National Institute of Genetics (NIG, Mishima, Shizuoka, Japan) in collaboration with Sasaki H at the National Cancer Center Research Institute (NCCRI, Tokyo, Japan), reported a novel mouse gene Gasdermin (*Gsdm*), which was identified in the process of investigating a causative gene of *Rim3* mouse mutant (Figure

2.A) [1, 12]. The mutant spontaneously arose in a mouse facility of NIG and features a generalized abnormality of the epidermis: thickened skin due to hyperkeratosis and hyperstratified prickle-cell layers (acanthosis), and few hair follicles (alopecia). It also has a thickened cornea, causing cataract (corneal opacity). The *Rim3* locus was mapped to a range of approximately 300kb, corresponding to three bacterial artificial chromosome (BAC) clones, on chromosome 11 by linkage analysis (Figure 2.B) [12].

Because of a lack of sequence information about the genomic region at that time, a causative gene discovery was attempted with a cDNA selection procedure in which skin cDNA and the BAC DNA were co-hybridized, and finally a sequence of unknown gene origin was isolated. Using the sequence as a probe, one novel gene was discovered and cloned from a mouse skin cDNA library, which was contained in one of the BAC clones, and northern blot analysis revealed that the gene was expressed in mouse stomach as well as in the skin (Figure 2.C). Because of this gastric and dermal expression, the novel gene was named Gasdermin (*Gsdm*).

Southern blot analysis revealed that it was also present in the human genome. At first, because of its expression in the skin and its localization in the *Rim3* locus, it was believed that the gene was the causal of *Rim3*; however, no mutation was identified in the gene of the *Rim3* mouse.

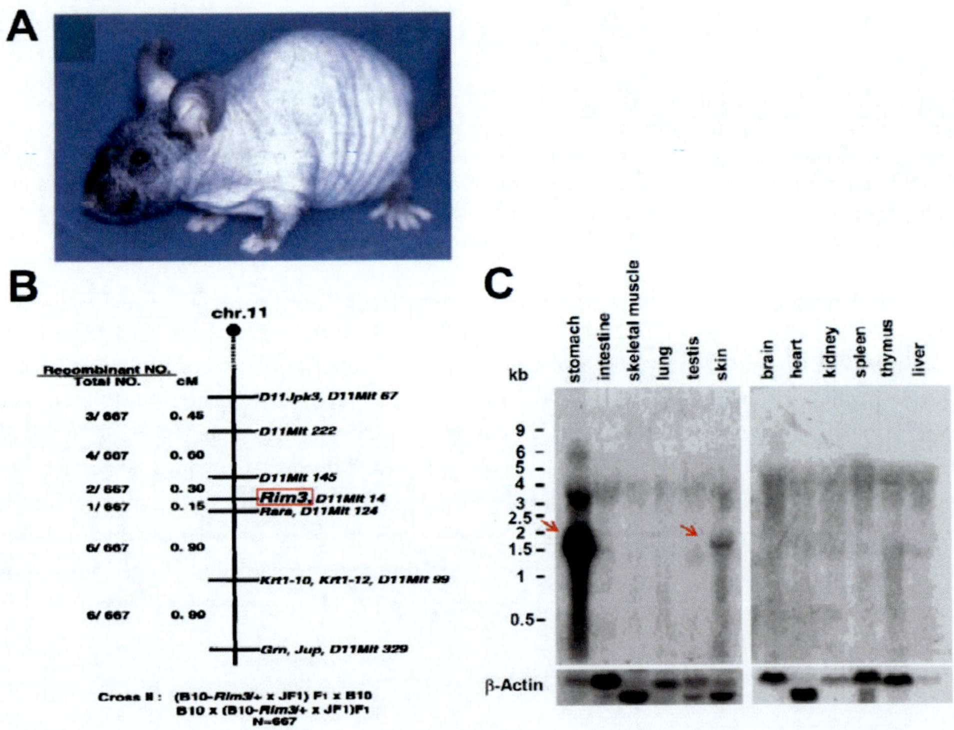

Figure 2. *Rim3* mutant mice and Gasdermin (*Gsdm*). (A) A mutant mouse *Rim3* showing skin abnormality. (B) *Rim3* locus mapped to chromosome 11 by linkage analysis. (C) *Gsdm* is expressed in the stomach and skin. (A, B: Adapted from Sato H, et al. A new mutation *Rim3* resembling *Re* (*den*) is mapped close to retinoic acid receptor alpha (*Rara*) gene on mouse chromosome 11. Mamm Genome 9 (1998), C: Adapted from Saeki N, et al. Gasdermin (*Gsdm*) localizing to mouse Chromosome 11 is predominantly expressed in upper gastrointestinal tract but significantly suppressed in human gastric cancer cells. Mamm Genome 11 (2000), with permission from Springer).

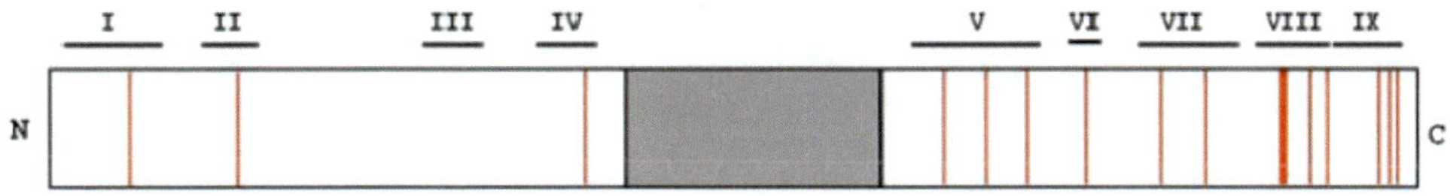

Figure 3. Structure of Gasdermin Family protein. The protein of the Gasdermin Family contains 9 conserved motifs. Red lines show leucines conserved in all Gasdermin Family members. The gray box representing the middle part is unique to each Gasdermin Family member. (Reprinted from Tamura M, et al. Members of a novel gene family, *Gsdm*, are expressed exclusively in the epithelium of the skin and gastrointestinal tract in a highly tissue-specific manner, Genomics 89 (2007), with permission from Elsevier).

In 2001, Saeki N and Sasaki H at the NCCRI identified 2 paralogs of *Gsdm* in the vicinity of the gene in chromosome 11 by searching the entire genomic sequence information of mouse released from Celera, and Shiroishi T and his colleagues at the NIG found a mutation in the one of the paralogs [13]. By comparative genomic analysis using sequence information deposited in the public data base, the NIG group noticed other genes similar to *Gsdm* in the mouse and human genomes and collectively defined those genes possessing the similarity as Gasdermin Family (Table 1, Figure 1) [4].

The family is classified into 4 groups, A, B, C and D. The human has one Group A gene (*GSDMA*) in chromosome 17q21 but the mouse has three: *Gsdma1* (originally named *Gsdm*), *Gsdma2* and *Gsdma3* (the causative gene for *Rim3*), in chromosome 11D. In the same manner, a human has one *GSDMC* in chromosome 8q24.2 and a mouse four *Gsdmc* in chromosome 15D1. Both have one Group D gene in the genome: chromosome 8q24.3 in human and chromosome 15D3 in mouse. Intriguingly, the Group B gene, *GSDMB*, is unique to humans (Table 1, Figure 1). The products of the family genes have 9 regions (motifs) in which conserved amino acids cluster (Figure 3). Both the N- and C-terminal sides of the proteins are relatively well conserved in the family, but the central portion is unique to each group. The 5th and 7th conserved regions contain a leucine-zipper motif, but whether the proteins have DNA binding ability is unknown.

Tissue-Specific Expression of Gasdermin Family Genes

Expression pattern analysis of the Gasdermin Family genes revealed tissue specificity of their expression (Figure 4 and Table 1). The pattern is thought to be conserved between human and mouse but it has yet to be ascertained sufficiently. Thus, systematic and comparative expression analyses on both human and mouse await investigation.

The Human Group A gene *GSDMA* is expressed mainly in the skin and stomach but also appear in the esophagus. In the mouse with 3 Group A genes, the gene-tissue relation is more distinct (Table 2). *Gsdma1* is expressed in the squamous epithelium of the skin, esophagus, forestomach and cornea. *Gsdma2* is expressed in the gastric gland. It is remarkable that *Gsdma3* is expressed in the sebaceous gland of the skin, possibly in the bulge [13], because *Rim3* caused by base-substitution in the gene shows hyperstratification of the epidermis, which could be a result from an abnormality in stem cell function. The epidermal stem cell is known to reside in the bulge located near the sebaceous gland, and *Gsdma3* in the sebaceous gland (possibly in the bulge) is down-regulated in the *Rim3* [13, 14]. *Gsdma3* is now also known as the causative gene for other mouse mutants showing skin abnormality (Table 3).

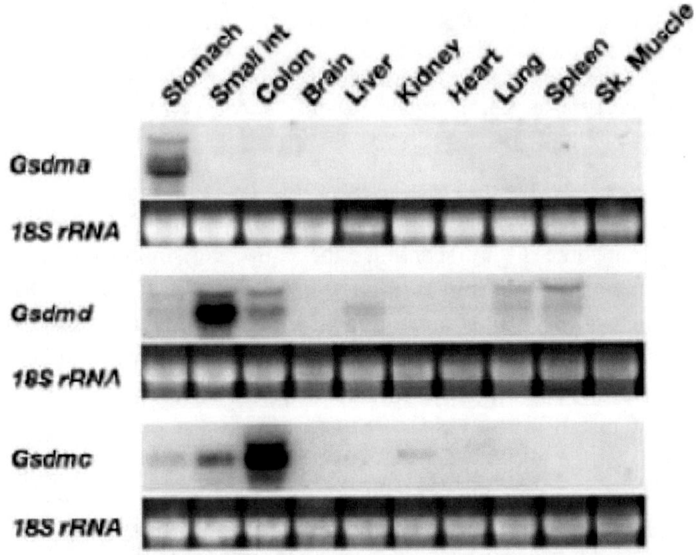

Figure 4. Major expression sites of Gasdermin Family genes in mice (RT-PCR). The mice have no ortholog of human *GSDMB* in their genome. (Reprinted from Tamura M, et al. Members of a novel gene family, *Gsdm*, are expressed exclusively in the epithelium of the skin and gastrointestinal tract in a highly tissue-specific manner, Genomics 89 (2007), with permission from Elsevier).

Table 2. Expression sites of *GSDMA/Gsdma*[1-4,13]

Type of Epithelium	Squamous					
	Skin*	Esophagus	Forestomach	Stomach(gland)	Hair Follicle	Other
Human						
GSDMA	+	+		+	?	Mammary Gland
Mouse						
Gsdma1	+	+	+	-	-	Cornea
Gsdma2	-	-	-	+	-	
Gsdma3 (*Rim3*)	+?	-	-	-	+	Sebaceous gland

*Interfollicular Epidermis

The Gasdermin Family genes are expressed in both the skin epithelium originating from the ectoderm and the digestive tract epithelium from the endoderm, which suggests that they may have a significant role in cell differentiation for realizing the function and/or structure of the epithelium.

Table 3. *Gsdma3* variation in mouse mutants with abnormal skin

Allele name	DNA change	Amino acid change	Location of mutation
Rim3	1124 G to A	348 A to T	Motif VII
Re-den	6bp duplication of 1314-1319	Duplication of 411-412 EA	Motif VIII
BSK	1113 A to G	344 Y to C	Motif VII
Rco2	1110 T to C	343 L to P	Motif VII
Dfl	Insertion of B2 element	Premature stop at 262	Absence of motif V-IX
Fgn	(no information)	278 T to P	Motif V

(Adapted from Tanaka S, et al. A new *Gsdma3* mutation affecting anagen phase of first hair cycle. Biochem Biophys Res Commun 359 (2007), with permission from Elsevier).

Distinct Expression Pattern of Gasdermin Family Genes in Epithelial Differentiation

Expression studies on the family genes in upper gastrointestinal tissues uncovered expression sites unique to each family gene, in relation to the differentiation state of the epithelial cells (Figure 5) [3]. *GSDMA* expression in the differentiated layer of the esophageal epithelium and in the pit region of the gastric epithelium suggests that it has a distinct function in mature epithelial cells. *GSDMB* is likely to be expressed in cells with proliferating ability, since it is expressed in the basal cell layer of the esophagus, where stem cells are harbored, and in the isthmus/neck region of the stomach, where rapidly amplifying precursors for superficial pit cells reside. *GSDMD* is expressed in differentiating cells and *GSDMC* seems to be expressed in both differentiating and differentiated cells in the esophagus. A distinct expression pattern was also observed for *Gsdmc* and *Gsdmd* in the small intestine and colon in mice (Figure 6) [4]. The distinct expression pattern of each family gene implies each has a distinct role in the process of epithelial differentiation.

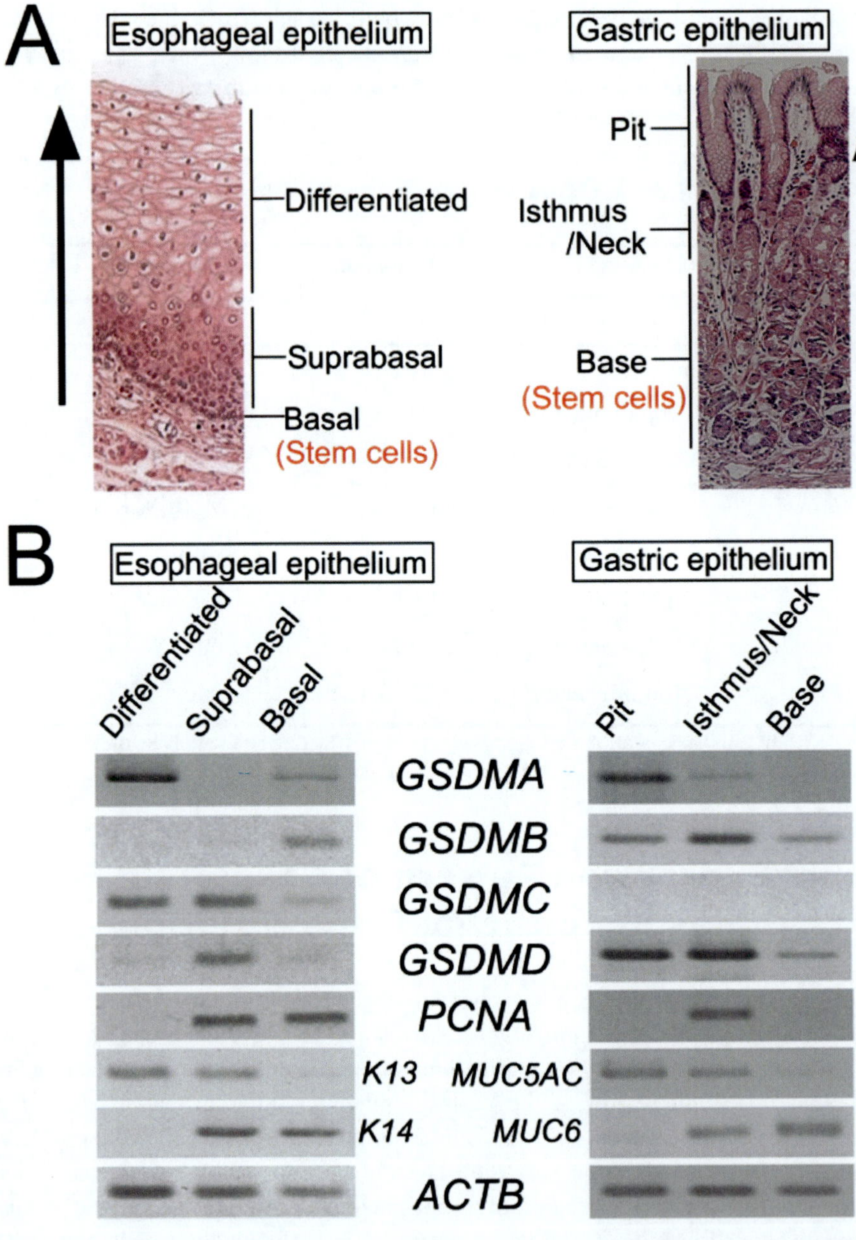

Figure 5. tract. (A) Schematic drawing shows structure of esophageal (left) and gastric (right) epithelium. Stem cells of the esophageal epithelium locate in the basal cell layer. Stem cells of the gastric epithelium are in the base, but there are rapidly amplifying cells, pit-cell precursors, in the isthmus of the gastric epithelium. Arrows indicate direction of migration of differentiating and differentiated cells. (B) RT-PCR analyses on microdissected specimens show distinct expression pattern of each family gene in the epithelia. (Adapted from Saeki N, et al. Distinctive expression and function of four *GSDM* family genes (*GSDMA-D*) in normal and malignant upper gastrointestinal epithelium. Genes Chromosomes Cancer 48 (2009), with permission from John Wiley and Sons)

Gasdermin-Related Genes

Besides the 4 gene groups of the Gasdermin Family, two other genes in the mouse and human genomes have similarity to *Gsdm*: *DFNA5* and *DFNB59* (classified as Gasdermin-related genes, the classification being originally introduced for this book). *DFNA5* is a causative gene of autosomal dominant deafness, whose expression was detected in several organs in human, including the placenta, heart, brain and kidney, and in the cochlea of the mouse [10, 15]. *DFNB59* is responsible for autosomal recessive deafness and the mouse counterpart is reported to be expressed in the inner ear and neurons of the auditory pathway as well as in other organs including the brain, eye, heart, lung, kidney, liver intestine, testis and skeletal muscle [11].

Unlike other superfamily genes, *DFNB59* possesses the putative nuclear localization signal and zinc-finger binding motif so it might be a transcriptional regulator.

The Superfamily Genes and Diseases

As mentioned previously, several mice with an aberration in *Gsdma3* were reported, all of which show a dermal phenotype; however, there has so far been no report for the superfamily genes as a causal of dermatological disorder in human. On the other hand, the superfamily genes were studied in the context of carcinogenesis.

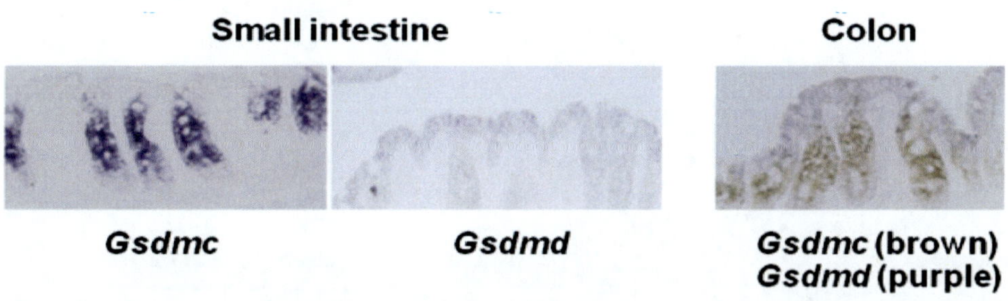

Figure 6. Expression of mouse *Gsdmc* and *Gsdmd* in the small intestine and colon. *Gsdmc* is expressed in the middle portion of the crypt but *Gsdmd* in the entire crypt. (Adapted from Tamura M, et al. Members of a novel gene family, *Gsdm*, are expressed exclusively in the epithelium of the skin and gastrointestinal tract in a highly tissue-specific manner, Genomics 89 (2007), with permission from Elsevier)

GSDMA has a strong tendency to be suppressed in esophageal and gastric cancers, whereas *GSDMB* is expressed in almost all cases of these cancers [1-3]. *GSDMB* is also expressed in about 80% of cases of hapatocellular carcinoma and uterine cervix cancer [16], and up-regulated in colorectal cancer with unknown frequency [7]. *GSDMC* is preferentially expressed in metastatic melanoma [8]. *GSDMD* expression was observed in the majority of gastric cancers [3]. *DFNA5* is down-regulated in breast cancer and its methylation status correlates to lymph node metastasis [17]. It is also suppressed by DNA methylation in gastric and colorectal cancers [18, 19].

Recently it was reported that *GSDMB* is associated with childhood asthma, which was revealed by a genome-wide association study and other succeeding study [20, 21], but how it is involved in the pathogenesis remains to be uncovered. As previously mentioned, *DFNA5* and *DFNB59* are known as a causal of human non-syndromic hearing impairment [10, 11].

Function of the Gasdermin Superfamily Genes

The expression pattern of the Gasdermin Family genes in the gastrointestinal tract and skin suggests that each gene has a unique function in the epithelium. *GSDMA* is the only gene in the family, whose function was investigated by a molecular biological approach. Saeki N and colleagues demonstrated a role in apoptosis induction in gastric pit cells, and that this phenomenon is regulated by TGF-β signalling (Figure 7) [2]. This molecular cascade is thought to be the machinery for the normal homeostatic turnover of gastric pit cells, by which senescent cells are eliminated from the pit. This function of *GSDMA* in apoptosis may be concordant with the phenotype of the *Rim3* mutant, in which hyperstratificaion of the prickle layer of the epidermis is observed. This phenotype may result from an abnormal turnover of epithelial cells. It is also concordant with the finding that *GSDMA* is frequently suppressed in esophageal and gastric cancers, in which it may function as a tumor suppressor [2, 3].

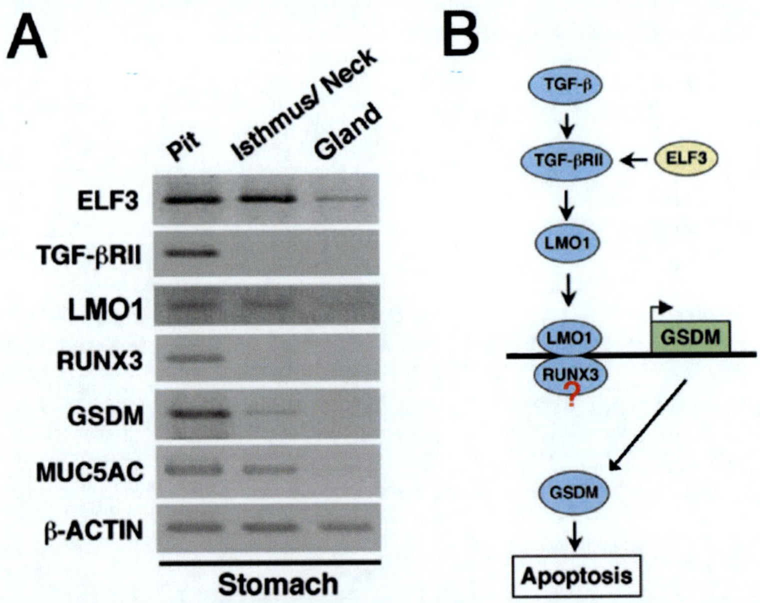

Figure 7. TGF-β/LMO1/GSDMA pathway for apoptosis of gastric pit cells. (A) RT-PCR analysis on microdissected specimens demonstrates that expressions of type II TGFβ receptor (*TGFβR II*), two transcription factors (*RUNX3*, *LMO1*) and *GSDM* (*GSDMA*) are co-localized to the pit region of the gastric epithelium. A transcription factor *ELF3*, which was reported to upregulate *TGFβR II* expression, is also expressed in the pit region. (B) A schematic diagram showing the TGF-β signalling for apoptosis of the pit cells of the gastric epithelium.

In addition to *GSDMA*, it was demonstrated by a colony formation assay in our study that *GSDMC* and *GSDMD* have cell-proliferation inhibition activity in a gastric cancer cell line (Figure 8) [3]. Interestingly, it was reported that *DFNA5* suppressed cell proliferation of colorectal cancer cells [19]. Therefore, the superfamily genes may have some common function and associated-molecules. It was reported that *GSDMB* has cell-proliferation promoting activity in HeLa cells (cervical cancer cell lines) [16], which is concordant with the fact that it is expressed in all the cases of esophageal and gastric cancers we examined [3]; however, it was not demonstrated in HepG2 cells (human hepatocellular carcinoma cell lines) [16]. Our study also showed that *GSDMB* has no promoting activity on cell proliferation in gastric cancer cells (MKN28 cell lines) (Figure 8) [3]. *GSDMC*, also designated as *MLZE* (Melanoma-derived Leucine Zipper, Extra-nuclear factor), was originally identified as a gene related to metastatic ability of mouse melanoma cells and its expression is unique to the vertical growth phase, a more advanced phase of tumor progression, of human melanoma [8]. Therefore, it is likely that it is involved in one or more of the processes of the invasion-metastasis cascade; local invasion, intravasation, circulation, adherence to endothelium, extravasation and proliferation [23]. In the esophagus, intestine and colon, *GSDMC* is expressed in the middle to upper region of the epithelium, where the epithelial cells are moving to the lumen side (Figures 5 and 6). In addition, our study showed no cell-growth promoting activity of the gene (Figure 8). *GSDMC* may have a function related to cell mobility. Although the gene product contains a functional nuclear localization signal and leucine-zipper motif, it shows cytoplasmic localization in melanoma cells. It may interact with cytoskeleton, but this remains unknown.

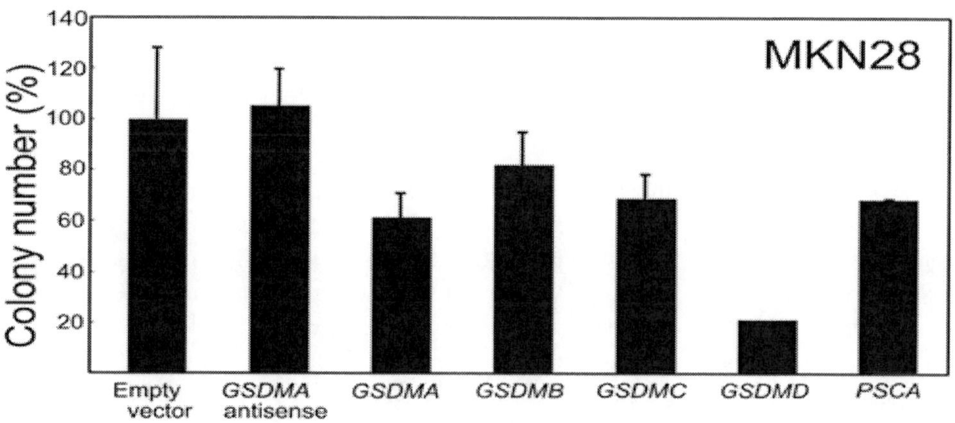

Figure 8. Colony formation assays demonstrated a cell-growth inhibition activity of the Gasdermin Family genes. Each gene in an expression vector was introduced into a gastric cancer cell line MKN28, and G418-resistant colonies were counted. In the family genes, *GSDMD* showed the strongest cell-growth inhibition activity. *PSCA* is a positive control whose cell-growth inhibition activity was shown previously [22]. (Adapted from Saeki N, et al. Distinctive expression and function of four *GSDM* family genes (*GSDMA-D*) in normal and malignant upper gastrointestinal epithelium. Genes Chromosomes Cancer 48 (2009), with permission from John Wiley and Sons).

Knock-out (KO) mice have been generated for some of the superfamily genes. Although *DFNA5* is expressed in multiple tissues, *Dfna5*-KO mouse showed a normal appearance by gross examination; however, an abnormality of the hair cells in the cochlea was demonstrated by electron-microscopic observation [24]. Targeting of *Gsdmd*, mainly expressed in the small intestine, resulted in no morphological abnormality in the digestive tract or other organs [25]; however, the mice might have minute changes in a function and/or in a microscopic appearance of the organs where the gene is expressed. It should be noted that *GSDMD* has the strongest cell-growth inhibition activity in a colony formation assay (Figure 8).

Phylogenetic Analysis of the Gasdermin Superfamily

The Gasdermin Superfamily genes are expressed in many tissues in a manner of tissue-type and differentiation-status specificities, which implies their important function in cell differentiation. Therefore, phylogenetic conservation of the genes in animals other than human and mouse is also expected; however, it is yet to be unveiled. So far, whole genome sequence information is available for phylogenetic analysis on the genome of several animals, some of which are well-known experimental animals: *Trichoplax adhaerens* (Placozoa), *Nematostella vectensis* (starlet sea anemone), *Branchiostoma floridae* (lancelet, amphyioxus), *Danio rerio* (zebrafih), *Gallus gallus*, *Ornithorhynchus anatinus* (platypus), *Mus musculus*, *Monodelphis domestica* (opossum), *Pan troglodytes* (chimpanzee) and *Homo sapiens*. A phylogenetical similarity analysis identified one sequence coding a hypothetical protein similar to human DFNB59, in the genome of *Trichoplax* and lancelet (Figure 9). As *Trichoplax* is a primitive form of the invertebrate, it is very likely that the superfamily genes diverged from this ancestor gene and now harbor also in the genome of other lower animals.

The authors are anticipating that a phylogenetic approach may disclose the biological significance of the superfamily. The tissue specific expression pattern of the family members in the alimentary tract may have some relation to differentiational regulation by Hox code. Amphioxus and *Ciona* (sea squirt) are phylogenetical ancestor animals for Chordates, and the whole genome sequencing projects revealed that they possess one cluster of *Hox* genes in their genome (Figure 10) [26, 27]. They possess an archetype of the digestive system featured by a distinct anatomical and functional regionalization (Figures 11 and 12) [28, 29], and a regionalized expression of *Fox* and *Pax* genes in the pharynx of *Ciona* was reported [30, 31]. To disclose the basic function of the superfamily genes in epithelial cell differentiation, an investigation conducted on the primitive form of the digestive tract in non-mammalians may be helpful. The relation between the expression site and the function of proteins resembles that between the two faces of a coin. The expression pattern, regulated in spatial and temporal manners, provides significant information for identification of the molecular function. Expression sites of the superfamily genes in the key animals, including hydra, sea urchin, sea squirt, amphioxus, *Danio rerio*, *Xenopus*, *Gallus*, mouse and human, will be a key of the identification.

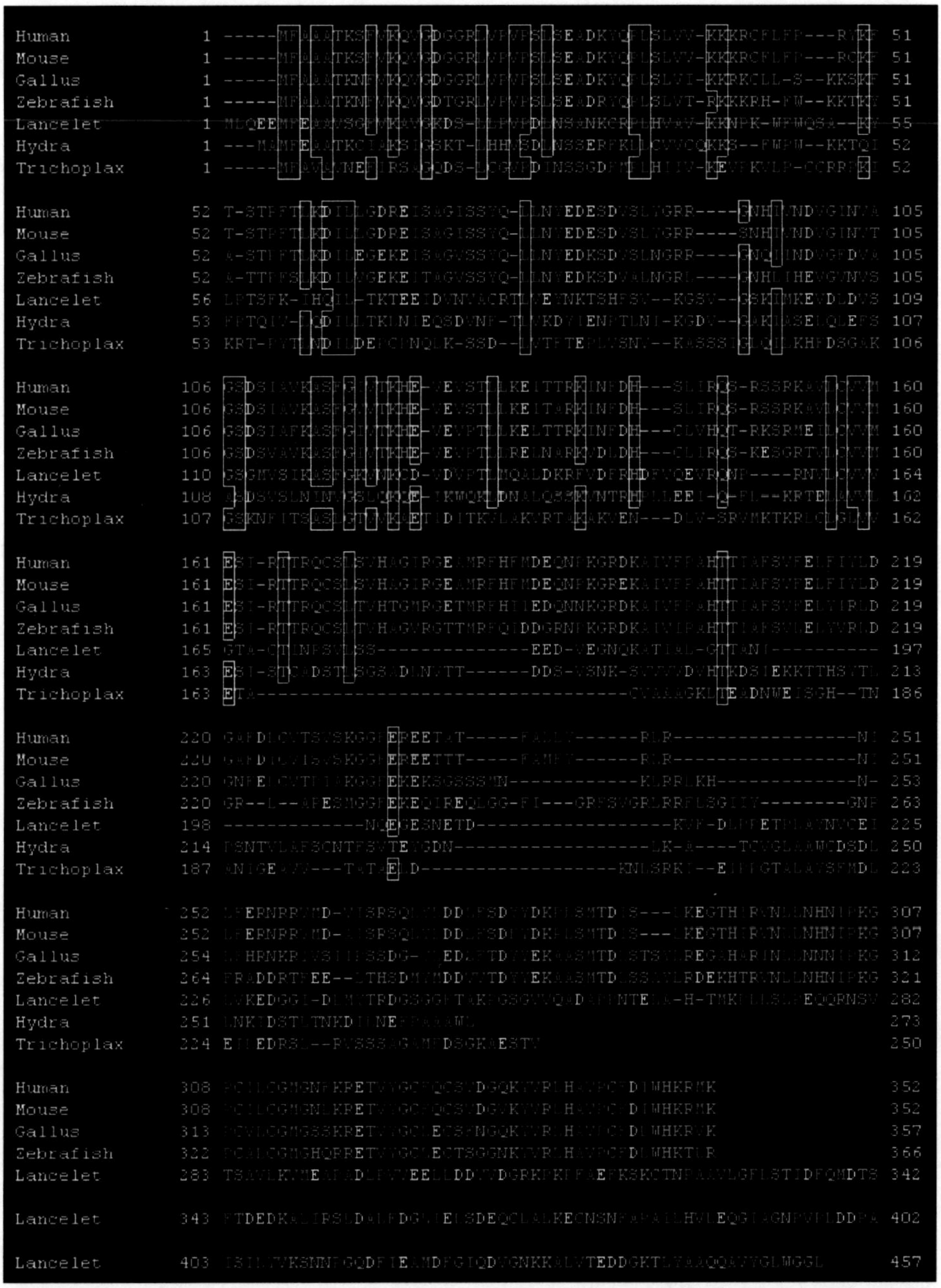

Figure 9. Amino acid sequences of hypothetical proteins, which show similarity to human and mouse DFNB59, encoded in the genome of other animals. The comparison analysis was performed using GENETYX Software (GENETYX Co.). Human; *Homo sapiens* DFMB59, Mouse; *Mus muscles* Dfnb5, Gallus; *Gallus gallus* Dfnb59, Zebrafish; *Danio rario* Dfnb59, Lancelet; *Branchiostoma floridae* hypothetical protein (XP_002587369.1), Trichoplax; *Trichoplax adherens* hypothetical protein (XP_002109134.1). N-terminal 165 amino acids of the Trichoplax hypothetical protein has the same basic structure as that of the Dfnb59 protein of human, mouse, *Gallus* and zebrafish, in which 31 amino acids are conserved between *Trichoplax* and human.

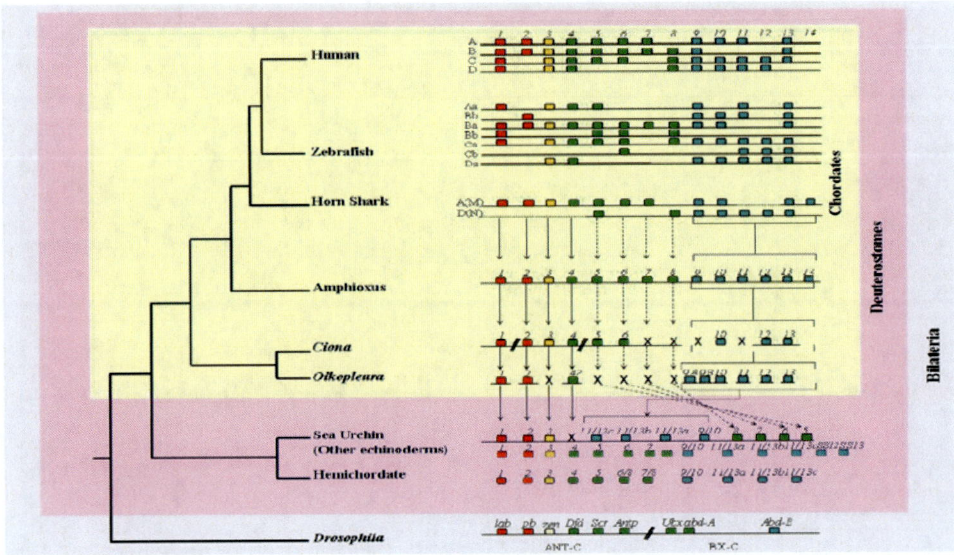

Figure 10. *Hox* gene cluster organization of deuterostomes. The organization of *Hox* gene cluster is conserved in animals, from sea urchin to human. (Reprinted from Monteiro AS, et al. Hox genes are not always Colinear. Int J Biol Sci 2 (2006), with permission from Ivyspring International Publisher).

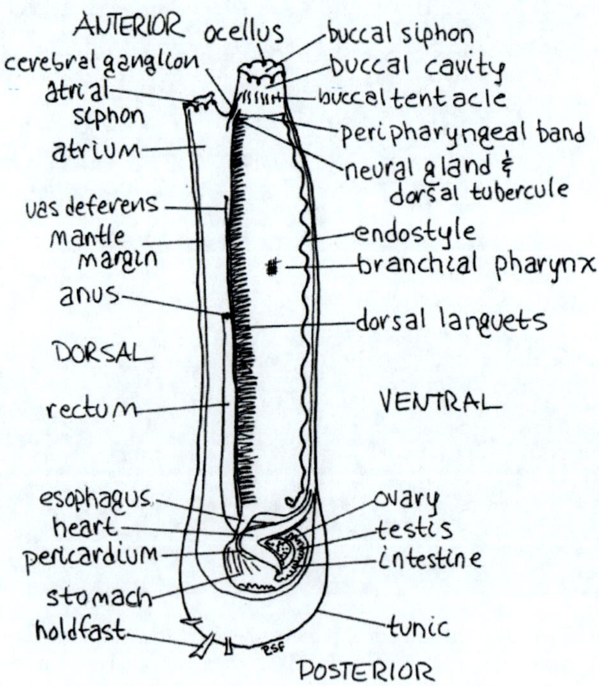

Figure 11. Internal structure of the sea squirt. The structure of the alimentary tract of *Ciona* is simple but obviously shows regionalization, i.e., consisting of esophagus, stomach and rectum. (Reprinted from "Invertebrate Anatomy OnLine, Ciona: http://webs.lander.edu/rsfox/invertebrates/ciona.html., with permission from Dr. RS Fox at Lander University).

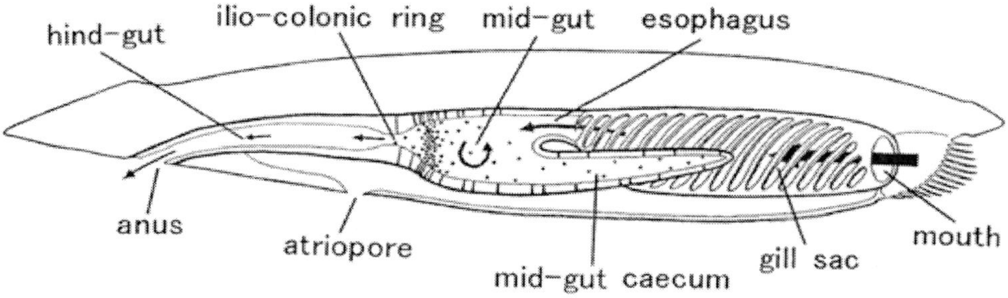

Figure 12. Digestive tract of the amphioxus. (Reprinted from Urata M, et al. Calcitonin-immunoreactive cells of the digestive tract of the amphioxus are distributed concentrically in a restricted region of the mid-gut. Acta Zoologica 82 (2001), with permission from John Wiley and Sons).

Additionally, it is notable that *Trichoplax* may possess the ancestor gene of the Gasdermin Superfamily. The body of *Trichoplax* consists of three layers of cells; the epitheloid layer (upper epithelium), bottom layer (lower epithelium) and fiber syncytium between them [32]. The epitheloid layer is made of ciliated cover cells. The bottom layer contains both cylinder cells with cilia for locomotion and gland cells which secret digestive enzyme for feeding. The fiber syncytium cells contain actin and a neurotransmitter-like substance. Namely, it has just 4 types of cells. Because of its simple structure, *Trichoplax* may be a practical model animal in the study of the epithelium as well as a principle function of the Gasdermin superfamily.

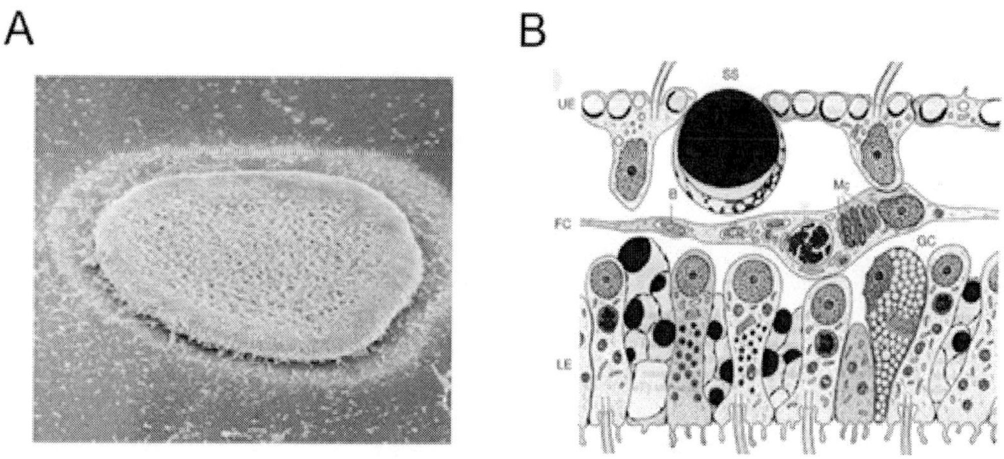

Figure 13. *Trichoplax* morphology. (A) Low-magnification scanning electron micrographs. It has a flat shape. (B) Diagram showing cross section. UE, upper epithelium; LE, lower epithelium; FC, contractile fiber cell; GC, gland cell; SS, shiny sphere; Mc, mitochondrial complex; B, (endosymbiotic?) bacterium in endoplasmic reticulum. (Adapted from Schierwater B. My favorite animal, *Trichoplax adhaerance*. BioEssays 27 (2005), with permission from John Wiley and Sons).

Conclusion and Future

As overviewed previously, many findings, which suggest the biological significance of the Gasdermin Superfamily genes, have already been obtained. However, their precise functions remain unrevealed. To explore them, a phylogenetic analysis may, the authors think, be helpful. The superfamily gene product is still hypothetical in some animals, i.e., their existence is not confirmed by molecular basis. Therefore, identification of the ortholog expression and cloning of the molecules are a primary requirement. Based on genome sequence information, it is likely that the orthologs and ancestor genes actually exist. The molecules will be investigated by molecular biological analyses and, at the same time, the bioinformatic approach can be applied to disclose: 1) the manner of molecular evolution of the proteins of the superfamily, 2) the manner of transition of the regulatory regions and regulatory molecules in phylogenesis, and 3) the overall phylogenetic relationship of the superfamily genes.

The functions of each protein encoded by the superfamily genes will be investigated by: 1) Bioinformatical estimation of the protein structure, 2) down-, over- and/or aberrant expression *in vivo* and *in vitro*, 3) gene expression profiling on the cells introduced with the superfamily genes or with siRNA for the genes, 4) a proteomics approach for identification of proteins modified by the superfamily gene products and 5) identification of binding proteins. Animal experimentation with genetic modification by transgenesis should prove fruitful. The study of the superfamily may lead us to;

1. Understanding the molecular mechanism of carcinogenesis:

   The majority of cancer originates from the epithelium. Since carcinogenesis is related to dysregulation of epithelial proliferation and differentiation, defining a normal regulatory pathway is a clue for understanding abnormal events. Each superfamily gene showed a characteristic expression pattern in normal and malignant epithelial cells and may be involved in critical processes of carcinogenesis.
2. Development of new therapeutics for cancers and other diseases:

   The diseases due to epithelial dysfunction include not only cancers but also other afflictions such as inflammatory bowel diseases. The molecules involved in the regulatory pathways for epithelial differentiation could be a direct target of pharmaceutical innovation. It is likely that the superfamily genes will be the direct target of the innovation or may lead us to crucial regulatory pathways for it, as they may be involved in essential steps of epithelial differentiation.
3. Artificial induction of differentiation of pluripotent stem cells:

   Because expression of the superfamily genes is regulated by molecular regulatory cascades for epithelial differentiation in organogenesis, information about regulatory cascades will lead to the development of new strategies for differentiation control of epithelial cells. The knowledge may enable us to produce many types of artificial epithelial tissues through manipulating pluripotent cells such as ES cells and iPS cells. Such information also has a potential to be applied to therapy for cancers, in which induction of cancer stem cells to differentiated cells is attempted.

Acknowledgements

This work was partly supported by Grant-in-Aid for Scientific Research (C) of Japan Society for the Promotion of Science.

References

[1] Saeki N, Kuwahara Y, Sasaki H, Satoh H, Shiroishi T. Gasdermin (*Gsdm*) localizing to mouse Chromosome 11 is predominantly expressed in upper gastrointestinal tract but significantly suppressed in human gastric cancer cells. *Mamm Genome,* 2000; 11: 718–24.

[2] Saeki N, Kim DH, Usui T, Aoyagi K, Tatsuta T, Aoki K, et al. *GASDERMIN*, suppressed frequently in gastric cancer, is a target of LMO1 in TGF-beta-dependent apoptotic signalling. *Oncogene,* 2007; 26: 6488–98.

[3] Saeki N, Usui T, Aoyagi K, Kim DH, Sato M, Mabuchi T, et al. Distinctive expression and function of four *GSDM* family genes (*GSDMA-D*) in normal and malignant upper gastrointestinal epithelium. *Genes Chromosomes Cancer,* 2009; 48: 261–71.

[4] Tamura M, Tanaka S, Fujii T, Aoki A, Komiyama H, Ezawa K, et al. Members of a novel gene family, *Gsdm*, are expressed exclusively in the epithelium of the skin and gastrointestinal tract in a highly tissue-specific manner. *Genomics*, 2007; 89: 618–29.

[5] Marchler-Bauer A, Anderson JB, Chitsaz F, Derbyshire MK, DeWeese-Scott C, Fong JH, et al. CDD: specific functional annotation with the Conserved Domain Database. *Nucleic Acids Res.* 2009; 37: D205–10.

[6] Yu Y, Zhang C, Zhou G, Wu S, Qu X, Wei H, et al. Gene expression profiling in human fetal liver and identification of tissue- and developmental-stage-specific genes through compiled expression profiles and efficient cloning of full-length cDNAs. *Genome Res.* 2001; 11: 1392–403.

[7] Carl-McGrath S, Schneider-Stock R, Ebert M, Röcken C. Differential expression and localisation of gasdermin-like (*GSDML*), a novel member of the cancer-associated GSDMDC protein family, in neoplastic and non-neoplastic gastric, hepatic, and colon tissues. *Pathology,* 2008; 40: 13–24.

[8] Watabe K, Ito A, Asada H, Endo Y, Kobayashi T, Nakamoto K, et al. Structure, expression and chromosome mapping of *MLZE*, a novel gene which is preferentially expressed in metastatic melanoma cells. *Jpn. J. Cancer Res.* 2001; 92; 140–51.

[9] Katoh M, Katoh M. Identification and characterization of human *DFNA5L,* mouse *Dfna5l*, and rat *Dfna5l* genes in silico. *Int. J. Oncol.* 2004; 25: 765–70.

[10] Van Laer L. Huizing EH, Verstreken M, van Zuijlen D, Wauters JG, Bossuyt PJ, et al. Nonsyndromic hearing impairment is associated with a mutation in *DFNA5*. *Nat. Genet.* 1998; 20: 194–7.

[11] Delmaghani S. del Castillo FJ, Michel V, Leibovici M, Aghaie A, Ron U, et al. Mutations in the gene encoding pejvakin, a newly identified protein of the afferent auditory pathway, cause DFNB59 auditory neuropathy. *Nat. Genet.* 2006; 38: 770–8.

[12] Sato H, Koide T, Masuya H, Wakana S, Sagai T, Umezawa A, et al. A new mutation *Rim3* resembling *Re*(*den*) is mapped close to retinoic acid receptor alpha (*Rara*) gene on mouse chromosome 11. *Mamm. Genome,* 1998; 9: 20–5.

[13] Tanaka S, Tamura M, Aoki A, Fujii T, Komiyama H, Sagai T, et al. A new *Gsdma3* mutation affecting anagen phase of first hair cycle. *Biochem. Biophys. Res. Commun.* 2007; 359: 902–7.

[14] Fuchs E, tumber T, Guasch G. Socializing with the neighbors: stem cells and their niche. *Cell*, 2004; 116: 769–78.

[15] Vrijens K, Van Camp G, Van Laer L Characterization of the murine *Dfna5* promoter and regulatory regions. *Gene,* 2009; 432: 82–90.

[16] Sun Q, Yang J, Xing G, Sun Q, Zhang L, He F. Expression of GSDML associates tumor progression in uterine cervix cancer. *Transl. Oncol.* 2008; 1: 73–83.

[17] Kim MS, Lebron C, Nagpal JK, Chae YK, Chang X, Huang Y, et al. Methylation of the *DFNA5* increases risk of lymph node metastasis in human breast cancer. *Biochem. Biophys. Res. Commun.* 2008; 370: 38–43.

[18] Akino K, Toyota M, Suzuki H, Imai T, Maruyama R, Kusano M, et al. Identification of *DFNA5* as a target of epigenetic inactivation in gastric cancer. *Cancer Sci.* 2007; 98: 88–95.

[19] Kim MS, Chang X, Yamashita K, Nagpal JK, Baek JH, Wu G, et al. Aberrant promoter methylation and tumor suppressive activity of the *DFNA5* gene in colorectal carcinoma. *Oncogene,* 2008; 27: 3624–34.

[20] Wu H, Romieu I, Sienra-Monge JJ, Li H, del Rio-Navarro BE, London SJ. Genetic variation in ORM1-like 3 (*ORMDL3*) and gasdermin-like (*GSDML*) and childhood asthma. *Allergy,* 2009; 64: 629–35.

[21] Moffatt MF, Kabesch M, Liang L, Dixon AL, Strachan D, Heath S, et al. Genetic variants regulating *ORMDL3* expression contribute to the risk of childhood asthma. *Nature*, 2007; 448: 470–3.

[22] Sakamoto H, Yoshimura K, Saeki N, Katai H, Shimoda T, Matsuno Y, et al. Genetic variation in *PSCA* is associated with susceptibility to diffuse-type gastric cancer. *Nat. Genet.* 2008; 40: 730–40.

[23] Fidler IJ. The pathogenesis of cancer metastasis: the 'seed and soil' hypothesis revisited. *Nat. Rev. Cancer,* 2003; 3: 453–8.

[24] Van Laer L. Pfister M, Thys S, Vrijens K, Mueller M, Umans L, et al. Mice lacking Dfna5 show a diverging number of cochlear fourth row outer hair cells. *Neurobiol. Dis.* 2005; 19: 386–99.

[25] Fujii T, Tamura M, Tanaka S, Kato Y, Yamamoto H, Mizushina Y, et al. Gasdermin D (*Gsdmd*) is dispensable for mouse intestinal epithelium development. *Genesis,* 2008; 46; 418–23.

[26] Monteiro AS, Ferrier DE. Hox genes are not always Colinear. *Int. J. Biol. Sci.* 2006; 2; 95–103.

[27] Amemiya CT, Prohaska SJ, Hill-Force A, Cook A, Wasserscheid J, Ferrier DE, et al. The amphioxus Hox cluster: characterization, comparative genomics, and evolution. *J. Exp. Zoo B Mol. Dev. Evol.* 2008; 310: 465–277.

[28] Urata M, Sasayama Y, Matada M, Kambegawa A, Suzuki M, Srivastav K. Calcitonin-immunoreactive cells of the digestive tract of the amphioxus are distributed concentrically in a restricted region of the mid-gut. *Acta Zoologica*, 2001; 82: 73–7.

[29] Fox R. invertebrate anatomy online. http://webs.lander.edu/rsfox/invertebrates/ciona.html.

[30] Ogasawara M, Satou Y. Expression of *FoxE* and *FoxQ* genes in the endostyle of *Ciona intestinalis*. *Dev. Genes Evol.* 2003; 213: 416–9.

[31] Ogasawara M, Wada H, Satoh M. Developmental expression of *Pax1/9* genes in urochordate and hemichordate gills: insight into function and evolution of the pharyngeal epithelium. *Development* 1999; 216: 2539–50.

[32] Schierwater B. My favorite animal, *Trichoplax adhaerance.* BioEssays 2005; 27: 1294–302.

In: Endothelium and Epithelium
Editors: J. Carrasco and M. Mota, pp. 213-229
ISBN 978-1-61470-874-2

Chapter X

Tight Junctions in upper Airway Epithelium

***Takashi Kojima*[1,*]*, Tetsuo Himi*[2]*, Hiroyuki Tsutsumi*[3] *and Norimasa Sawada*[1]**

[1]Department of Pathology, Sapporo Medical University School of Medicine, Sapporo 060-8556, Japan

[2]Department of Otolaryngology, Sapporo Medical University School of Medicine, Sapporo 060-8556, Japan

[3]Department of Pediatrics, Sapporo Medical University School of Medicine, Sapporo 060-8556, Japan

Abstract

The mucosal barrier of the upper respiratory tract including the nasal cavity, which is the first site of exposure to inhaled antigens, plays an important role in host defense in terms of innate immunity and is regulated in large part by tight junctions of epithelial cells. Tight junction molecules are expressed in both M cells (membranous or microfold cells) and dendritic cells (DCs) as well as epithelial cells of upper airway. Various antigens may be sampled, transported, and released to lymphocytes through the cells while they maintain the integrity of the airway epithelial barrier. On the other hand, epithelial-derived thymic stromal lymphopoietin (TSLP) triggers DC-mediated Th2-type inflammatory responses and is a master switch for allergic inflammatory diseases including allergic rhinitis. TSLP enhances the tight junctional barrier function of human nasal epithelial cells in vitro. Furthermore, we established a model of respiratory syncytial virus (RSV)-infection using human nasal epithelial cells in vitro, showing that tight junction molecules were markedly increased together with proinflammatory cytokine release. This review summarizes the recent progress in our understanding of the regulation of tight junctions in the upper airway epithelium under normal, allergic and RSV-infected condition.

[*]Tel: +81-11-611-2111 (Ext. 2701) , Fax: +81-11-613-5665, E-mail: ktakashi@sapmed.ac.jp

Keywords: Epithelial cells, tight junction, M cells, dendritic cells, TLR, TSLP, RSV

The epithelium in the upper respiratory consists of pseudostratified ciliated columnar epithelial cells, including M cells (membranous or microfold cells), which are specialized for antigen uptake, and form a continuous barrier against a wide variety of exogenous antigens [1-5], and dendritic cells (DCs), which take up transported antigens via M cells and present antigens for CD4+T cells, while they maintain the integrity of the airway epithelial barrier [6-8]. The epithelium plays a crucial role as an interface of adaptive responses and innate responses via tight junctions to prevent invasion of inhaled environmental agents such as allergens and pathogens (Figure1).

Tight Junctions in Epithelium

The airway epithelium of the human upper respiratory mucosa acts as the first physical barrier that protects against inhaled substances and pathogens [9,10]. The epithelium is a highly regulated and impermeable barrier exclusively formed by tight junctions [9,10].

Tight junctions, the most apically located of the intercellular junctional complexes, inhibit solute and water flow through the paracellular space (termed the "barrier" function (Figure 2.A)[11,12]. They also separate the apical from the basolateral cell surface domains to establish cell polarity (termed the "fence" function)(Figure 2.A)[13,14]. Recent evidence suggests that tight junctions also participate in signal transduction mechanisms that regulate epithelial cell proliferation, gene expression, differentiation and morphogenesis (Figure 2.A)[15].

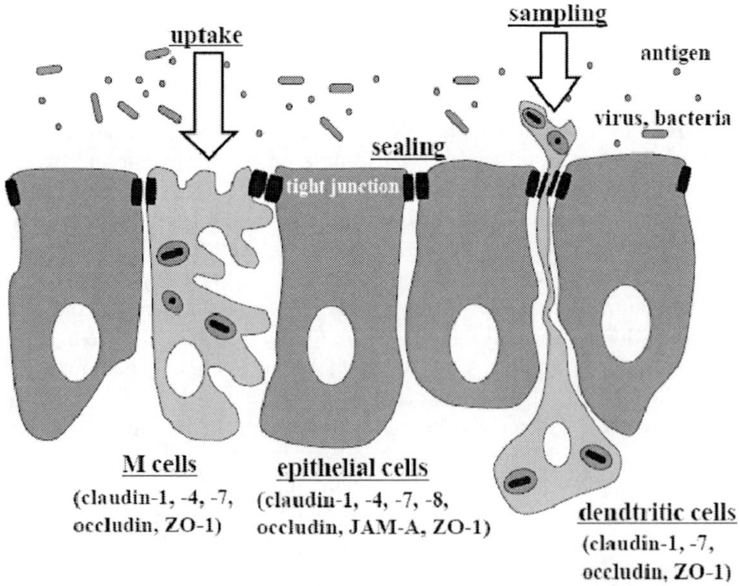

Figure 1. Schema of putative sealing intercellular spaces by tight junction molecules in the upper airway epithelium.

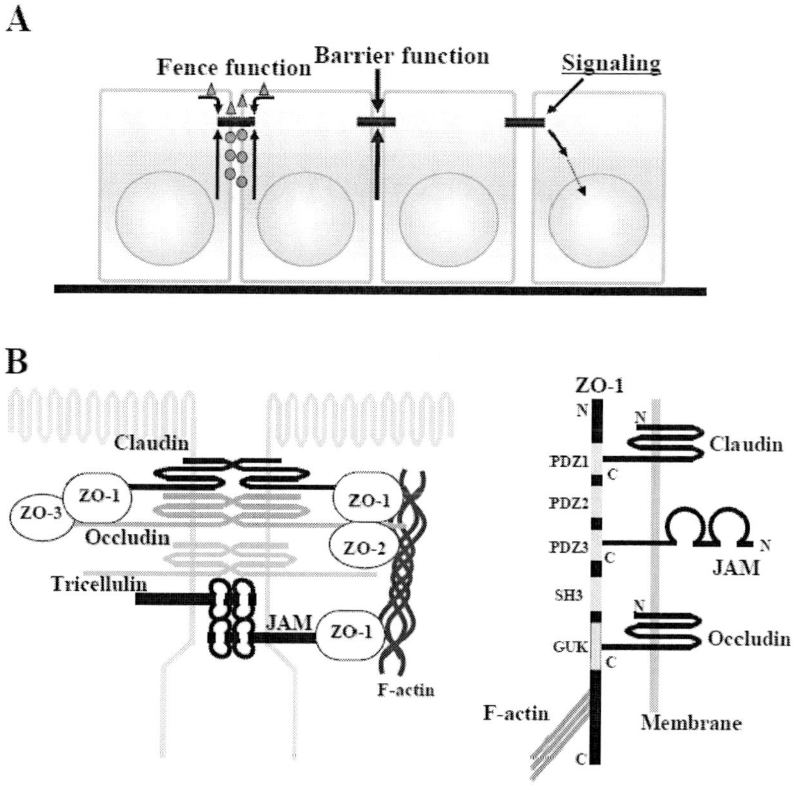

Figure 2. Function and molecular components of tight junctions. A: Three major tight junctional functions (termed ''fence'', ''barrier'', ''signaling''). B: The major molecular components and the organization of tight junction proteins. C: The binding of proteins claudin, occludin and JAM to ZO-1.

Tight junctions are formed by not only the integral membrane proteins claudins, occludin, and JAMs, but also many peripheral membrane proteins, including the scaffold PDZ-expression proteins Zonula occludens (ZO)-1, ZO-2, ZO-3, multi-PDZ domain protein-1 (MUPP1) and membrane-associated guanylate kinase with inverted orientation-1 (MAGI)-1, MAGI-2, MAGI-3, and cell polarity molecules ASIP/PAR-3, PAR-6, PALS-1 and PALS-1 associated tight junction (PATJ) and the non-PDZ-expressing proteins, Cingulin, Symplekin, ZONAB, GEF-H1, aPKC, PP2A, Rab3b, Rab13, PTEN and 7H6 (Figure 2.B)[16-18]. Zonula occludens-1 (ZO-1), ZO-2 and ZO-3 are members of the membrane associated guanylate kinase (MAGUK) family of proteins displaying a characteristic multidomain structure comprised of SH3, guanylate kinase-like (GUK) and multiple PDZ (PSD95-Dlg-ZO1) domains [19]. ZO-1 and ZO-2 are also closely associated with polymerization of claudins [20]. More recently, tricellulin was identified at tricellular contacts where there are three epithelial cells and shown to have a barrier function [21].

The claudin family, consisting of over 24 members, is solely responsible for forming tight junction strands and shows tissue- and cell-specific expression of individual members [17]. Several lines of evidence point to claudins as the basis for the selective size, charge, and conductance properties of the paracellular pathway [22]. The claudins have two extracellular loop domains and the first extracellular loop influences the paracellular charge selectivity and the second extracellular loop is the receptor for a bacterial toxin [22].

Occludin, the first-discovered integral membrane protein of tight junctions, is most ubiquitously expressed at the apicalmost basolateral membranes, and is the most reliable immunohistochemical marker for tight junctions [17, 23]. Overexpression of occludin increases the barrier function, indicated as an increase in transepithelial electric resistance (TER) increase in mammalian epithelial cells [24, 25].

However, TJ strands can be formed without occludin in some cell types, including occludin-deficient embryonic stem cells [26, 27]. Moreover, an occludin-deficient mouse model does not display a perturbation of epithelial barrier function, although a complex pathophysiological phenotype is observed with growth retardation, chronic inflammation and hyperplasia of the gastric epithelium, calcification in the brain, testicular atrophy, loss of cytoplasmic granules in striated duct cells of the salivary gland, and thinning of the compact bone [29].

JAMs (JAM-A, -B, -C, -4) are immunoglobulin superfamily proteins expressed at cell junctions in epithelial and endothelial cells as well as on the surfaces of leukocytes, platelets, and erythrocytes [30]. They are important for a variety of cellular processes, including tight junction assembly, leukocyte transmigration, platelet activation, angiogenesis and adenovirus binding.

The integral proteins, claudin, occludin and JAM bind to the domains of scaffold proteins ZO-1, PDZ1, GUK, and PDZ3, respectively (Figure 2.C)[16].

Expression and Localization of Tight Junction Molecules and Its Function in the Epithelium of the upper Respiratory *In Vivo* and *In Vitro*

We investigated the expression and localization of tight junction molecules and its function in the epithelium of the upper respiratory in vivo and in vitro (Figure 3, 4) [31-37].

In, human nasal mucosa in vivo in which was observed many cilia on the surface, mRNAs of occludin, JAM-A, ZO-1, ZO-2, claudin-1, -4, -7, -8, -12, -13, -14 and tricellulin are detected [31, 34]. Occludin, JAM-A and ZO-1 were found in the uppermost layer. Claudin-1 was observed in the uppermost and basal layers in the epithelium. Claudin-4 and -7 were observed throughout the epithelium. In freeze-fracture replicas, continuous lines of tight junction strands formed well-developed networks on the subapical membranes.

It is known that the proliferation and storage of epithelial cells in primary cultures are very limited. We introduced the catalytic component of telomerase, the hTERT gene, into primary cultured human nasal epithelial cells [32]. The ectopic expression of hTERT in the epithelial cells resulted in greater growth potential and a longer lifespan of the cells. The cells had a small cobblestone appearance in phase-contrast images. The cilia-like structures, a differentiation marker of nasal epithelial cells, were observed on the surface of hTERT-transfected cells (Figure 1). The properties of the passaged hTERT-transfected cells were similar to those of the cells in primary cultures [32].

We investigated tight junctions in hTERT-HNECs compared to primary cultured cells [32, 33]. In both hTERT-HNECs and primary cultured cells, mRNAs of claudin-1, -2, -4, -5, -6, -7, -8, -9, and -12, occludin, and JAM-A was detected. Claudin-1, -4, occludin and JAM-A was observed at cell borders cells. The continuous lines of tight junction strands formed well-developed networks on the subapical membranes. The hTERT-transfected human nasal epithelial cells (hTERT-HNECs) can be used as an indispensable and stable model for studying regulation of tight junctions in human nasal epithelium.

To evaluate the barrier function of tight junctions, the nasal mucosa in vivo was incubated with rhodamine-dextran (molecular weight, 1,000 Da) dissolved in PBS for 15 min on ice. The dye leaked from the surface in the epithelium but it did not leak from the surface regions in the nasal epithelium [31]. In HNECs in vitro, the maximum values of transepithelial resistance (TER), which indicates barrier function of tight junctions, were 80 ± 20 Ωcm^2 [32, 33].

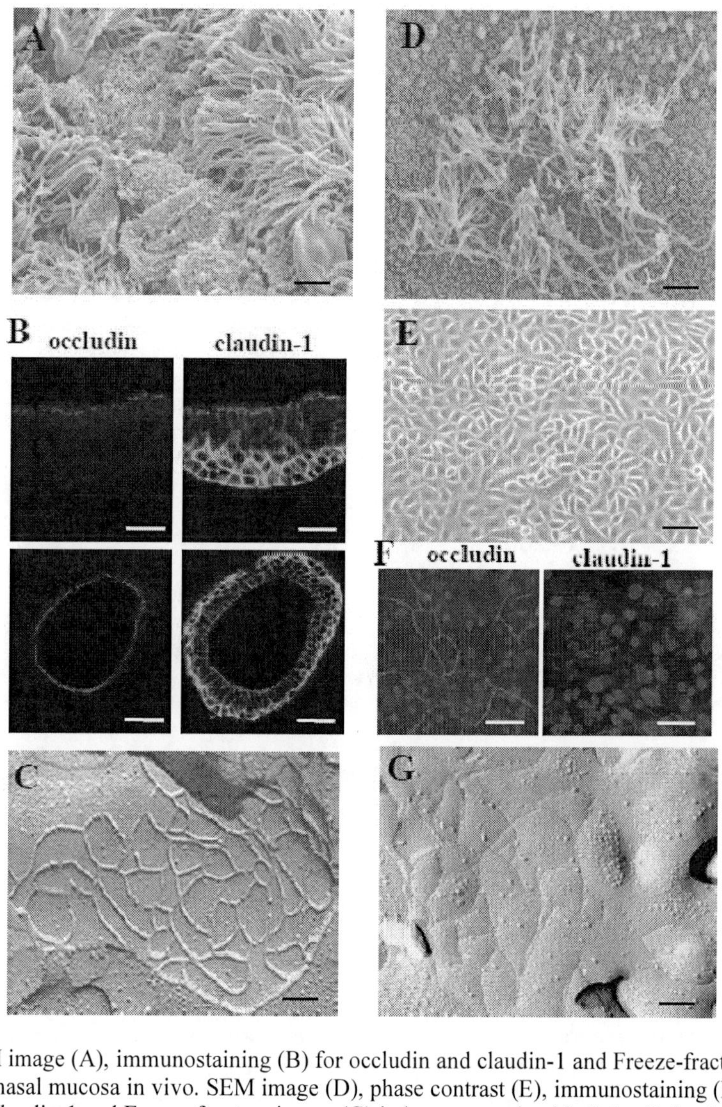

Figure 3. SEM image (A), immunostaining (B) for occludin and claudin-1 and Freeze-fracture image (C) in human nasal mucosa in vivo. SEM image (D), phase contrast (E), immunostaining (F) for occludin and claudin-1 and Freeze-fracture image (G) in human nasal epithelial cells in vitro (hTERT-HNECs). Scale bars: A=20 μm,

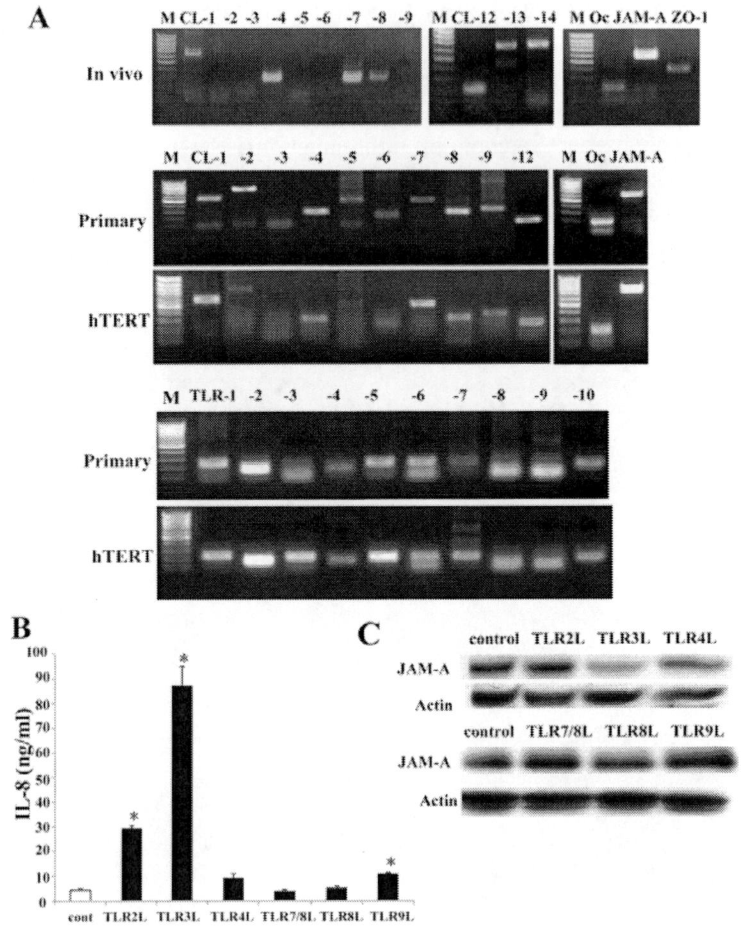

Figure 4. RT-PCR (A) for mRNAs of tight junction molecules in human nasal mucosa in vivo and two types of human nasal epithelial cells in vitro (primary and hTERT-HNECs). M, 100-bp ladder DNA marker; Oc, occludin; CL, claudin. ELISA (B) for IL-8 production and Western blotting (C) for JAM-A in human nasal epithelial cells in vitro (hTERT-HNECs) after treatment with TLR ligands for 24 h.

Tight Junctions of M Cells in the Epithelium of the upper Respiratory

M-cells (membranous or microfold cells) are known as specialized epithelial cells of the follicle-associated epithelium (FAE) and the role seems to be the rapid uptake of particular antigens and microorganisms to the immune cells of the lymphoid follicle to induce an effective immune response [38].

The structure and functional peculiarities of M-cells are observed in different species and at the different sites of lymphoid tissue along the digestive and respiratory tracts [39]. The respiratory M cells act as a nasopharynx-associated lymphoid tissue in the upper respiratory tract [3]. The adenoidal epithelium, including M cells, which are specialized for antigen uptake, forms a continuous barrier against a wide variety of exogenous antigens [40]. Several markers such as lectin histochemistry and immunoreactivity to vimentin, cytokeratins and annexin-V have been proposed to identify M cells in the mouse, rat, hamster, rabbit and pig [38, 39, 41-43]. In the nasopharyngeal tonsil of the horse, lectin GS-1 B4 (Griffonia simplicifolia 1 isolectin B4) has been used as a marker for M cells [44]. It was reported that cytokeratin20 (Ck20) could serve as an M cell marker for rabbit palatine tonsils [45]. Furthermore, claudin-1, -3 and ZO-1 are detected in M cells of mouse intestinal follicle-associated epithelium (FAE)[46]. In humans, clusterin is expressed in M cells and follicular dendritic cells at inductive sites of human mucosa-associated lymphoid tissue [47]. Class II beta-tubulin is a specific histochemical marker for human tonsillar M cells and follicular dendritic cells [48]. However, universal markers for human M cells have not yet been established. We identified M-like cells using an anti-Ck20 antibody in human adenoidal tissues in vivo and in vitro and investigated the expression of tight junctions (Figure 5)[5]. In human adenoidal epithelium in vivo, some M-like cells, characterized by irregular microvilli, were observed on the surface. Some Ck20-positive cells were randomly observed in the epithelium and appeared as pocket-like structures. In both Ck20-positive and –negative cells of the adenoidal epithelium in vivo, occludin, ZO-1, claudin-1 and -7 were observed. In the primary cultures, Ck20-positive cells took up fluorescent microparticles in vitro. In both Ck20-positive and –negative cells of primary cultures in vitro, occludin, ZO-1, claudin-1 and -7 were clearly observed at cell borders. These findings indicated that M cells in the adenoid could take up antigens while maintaining a continuous barrier.

Tight Junctions of Dendritic Cells in the Epithelium of the upper Respiratory

This function requires DCs, professional antigen-presenting cells that act as peripheral sentinels specializing in the uptake, processing and presentation of antigenic material. Rescigno et al. discovered a new mechanism for pathogen uptake in the mucosa by which DCs open the tight junctions between epithelial cells and send dendrites outside the epithelium to directly sample the pathogen. DCs express tight-junction proteins such as occludin, claudin-1 and ZO-1 to preserve the integrity of the epithelial barrier [49]. The epithelial DC population expresses high levels of the Langerhans cell (LC) marker langerin and the tight junction proteins claudin-1, -7, and ZO-2 [50]. Claudin-1 is detected in murine $CD207^+$ LCs residing in the epidermis but not in other skin DCs [51]. In human THP-1 monocytes, mRNAs of occludin, tricellulin, JAM-A, ZO-1, ZO-2 and claudin -4, -7, -8, and -9 can be detected. In mature DCs that have dendrites elongated by treatment with IL-4, GM-CSF, TNF-α, and ionomycin, mRNA and protein of JAM-A are significantly increased compared to monocytes [52]. We previously reported that in mouse XS52 DCs, claudin-1, -3, -4, -6, -7, -8 and occludin are detected and claudin-7 is induced via an NF-κB pathway by thymic stromal lymphopoietin (TSLP) and ligands of Toll-like receptor 2 (TLR2), TLR4 or TLR7/8 [53]. In the human nasal mucosa of allergic rhinitis, HLA-DR- and CD11c-positive DCs express tight junction protein claudin-1 and penetrate beyond the apicalmost tight junction protein occludin to minimize the increase in permeability of the epithelial barrier (Figure 5) [31, 54].

TLR3 Ligand Reduced Tight Junctions in the Epithelium of the upper Respiratory

TLRs are a component of the innate immune system [55, 56]. They enable the host to recognize a large number of pathogen-associated molecular patterns such as those of bacterial lipopolysaccharides (LPS), viral RNA, CpG-containing DNA, and flagellin, among others [57]. TLRs are also expressed in the epithelium of the upper respiratory, and may play a vital role in the immunological outcomes in these tissues, which produce proinflammatory cytokines and chemokines upon ligation [58]. In human nasal epithelium in vivo and in vitro, mRNAs for all 10 known human TLRs are detected in vivo and vitro (Figure 4)[59]. TLR3 recognizes viral double-stranded (dsRNA) and its synthetic analogue polyinosinic-polycytidylic acid [poly(I:C)] and stimulates innate immune responses [60]. In primary cultures of human adenoid epithelial cells that expressed mRNAs of TLR1, 2, 3, 4, 6, 7 and 10, stimulation by the TLR3 ligand poly(I;C), induced production of not only TNFα and IL-8 but also reduced JAM-A expression (Figure 4). The changes were regulated via distinct signaling transduction pathways [37]. The control of TLR3-mediated signaling pathways in human nasal epithelium may be important not only in infection by viral dsRNA but also in autoimmune diseases caused by endogenous dsRNA released from necrotic cells.

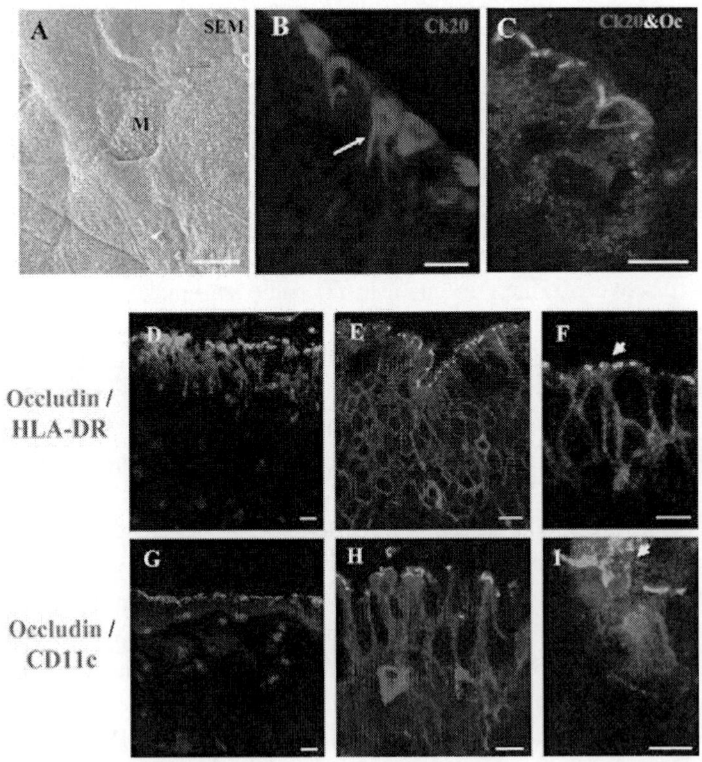

Figure 5. SEM image (A) and immunostaining for Ck20 (B), Ck20 and occludin (C) in human adenoidal epithelium in vivo. Immunostaining for occludin and HLA-DR (D, E, F) and occludin and CD11c (G, H, I) in human nasal epithelium in vivo. M: M-like cell. Scale bars = A: 5 μm; B-E: 20 μm.

TSLP Induced Tight Junctions in the Epithelium of the upper Respiratory

The epithelial-derived factor TSLP is an IL-7-like cytokine that potently induces deregulation of Th2 responses, a hallmark feature of allergic inflammatory diseases such as asthma, atopic dermatitis and allergic rhinitis [35, 61-63]. TSLP-stimulated CD11c^{+} DCs induce naïve CD4^{+} T cells to differentiate into Th2 cells that produce IL-4, IL-5, IL-13 and TNF-α [62]. We found high expression of TSLP in epithelium from patients with allergic rhinitis with recruitment and infiltration of CD11c^{+}DCs (Figure 6)[35]. In vitro, TSLP was significantly produced in HNECs by treatment with a TLR2 ligand Pam$_3$Cys-Ser-(Lys)$_4$ and a mixture of IL-1β and TNF-α. Treatment with TSLP rapidly enhanced the barrier function of cultured HNECs together with an increase of tight junction proteins claudin-1, -4, -7, and occluding (Figure 6). The nasal epithelial-derived TSLP not only activates DCs but also preserves the epithelial barrier via upregulation of tight junction proteins to regulate antigen sensitization during the early stage of allergic rhinitis.

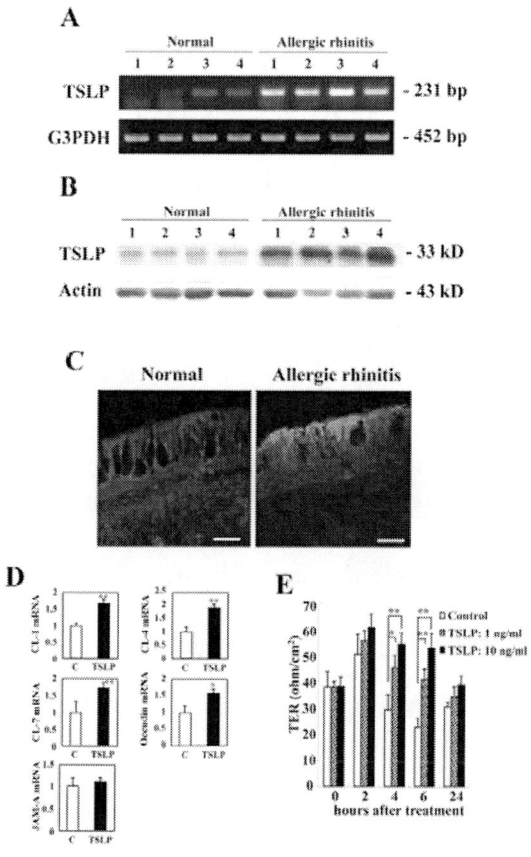

Figure 6. RT-PCR (A), Western blotting (B), and immunohistochemistry (C) for TSLP in nasal mucosa from normal controls and patients with allergic rhinitis. Real-time PCR (D) for tight junction molecules, and barrier function measured as TER (E) in human nasal epithelial cells in vitro (hTERT-HNECs) after treatment with TSLP.

RSV Induced Tight Junctions in the Epithelium of the upper Respiratory

Respiratory syncytial virus (RSV) is the major cause of bronchitis, asthma and severe lower respiratory tract disease in infants and young children [64]. The airway epithelium, which has a well-developed barrier regulated by tight junctions, is the first line of defense during respiratory virus infection. Moreover, it is also known that tight junctions include targets or receptors of viruses and bacteria such as claudin-1 and occludin as coreceptors of HCV, JAM as a reovirus receptor, CAR as a coxsackie and adenovirus receptor, and some claudins as CPE receptors [65]. To investigate the detailed mechanisms of replication and budding of RSV in human nasal epithelial cells (HNECs) and the epithelial cell responses including tight junctions, we established an RSV-infected model using hTERT-transfceted HNECs (Figure 7)[66]. When the HNECs were infected with RSV at MOI 1, the mRNA and protein of RSV/G-protein were markedly increased from 24 h. At 24 h after infection by RSV, the expression of RSV/G- and /F-proteins was detected in most cells by immunocytochemistry using specific antibodies. During RSV infection, it is thought that there is a close physical interaction between F-actin and the virus, which is indicated by both the virus filaments and inclusion bodies [67]. In the RSV-infected HNECs, many virus filaments were observed on the cell surface by SEM and inclusion bodies were also detected at submembranes by TEM (Figure 7). We found that the expression, structure and barrier function of tight junction molecules claudin-4 and occludin were markedly induced together with production of proinflammatory cytokines IL-8 and TNFα in HNECs after RSV infection, and the induction of tight junction molecules possibly contributed to budding of RSV (Figure 7)[66]. However, knockdown of claudin-4 and occludin by siRNAs did not affect of replication of RSV. The replication and budding of RSV and the epithelial cell responses in HNECs were regulated via a PKCδ/HIF-1α/NF-κB pathway [66]. The control of this pathway in HNECs may be useful not only for prevention of replication and budding of RSV but also in therapy for RSV-induced respiratory pathogenesis.

Concluding Remarks

This review summarizes the recent progress in our understanding of the regulation of tight junctions in the upper airway epithelium under normal, allergic and RSV-infected condition. In HNECs in vitro using our culture systems, tight junction molecules and the barrier function are upregulated by various stimuli (Table 1). The epithelial tight junction barrier of the upper airway epithelium, including M cells and DCs, is stably maintained via the regulation of tight junction molecules expressed in the epithelial cells, M cells and DCs, various antigens may be sampled, transported, and released to lymphocytes through the cells while maintaining the integrity of the epithelial barrier (Figure 1). Claudin-4 is known as a rodent M cell-specific gene [68].

The high transcytotic ability of M cells is an attractive target for mucosally delivered vaccines and therapeutics. It is thought that claudin-4-targeted protein incorporated into nanoparticles can mediate M cell targeted delivery [69].

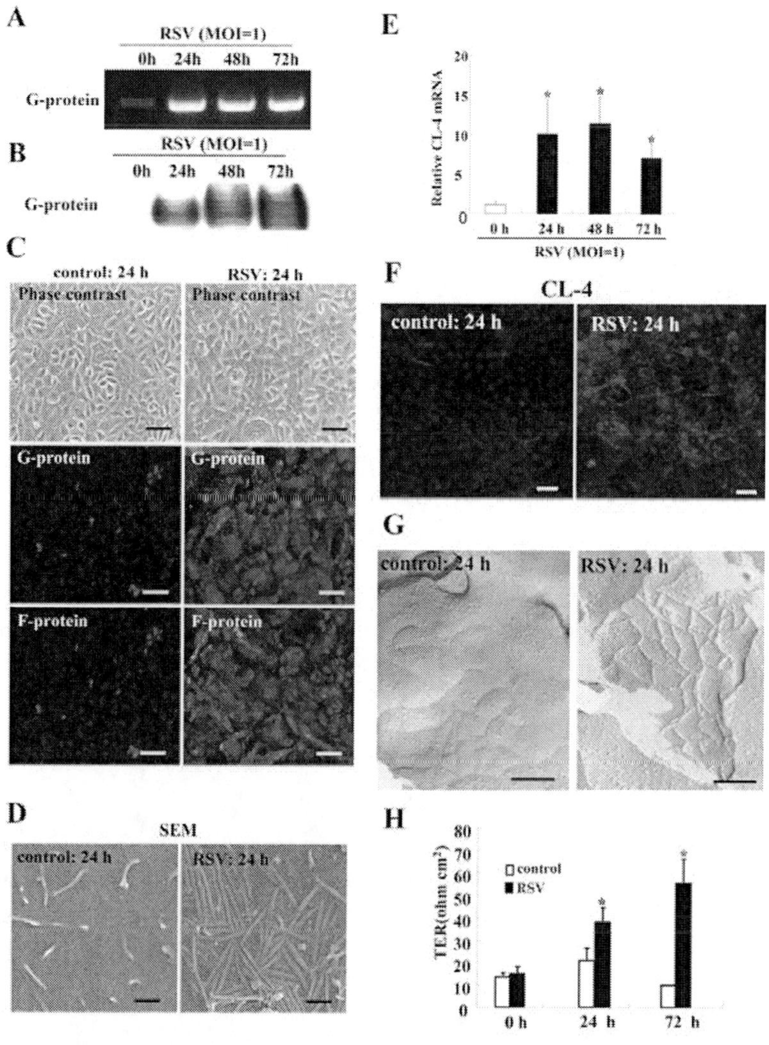

Figure 7. RT-PCR (A) and Western blotting (B) for RSV/G-protein, real-time PCR (E) for claudin-4 and barrier function measured as TER (H) in human nasal epithelial cells in vitro (hTERT-HNECs) from 24 h to 72 h after infection with RSV. Phase contrast (C) and immunostaining (C) for RSV/G- and F-proteins, SEM image (D), immunostaining (F) for claudin-4 and freeze-fracture image (G) in human nasal epithelial cells in vitro (hTERT-HNECs) at 24 h after infection with RSV. C=Black bar: 40 μM, white bar: 20 μM, SEM bars: 400 nm.

Tricellular tight junctions form at the convergence of bicellular tight junctions where three epithelial cells meet in polarized epithelia, and tricellulin was the first marker of the tricellular tight junction identified in epithelial cells [21]. In various DCs, including human THP-1 cells, mouse XS52 cells and epidermal LCs, tricellulin is also detected [52, 53]. The tricellular tight junction may a good penetration point for DCs into the epithelium [70]. Thus, further study of tricellulin may be important to understand the regulation of the penetration of DCs into the epithelium. On the other hand, it is thought that not only TSLP but also epithelial cell-derived cytokines, IL-25, and IL-33 are also critical regulators of innate and adaptive immune responses associated with Th2 cytokine-mediated inflammation [71].

Table 1. Changes of tight junction proteins and barrier function in HNECs in vitro

Treatments		Tight junction proteins	Barrier function
FBS		CL-1↑; CL-4↑	upregulation
growth factor	TGF-β	CL-4↑	no change
PKC activator	TPA	CL-1↑; Oc↑; ZO-1↑; ZO-2↑	upregulation
PPARγ ligands	Rosiglitazone	CL-1↑; CL-4↑; Oc↑; TRIC↑	upregulation
	Troglitazone	CL-1↑; CL-4↑; Oc↑	
TLR3 ligand	Poly I:C	JAM-A↓	no change
cytokine	TSLP	CL-1↑; CL-4↑; CL-7↑; Oc↑	upregulation
virus	RSV	CL-4↑; Oc↑	upregulation

In addition, IL-33/its receptor ST2 pathway is also necessary not only for the development of the allergic inflammatory response but also for its maintenance [72]. We recently found elevated serum IL-33 and high expression of IL-33 and its receptor ST2 in the nasal epithelium of patients with allergic rhinitis. The IL-33/ST2 pathway may play an important role in the development of allergic rhinitis although its effects on epithelial barrier remain unclear. These studies should provide new insights not only into pathological conditions but also in the context of new vaccines and therapeutics against infectious and inflammatory mucosal diseases in upper airway.

Acknowledgments

This work was supported by Grants-in-Aid from the National Project "Knowledge Cluster Initiative" (2nd stage, "Sapporo Biocluster Bio-S"), Program for developing the supporting system for upgrading education and research, the Ministry of Education, Culture, Sports Science, and Technology, and the Ministry of Health, Labour and Welfare of Japan.

References

[1] Holgate ST. The airway epithelium is central to the pathogenesis of asthma. *Allergol. Int*. 2008;57:1-10.

[2] Fujimura Y. Evidence of M cells as portals of entry for antigens in the nasopharyngeal lymphoid tissue of humans. *Virchows Arch.* 2000;436:560-566.

[3] Kim DY, Sato A, Fukuyama S, Sagara H, Nagatake T, Kong IG, Goda K, Nochi T, Kunisawa J, Sato S, Yokota Y, Lee CH, Kiyono H. The airway antigen sampling system: respiratory m cells as an alternative gateway for inhaled antigens. *J. Immunol.* 2011;186:4253-4262.

[4] Nawijn MC, Hackett TL, Postma DS, van Oosterhout AJ, Heijink IH. E-cadherin: gatekeeper of airway mucosa and allergic sensitization. *Trends Immunol.* 2011.

[5] Takano K, Kojima T, Ogasawara N, Go M, Kikuchi S, Ninomiya T, Kurose M, Koizumi J, Kamekura R, Murata M, Tanaka S, Chiba H, Himi T, Sawada N. Expression

of tight junction proteins in epithelium including Ck20-positive M-like cells of human adenoids in vivo and in vitro. *J. Mol. Histol.* 2008;39:265-273.

[6] Steinman RM, Pack M, Inaba K. Dendritic cells in the T-cell areas of lymphoid organs. *Immunol. Rev.* 1997;156:25-37.

[7] Yamanaka T, Helgeland L, Farstad IN, Fukushima H, Midtvedt T, Brandtzaeg P. Microbial colonization drives lymphocyte accumulation and differentiation in the follicle-associated epithelium of Peyer's patches. *J. Immunol.* 2003;170:816-822.

[8] Hammad H, Lambrecht BN. Dendritic cells and airway epithelial cells at the interface between innate and adaptive immune responses. *Allergy*. 2011; 66:579-87.

[9] Holgate ST. Epithelium dysfunction in asthma. *J. Allergy Clin. Immunol.* 2007;120:1233-1246.

[10] Schleimer RP, Kato A, Kern R, Kuperman D, Avila PC. Epithelium: At the interface of innate and adaptive immune responses. *J. Allergy Clin. Immunol.* 2007;120:1279-1284.

[11] Gumbiner BM. Breaking through the tight junction barrier. *J. Cell Biol.* 1993;123:1631-1633.

[12] Schneeberger E.E. and Lynch R.D. Structure, function, and regulation of cellular tight junctions. *Am. J. Physiol.* 1992; 262: L647-L661.

[13] van Meer G., Simon K. The function of tight junctions in maintaining differences in lipid composition between the apical and basolateral cell surface domains of MDCK cells. *EMBO J.* 1986; 5:1455-1464.

[14] Cereijido M, Valdés J, Shoshani L, Contreras RG.. Role of tight junctions in establishing and maintaining cell polarity. *Annu. Rev. Physiol.* 1998: 60: 161-177.

[15] Balda MS, Matter K. Tight junctions and the regulation of gene expression. *Biochim. Biophys. Acta, 2009*;1788:761-767.

[16] Schneeberger EE, Lynch RD. The tight junction: a multifunctional complex. *Am. J. Physiol. Cell Physiol.* 2004;286:1213-1228.

[17] Tsukita S, Furuse M, Itoh M. Multifunctional strands in tight junctions. *Nat. Rev. Mol. Cell Biol.* 2001;2:285-293.

[18] Sawada N, Murata M, Kikuchi K, Osanai M, Tobioka H, Kojima T, Chiba H. Tight junctions and human diseases. *Med. Electron Microsc.* 2003;36:147-156.

[19] Anderson JM. Cell signaling: MAGUK magic. *Curr. Biol.* 1996; 6: 382-384.

[20] Umeda K, Ikenouchi J, Katahira-Tayama S, Furuse K, Sasaki H, Nakayama M, Matsui T, Tsukita S, Furuse M, Tsukita S. ZO-1 and ZO-2 independently determine where claudins are polymerized in tight-junction strand formation. *Cell*, 2006; 126: 741-754.

[21] Ikenouchi J, Furuse M, Furuse K, Sasaki H, Tsukita S. Tricellulin constitutes a novel barrier at tricellular contacts of epithelial cells. *J. Cell Biol.* 2005;171:939-945.

[22] Van Itallie CM, Anderson JM. Claudins and epithelial paracellular transport. *Annu. Rev. Phys*iol. 2006; 68: 403-429.

[23] Furuse M, Hirase T, Itoh M, Nagafuchi A, Yonemura S, Tsukita S, Tsukita S. Occludin: a novel integral membrane protein localizing at tight junctions. *J. Cell Biol.* 1993 ; 123 : 1777-1788.

[24] Balda MS, Whitney JA, Flores C, Gonzalez S, Cereijido M, Matter K. Functional dissociation of paracellular permeability and transepithelial electrical resistance and disruption of the apical-basolateral intramembrane diffusion barrier by expression of a mutant tight junction membrane protein. *J. Cell Biol.* 1996; 134: 1031-1049.

[25] McCarthy KM, Skare IB, Stankewich MC, Furuse M, Tsukita S, Rogers RA, Lynch RD, Schneeberger EE. Occludin is a functional component of the tight junction. *J. Cell Sci.* 1996; 109: 2287-2298.

[26] Hirase T, Staddon J.M, Saitou M, Ando-Akatsuka Y, Itoh M, Furuse M, Fujimoto K, Tsukita S, Rubin LL. Occludin as a possible determinant of tight junction permeability in endothelial cells. *J. Cell Sci.* 1997; 110: 1603-1613.

[27] Saitou M, Fujimoto K, Doi Y, Itoh M, Fujimoto T, Furuse M, Takano H, Noda T, Tsukita S. Occludin-deficient embryonic stem cells can differentiate into polarized epithelial cells bearing tight junctions. *J. Cell Biol.* 1998; 141: 397-408.

[28] Saitou M, Furuse M, Sasaki H, Schulzke JD, Fromm M, Takano H, Noda T, Tsukita S. Complex phenotype of mice lacking occludin, a component of tight junction strands. *Mol. Biol. Cell* 2000; 11: 4131-4142.

[29] Martin-Padura I, Lostaglio S, Schneemann M, Williams L, Romano M, Fruscella P, Panzeri C, Stoppacciaro A, Ruco L, Villa A, Simmons D, Dejana E. Junctional adhesion molecule, a novel member of the immunoglobulin superfamily that distributes at intercellular junctions and modulates monocyte transmigration. *J. Cell Biol.* 1998; 142: 117-127.

[30] Takano K, Kojima T, Go M, Murata M, Ichimiya S, Himi T, Sawada N. HLA-DR- and CD11c-positive dendritic cells penetrate beyond well-developed epithelial tight junctions in human nasal mucosa of allergic rhinitis. *J. Histochem. Cytochem.* 2005;53:611-619.

[31] Kurose M, Kojima T, Koizumi JI, Kamekura R, Ninomiya T, Murata M, Ichimiya S, Osanai M, Chiba H, Himi T, Sawada N. Induction of claudins in passaged hTERT-transfected human nasal epithelial cells with an extended life span. *Cell Tissue Res.* 2007;330:63-74.

[32] Koizumi J, Kojima T, Ogasawara N, Kamekura R, Kurose M, Go M, Harimaya A, Murata M, Osanai M, Chiba H, Himi T, Sawada N. Protein kinase C enhances tight junction barrier function of human nasal epithelial cells in primary culture by transcriptional regulation. *Mol. Pharmacol.* 2008;74:432-442.

[33] Ohkuni T, Kojima T, Ogasawara N, Masaki T, Ninomiya T, Kikuchi S, Go M, Takano K, Himi T, Sawada N. Expression and localization of tricellulin in human nasal epithelial cells in vivo and in vitro. *Med. Mol. Morphol.* 2009;42:204-211.

[34] Kamekura R, Kojima T, Koizumi J, Ogasawara N, Kurose M, Go M, Harimaya A, Murata M, Tanaka S, Chiba H, Himi T, Sawada N. Thymic stromal lymphopoietin enhances tight-junction barrier function of human nasal epithelial cells. *Cell Tissue Res.* 2009;338:283-293.

[35] Ogasawara N, Kojima T, Go M, Ohkuni T, Koizumi JI, Kamekura R, Masaki T, Murata M, Tanaka S, Fuchimoto J, Himi T, Sawada N. PPARgamma agonists upregulate the barrier function of tight junctions via a PKC pathway in human nasal epithelial cells. *Pharmacol. Res.* 2010;61:489-498.

[36] Ohkuni T, Kojima T, Ogasawara N, Masaki T, Fuchimoto J, Kamekura R, Koizumi JI, Ichimiya S, Murata M, Tanaka S, Himi T, Sawada N. Poly(I:C) reduces expression of JAM-A and induces secretion of IL-8 and TNF-α via distinct NF-κB pathways in human nasal epithelial cells.*Toxicol. Appl. Pharmacol.* 2011; 250:29–38.

[37] Mach J, Hshieh T, Hsieh D, Grubbs N, Chervonsky A. Development of intestinal M cells. *Immunol. Rev.* 2005;206:177-189.

[38] Gebert A. The role of M cells in the protection of mucosal membranes. *Histochem. Cell Biol.* 1997;108:455-470.

[39] Van Kempen MJ, Rijkers GT, Van Cauwenberge PB. The immune response in adenoids and tonsils. *Int. Arch. Allergy Immunol.* 2000;122:8-19.

[40] Giannasca PJ, Boden JA, Monath TP. Targeted delivery of antigen to hamster nasal lymphoid tissue with M-cell-directed lectins. *Infect. Immun.* 1997;65:4288-4298.

[41] Kucharzik T, Lugering N, Schmid KW, Schmidt MA, Stoll R, Domschke W. Human intestinal M cells exhibit enterocyte-like intermediate filaments. *Gut,* 1998;42:54-62.

[42] Verbrugghe P, Waelput W, Dieriks B, Waeytens A, Vandesompele J, Cuvelier CA. Murine M cells express annexin V specifically. *J. Pathol.* 2006;209:240-249.

[43] Kumar P, Timoney JF. Light and electron microscope studies on the nasopharynx and nasopharyngeal tonsil of the horse. *Anat. Histol. Embryol.* 2001;30:77-84.

[44] Carapelli A, Regoli M, Nicoletti C, Ermini L, Fonzi L, Bertelli E. Rabbit tonsil-associated M-cells express cytokeratin 20 and take up particulate antigen. *J. Histochem. Cytochem.* 2004;52:1323-1332.

[45] Clark MA, Hirst BH. Expression of junction-associated proteins differentiates mouse intestinal M cells from enterocytes. *Histochem. Cell Biol.* 2002;118:137-147.

[46] Verbrugghe P, Kujala P, Waelput W, Peters PJ, Cuvelier CA. Clusterin in human gut-associated lymphoid tissue, tonsils, and adenoids: localization to M cells and follicular dendritic cells. *Histochem. Cell Biol.* 2008;129:311-320.

[47] Lee JH, Kong SK, Wu ZS, Wu Q, Choe J, Cho HY, Ha SY, Kee SH, Kim YS. Class II beta-tubulin is a novel marker for human tonsillar M cells and follicular dendritic cells. *J. Oral. Pathol. Med.* 2010; 39:533-539.

[48] Rescigno M, Urbano M, Valzasina B, Francolini M, Rotta G, Bonasio R, Granucci F, Kraehenbuhl JP, Ricciardi-Castagnoli P. Dendritic cells express tight junction proteins and penetrate gut epithelial monolayers to sample bacteria. *Nat. Immunol.* 2001;2:361-367.

[49] Sung SS, Fu SM, Rose Jr CE, Gaskin F, Ju ST, Beaty SR. A major lung CD103 (αE)-β7 integrin-positive epithelial dendritic cell population expressing Langerin and tight junction proteins. *J. Immunol.* 2006;176:2161-2172.

[50] Zimmerli SC, Hauser C. Langerhans cells and lymph node dendritic cells express the tight junction component claudin-1. *J. Invest. Dermatol.* 2007;127:2381-2390.

[51] Ogasawara N, Kojima T, Go M, Fuchimoto J, Kamekura R, Koizumi J, Ohkuni T, Masaki T, Murata M, Tanaka S, Ichimiya S, Himi T, Sawada N. Induction of JAM-A during differentiation of human THP-1 dendritic cells. *Biochem. Biophys. Res. Commun.* 2009;389:543-549.

[52] Kamekura R, Kojima T, Takashima A, Koizumi J, Ogasawara N, Go M, Takano K, Murata M, Tanaka S, Ichimiya S, Himi T, Sawada N. Thymic stromal lymphopoietin induces tight junction protein claudin-7 via NF-kappaB in dendritic cells. *Histochem. Cell Biol.* 2010;133:339-348.

[53] Ogasawara N, Kojima T, Go M, Takano KI, Kamekura R, Ohkuni T, Koizumi JI, Masaki T, Fuchimoto J, Obata K, Kurose M, Shintani T, Sawada N, Himi T. Epithelial barrier and antigen uptake in lymphoepithelium of human adenoids. *Acta Otolaryngol.* 2011; 131: 116-123.

[54] Iwasaki A, Medzhitov R. Toll-like receptor control of the adaptive immune responses. *Nat. Immunol.* 2004;5:987-995.
[55] Takeda K, Akira S. Toll-like receptors in innate immunity. Int Immunol 2005;17:1-14.
[56] Kawai T, Akira S. TLR signaling. *Cell Death Differ*, 2006;13:816-825.
[57] Lange MJ, Lasiter JC, Misfeldt ML. Toll-like receptors in tonsillar epithelial cells. *Int. J. Pediatr. Otorhinolaryngol.* 2009;73:613-621.
[58] Lesmeister MJ, Bothwell MR, Misfeldt ML. Toll-like receptor expression in the human nasopharyngeal tonsil (adenoid) and palantine tonsils: a preliminary report. *Int. J. Pediatr. Otorhinolaryngol.* 2006;70:987-992.
[59] Alexopoulou L, Holt AC, Medzhitov R, Flavell RA. Recognition of double-stranded RNA and activation of NF-kappaB by Toll-like receptor 3. *Nature*, 2001;413:732-738.
[60] Holgate ST. Epithelium dysfunction in asthma. *J. Allergy Clin. Immunol.* 2007;120:1233-1246.
[61] Soumelis V, Reche PA, Kanzler H, Yuan W, Edward G, Homey B, Gilliet M, Ho S, Antonenko S, Lauerma A, Smith K, Gorman D, Zurawski S, Abrams J, Menon S, McClanahan T, de Waal-Malefyt Rd R, Bazan F, Kastelein RA, Liu YJ. Human epithelial cells trigger dendritic cell-mediated allergic inflammation by producing TSLP. *Nature Immunol.* 2002;3:673-680.
[62] Ying S, O'Connor B, Ratoff J, Meng Q, Mallett K, Cousins D, Robinson D, Zhang G, Zhao J, Lee TH, Corrigan C. Thymic stromal lymphopoietin expression is increased in asthmatic airways and correlates with expression of Th2-attracting chemokines and disease severity. *J. Immunol.* 2005;174:8183-8190.
[63] Bitko V, Barik S. Persistent activation of RelA by respiratory syncytial virus involves protein kinase C, underphosphorylated IkBb, and sequestration of protein phosphatase 2A by the viral phosphoprotein. *J. Virol.* 1998; 72: 5610-5618.
[64] Guttman JA, Finlay BB. Tight junctions as targets of infectious agents. *Biochim. Biophys. Acta.* 2009;1788:832-841.
[65] Masaki T, Kojima T, Okabayashi T, Ogasawara N, Ohkuni T, Obata K, Takasawa A, Murata M, Tanaka S, Hirakawa S, Fuchimoto J, Ninomiya T, Fujii N, Tsutsumi H, Himi T, Sawada N. An NF-κB signaling pathway via PKCδ regulates replication of respiratory syncytial virus in polarized normal human nasal epithelial cells. *Mol. Biol. Cell,* 2011 (in press).
[66] Jeffree CE, Brown G, Aitken J, Su-Yin DY, Tan BH, Sugrue RJ. Ultrastructural analysis of the interaction between F-actin and respiratory syncytial virus during virus assembly. *Virology.* 2007;369:309-323.
[67] Wang J, Lopez-Fraga M, Rynko A, Lo DD. TNFR and LTbetaR agonists induce follicle-associated epithelium and M cell specific genes in rat and human intestinal epithelial cells. *Cytokine,* 2009;47:69-76.
[68] Rajapaksa TE, Stover-Hamer M, Fernandez X, Eckelhoefer HA, Lo DD. Claudin 4-targeted protein incorporated into PLGA nanoparticles can mediate M cell targeted delivery. *J. Control Release,* 2010;142:196-205.
[69] Kubo A, Nagao K, Yokouchi M, Sasaki H, Amagai M. External antigen uptake by Langerhans cells with reorganization of epidermal tight junction barriers. *J. Exp. Med.* 2009;206:2937-2946.

[70] Saenz SA, Taylor BC, Artis D. Welcome to the neighborhood: epithelial cell-derived cytokines license innate and adaptive immune responses at mucosal sites. *Immunol. Rev.* 2008;226:172-190.

[71] Kearley J, Buckland KF, Mathie SA, Lloyd CM. Resolution of allergic inflammation and airway hyperreactivity is dependent upon disruption of the T1/ST2-IL-33 pathway. *Am. J. Respir. Crit. Care Med.* 2009;179:772-781.

In: Endothelium and Epithelium
Editors: J. Carrasco and M. Mota, pp. 231-245
ISBN 978-1-61470-874-2

Chapter XI

β-Catenin Mediates FGF-7-Induced Ocular Surface Squamous Neoplasia in the Transgenic Mice

Chia-Yang Liu**, Yujin Zhang, Mindy Call and Winston W.-Y. Kao

Crawley Vision Research Center/Department of Ophthalmology, College of Medicine, University of Cincinnati, Cincinnati Ohio, 45267-0838, U.S.A

Abstract

We have recently reported that excess fibroblast growth factor-7 (FGF-7) in *Krt12*$^{rtTA/rtTA}$*/tetO-FGF-7* double transgenic mice induced by doxcycline (Dox) resulted in a corneal epithelial tumor resembling human ocular surface squamous neoplasia (OSSN) (Chikama et al. Am. J. Pathol. 172:638-649).

Here, we further investigate the role of β-catenin in mediating FGF-7 induced corneal tumorigenesis. Pannus surgically removed from OSSN patients all had nuclear β-catenin and exhibited excess FGF-7. *In vitro*, FGF-7 induces stabilization and nuclear translocation of β-catenin in human corneal epithelial cells.

In vivo, conditional knock-out of β-catenin gene (*Ctnnb1*) did not have significant effects on corneal morphogenesis or homeostasis, but abolished OSSN formation in Dox-treated *Krt12*$^{rtTA/rtTA}$*/tetO-FGF-7/tetO-Cre/Ctnnb1*$^{lox(E2-6)/lox(E2-6)}$ mice. These results suggest that β-catenin plays a pivotal role in mediating the formation of OSSN. Thus, the administration of inhibitor(s) that prevents activation of β-catenin may be beneficial for treating OSSN.

*E-mail: Liucg@uc.edu/ Phone: 513-558-0847/ Fax: 513-558-6021

Introducton

Wnt signaling pathways (canonical and non-canonical) play critical roles in the life of an animal by regulating the spatial-temporal expression of genes during embryonic development and homeostasis in adults [1, 2]. The disruption of the signaling pathways leads to congenital defects of neonate and diseases in adults [3, 4]. The gateway of the canonical Wnt pathway is mediated via β-catenin which is a multifunctional protein. β-catenin is a component of adherent complex of cadherin, which mediates cell-cell adhesion, and can serve as a transcription factors upon the activation of Wnts-dependent signaling cascade [5] and Wnt-independent signal transduction pathways such as fibroblast growth factors (FGFs) [6, 7], epidermal growth factor (EGF) [8], insulin-like growth factor [9, 10], integrin-linked kinase [11], nuclear factor-κB (NF-κB) [12], PTEN [13], FP(B) prostanoid receptor [14], nuclear hormone receptors [15-18], and oxidative stress [19]. In the absence of Wnt ligand, cytoplasmic β-catenin molecules are phosphorylated at the N-terminus by GSK3β (glycogen synthase kinase 3β) and become labile and are degraded via ubiquitin pathway. In contrast, when Wnt binds to its receptors Frizzled and Lrp5/6, the assembly of proteins responsible for this phosphorylation state is abolished. Consequently, the unphosphorylated β-catenin molecules are stable and accumulate in cytoplasm, and assume nuclear translocation, where it serves as a transcription factor regulating expression of target genes. Once in the nucleus, β-catenin associates with transcription factors of the high-mobility-group (HMG) box T cell factor/leukemia enhancing factor (Tcf/Lef) family, which stimulate the expression of downstream genes including c-*myc* [20], *cyclin D1* [21-22], c-*jun*, *Tcf-1*, *Lef-1*, *Conductin/Axin2*, and the metalloproteinase matrilysin (*MMP-7*) [23] and resulting in dramatic alterations in cell proliferation, differentiation, and/or migration.

Transparent cornea consists of three tissue layers, i.e., epithelium, stroma and endothelium. Corneal epithelium is a stratified but non-keratinized tissue. The maintenance of transparent corneas requires continued replenishment of corneal epithelial cells that have a half life of about 2-3 weeks by limbal stem cells. Morphogenesis of the corneal epithelium is part of a differentiation program of the ocular surface ectoderm which in addition to the cornea also gives rise to epithelia of lens, conjunctiva, eyelid epidermis, lacrimal gland, and Meibomian gland [24]. Corneal epithelium is distinct from other epithelia by the unique expression of Keratin 12 (K12) which pairs with K3 to form the intermediate filament of corneal epithelial cells and signifies the corneal type-epithelium differentiation. Recent studies demonstrate that modulation of Wnt/β-catenin signaling pathways plays an important role in corneal epithelial morphogenesis, homeostasis, and wound healing [25-27]. For example, suppression of canonical Wnt signaling by DKK2 is essential for peridermal epithelium of ocular surface ectoderm to assume corneal type-epithelium differentiation.

Human ocular surface squamous neoplasia is the most common ocular surface pre-cancerous and cancerous lesion previously known by various names such as conjunctival intraepithelial neoplasia, corneal intraepithelial neoplasia (CIN), or both together (CCIN) [28]. Clinically, OSSN manifests in different grades ranging from simple dysplasia to squamous cell carcinoma. Because of the high incidence of OSSN in the limbal area, where the corneal epithelial stem cells reside, the limbal transition zone/stem cell theory has been proposed for the development of CIN [28]. It has been suggested that the slow cycling limbal stem cells may become hyper-proliferative by several factors e.g., carcinogen, irradiation, and the phorbol ester tumor promoter, 12-O-tetradecanoylphorbol 13-acetate (TPA) [29]. Alterations within the limbal area can cause abnormal maturation of the conjunctival and corneal epithelium and lead to the formation of CIN. Nevertheless, the etiology and pathogenesis of CIN and ocular surface carcinoma remain elusive. We have recently developed a *Krt12*$^{rtTA/rtTA}$/*tet-O-FGF-7* double transgenic mouse model in which over-expression of FGF-7 by Dox induction results in a phenotype resembling OSSN in human [30]. In this tumor model, we found nuclear localization of β-catenin in hyperplastic corneal epithelial cells of the Dox-treated *Krt12*$^{rtTA/rtTA}$/*tet-O-FGF-7* mice. This suggests that excess levels of FGF-7 signaling might activate β-catenin in a Wnt-dependent or -independent manner.

To further understand the relationship between FGF-7 and Wnt signaling pathway in our mouse model, we examined conditional loss-of-function mutation of β-catenin in the corneal epithelium to further examine the molecular and cellular mechanisms of β-catenin in corneal morphogenesis, homeostasis, and oncogenesis. Genetically modified *Ctnnb1*$^{lox(E2-6)/lox(E2-6)}$ mice harbored loxP sites flanking exons 2-6 of *Ctnnb1* allele, which encode the N-terminal domain and Armadillo repeats 1-4 of the β-catenin protein [31]. Excision of the floxed exons result in a frameshift mutation and production of a null-allele. We reported herein that loss of β-catenin did not alter corneal epithelium morphpgenesis and homeostasis but prevented corneal intra-epithelial neoplasia induced by excess FGF-7.

Results

Over-Expression of FGF-7 and Nuclear β-Catenin in Human OSSN

We have recently reported that excess FGF-7 in *Krt12*$^{rtTA/rtTA}$/*tet-O-FGF-7* double transgenic mice induced by Dox resulted in a corneal papillomatous squamous cell carcinoma with neovascularization resembling human OSSN. In this animal model, β-catenin was found in the nucleus of the corneal epithelium [30].

To investigate whether FGF-7 over-expression and β-catenin nuclear translocation associates with human OSSN, we have examined human ocular specimens from the eye pathology laboratory by immunohistochemistry. From eight human OSSN samples, all showed nuclear β-catenin staining and were positive for FGF7 expression as compared to a normal eye bank cornea, which has neither nuclear β-catenin staining nor detectable FGF-7 ([32], Figure 1B-E, Table 1).

FGF-7 Activated β-Catenin in Corneal Epithelial Cells

To study how FGF-7 activates the β-catenin in corneal epithelial cells, a telomerase-immortalized human corneal epithelial cell line (HTCE) [33] was starved for 12hr and treated with or without FGF-7 under serum-free culture condition. Total HTCE cell numbers increased by 2.35 and 4.0 fold at 24 hr and 48 hr after FGF-7 treatment, respectively, while decreased at 48hr in the absence of FGF-7 under serum-free condition (Figure 1F-G).

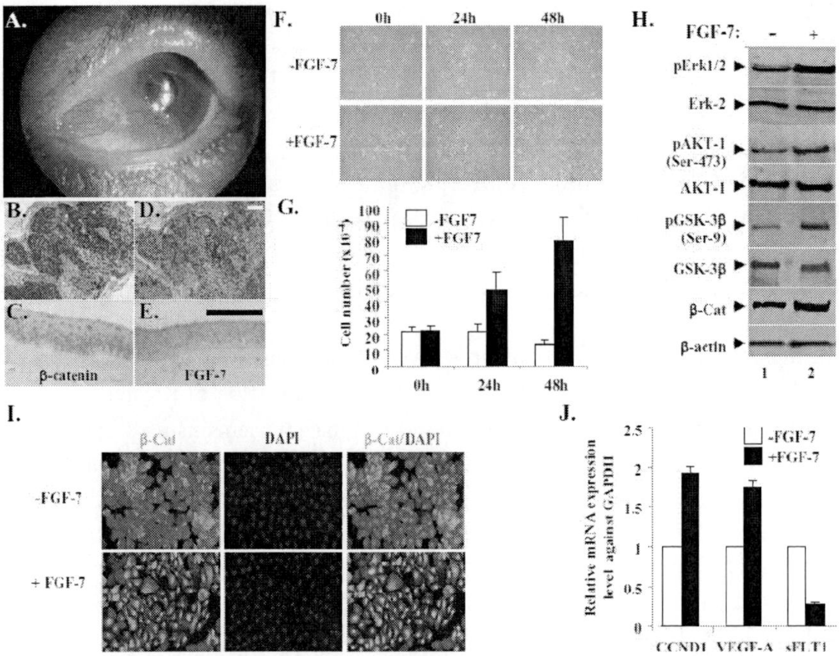

Figure 1. FGF-7/β-catenin signaling associated with human OSSN. (A), A photograph taken from a human OSSN patient (see Table S1, case 4) who presented a leukoplakic pannus-like lesion with engorged vessels involving the area from the 5-o'clock to 10-o'clock position with a 4-mm extension onto the corneal side. (B-E), Immunohistochemical staining with anti-β-catenin (B, C) or anti-hFGF-7 (D, E) to the OSSN sample (B, D) and normal eye bank cornea (C, E) were performed. It is clear that normal corneal epithelium showed membranous β-catenin (C) and little FGF-7 (E) expression. In the contrast, OSSN showed abundant nuclear β-catenin (B) and FGF-7 expression (D). (F, G), HTCE cells were starved in KSFM basal medium for 12hr and treated with or without FGF-7 (50 ng/ml).Total cell numbers were counted by hemocytometer at 0, 24, and 48 hr after FGF-7 treatment. Representative photo (F) and quantitative histograms (G) showed that FGF-7 enhanced HTCE cell growth. (H), Total protein lysates from non-treated (lane 1) or FGF-7-treated (lane 2) cells were subjected to western blot analysis with anti-ERK2, anti-pERK1/2, anti-AKT-1, anti- pAKT-1 (ser473) ,anti-GSK3β,anti-pGSK3β (Ser-9), and anti-β-catenin antibodies respectively. Note that FGF-7 increased the level of phosphorylation of ERKs, AKT-1, and GSK-3β. The phosphorylation of GSK-3β lead to β-catenin stabilization. (I), HTCE cells treated with (top) or without (bottom) FGF-7. Immunofluorescent staining with anti-β-catenin antibody (green) and counter stained with DAPI (blue). Note that β-catenin was located at the cell junction and cytoplasm but was translocated into nucleus by FGF-7. (J), qRT-PCR analysis showed FGF-7 up-regulated growth promoting gene, CCND1, and pro-angiogenic factors FLT1, but down-regulated anti-angiogenic factor, sFLT1 gene expression. These experiments implicated that FGF-7 triggered phosphorylation of ERK1/2, AKT-1, GSK-3β and nuclear translocation of β-catenin to enhance proliferation, up-regulated pro-angiogenic but down-regulated anti-angiogenic factor in corneal epithelial cells. Scale bars in D and E: 100 μm.

Biochemical analysis showed that exogenous FGF-7 (50ng/ml) was able to induce up-regulation of phospho-ERK1/2, phospo-AKT-1(Ser-473), and phospho-GSK-3β (Ser-9) in 30 min (Figures 1H) as well as increasing the expression level of β-catenin and its nuclear translocation (Figures 1I). In addition, real time RT-PCR revealed that the expression level of CCDN1 (cyclin D) and VEGF-A, known to be downstream targets of β-catenin, were increased whereas the soluble VEGFR1/FLT1 expression was decreased by two-fold in FGF-7-treated HTCE cells (Figure 1J).

Table 1. Summary of FGF-7 and β–cetenin immunohistochemical staining from human OSSN patients

Case number	Nuclear β-Catenin	Expression of FGF-7	Diagnosis	Gender/Age
1	++	++	Conjunctiva SCC and extend to orbital cavity; SCC of right forearm(distance metastasis)	Female/89
2	+++	++	Conjunctiva and lid SCC, OD	Male/60
3	+++	+++	SCC of cornea and conjunctiva, OD.	Male/83
4	+++	+++	CIN of conjunctiva and cornea	Female/74
5	++	+	Conjunctival tumor, OS	Male/81
6	++	+	Malignant SCC of orbit and conjunctiva, OD	Male/69
7	++	++	Conjunctival CIN, OD	Male/63
8	++	++	Conjunctival tumor	Male/97
9	-	-	Normal eye bank cornea	Male/74

+++: Strong positive, ++: intermediate positive, +; weak positive, -: negative staining
SCC, squamous cell carcinoma; CIN, corneal and/or conjunctiva intraepithelial neoplasia; OS, left eye; OD, right eye.

FGF-7 Activated β-Catenin in Corneal Epithelial Cells Via Wnt Independent Manner

To examine whether FGF-7 over-expression stimulate Wnt ligands expression, which in turn causes the nuclear translocation of β-catenin, we surveyed the expression of Wnt ligand genes including *Wnt-4*, *Wnt-5a*, *Wnt-7a*, *Wnt-8a*, *Wnt-9a*, *Wnt-9b*, *Wnt-10a*, and *Wnt-16a* in corneas of *Krt12*$^{rtTA/rtTA}$/*tetO-FGF-7* double transgenic mice administered with or without Dox. No significant difference of mRNA expression level was found. Thus, our data argued against the notion that β-catenin activation was due to the up-regulation of Wnt genes by over-expression of FGF-7 in Dox-induced *Krt12*$^{rtTA/rtTA}$/*tetO-FGF-7* double transgenic mice (Figure 2).

β-Catenin Is Required for FGF-7-Induced Corneal Epithelial Neoplasia

Since our data implicated a role for β-catenin signaling in mediating FGF-7-induced neoplasia in corneal epithelial cells, we then asked if knocking out *Ctnnb1* gene can abolish corneal neoplasia formation using *Krt12*$^{rtTA/rtTA}$/*tetO-FGF-7* mouse model treated with Dox. By combining *Krt12*$^{rtTA/rtTA}$, *tetO-FGF-7, tetO-Cre,* and *Ctnnb1*$^{lox(E2-6)}$ genotypes together, our results demonstrated that Dox treatment induced corneal epithelial hyperplasia in *Krt12*$^{rtTA/rtTA}$/*tetO-FGF-7/tetO-Cre/Ctnnb1*$^{lox(E2-6)/Wt}$ quadruple transgenic mice, but were surprisingly prevented in *Krt12*$^{rtTA/rtTA}$/*tetO-FGF-7/tetO-Cre/Ctnnb1*$^{lox(E2-6)/lox(E2-6)}$ quadruple transgenic mice induced with Dox from post-natal day 1 (P1) to P21 (Figure 3).

Primer name and sequence	PCR product (base pair)
mWnt4-F 5'GGCGTAGCCTTCTCACAGTC mWnt4-R 5'TACAGCACGCCAGCACGTCT	190
mWnt5a-F 5'AACAATGAAGCAGGCCGTAG mWnt5a-R 5'CGCGCTATCATACTTCTCCT	155
mWnt7a-F 5'GGACGCTCATGAACTTACAC mWnt7a-R 5'AGCCTAGCTCTCGGAACTGT	130
mWnt8a-F 5'CAGACTCTTCGTGGACAGTT mWnt8a-R 5'GCTTCCTGAGATGCCATGAC	135
mWnt9a-F 5'CCGAGTGGACTTCCACAACA mWnt9a-R 5'ACCTTGAGCGAGGTCTCATA	179
mWn9b-F 5'GGCATCAAGGCTGTGAAGAG mWnt9b-R 5'GTGTCATAGCGTAGCTTCAG	139
mWnt10a-F 5'GCAGGCGGTGATGGAGAACA mWnt10-R 5'CCTACTGTGCGGAACTCAGG	110
mWnt16-F 5'ACGGCATGTGGTTCAGCAGA mWnt16-R 5'CAGCGGCAGTCCACAGACAT	145
mGapdh-F 5'AAGGTGGTGAAGCAGGCATC mGapdh-R 5'TCTTACTCCTTGGAGGCCAT	150

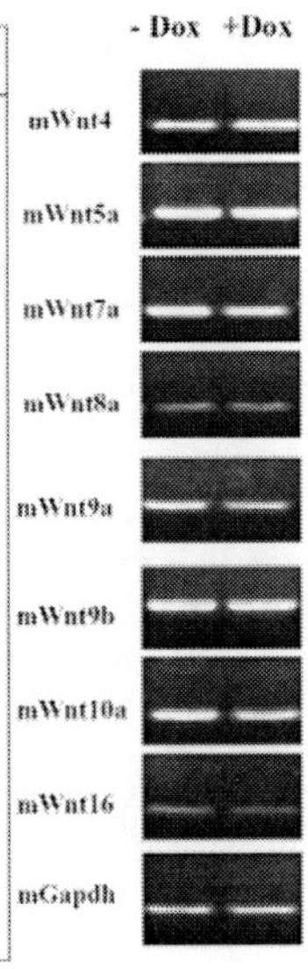

Figure 2. Wnt expressions in corneas determined by RT-PCR. Three-week-old *Krt12*$^{rtTA/rtTA}$/*tetO-FGF-7* mice administered with or without Dox for 14 days. Total RNAs isolated from 10 corneas of each group and subjected to RT-PCR analysis with primer pairs specific for the individual mouse Wnt genes as indicated (right panel). RT-PCR-amplified products were analyzed on a 2% agarose/TAE gel electrophoresis and visualized with ethidium bromide. No significant changes of all the Wnt genes analyzed, suggesting that FGF-7 over-expression induced by Dox did not up-regulate Wnt expression.

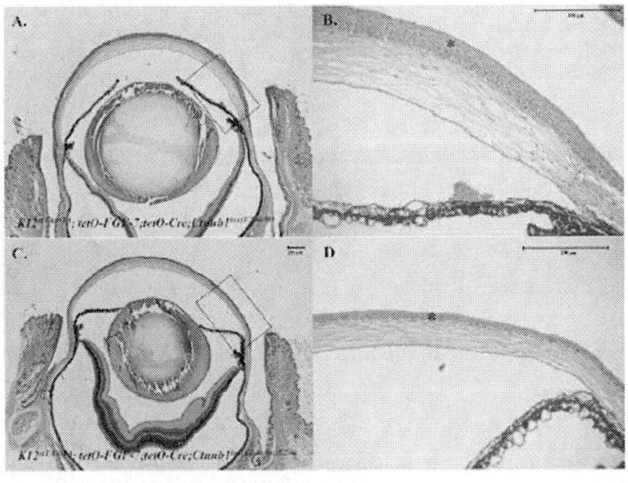

Figure 3. *Ctnnb1* is required for corneal epithelial hyperplasia induced by FGF-7. Hematoxyline-Eosin staining of paraffin sections of *Krt12*$^{rtTA/rtTA}$/*tetO-FGF-7/tetO-Cre/Ctnnb1*$^{lox(E2-6)/Wt}$ (A, B), *and Krt12*$^{rtTA/rtTA}$/*tetO-FGF-7/tetO-Cre/Ctnnb1* $^{lox(E2-6)/lox(E2-6).}$ (C, D) offsprings. Notice that Dox-induction caused corneal epithelial hyperplasia in *Ctnnb1*$^{f/W\ lox(E2-6)/Wt}$ (A, B) but not in *Ctnnb1* $^{lox(E2-6)/lox(E2-6)}$ (C, D) background. It seemed that one allele of *Ctnnb1* was sufficient to respond to FGF-7 signaling and increased the corneal epithelial stratification to approximately 10 layers (B, asterisk) while *Ctnnb1* null is resistant to FGF-7-induced hyperplasia (D, asterisk).

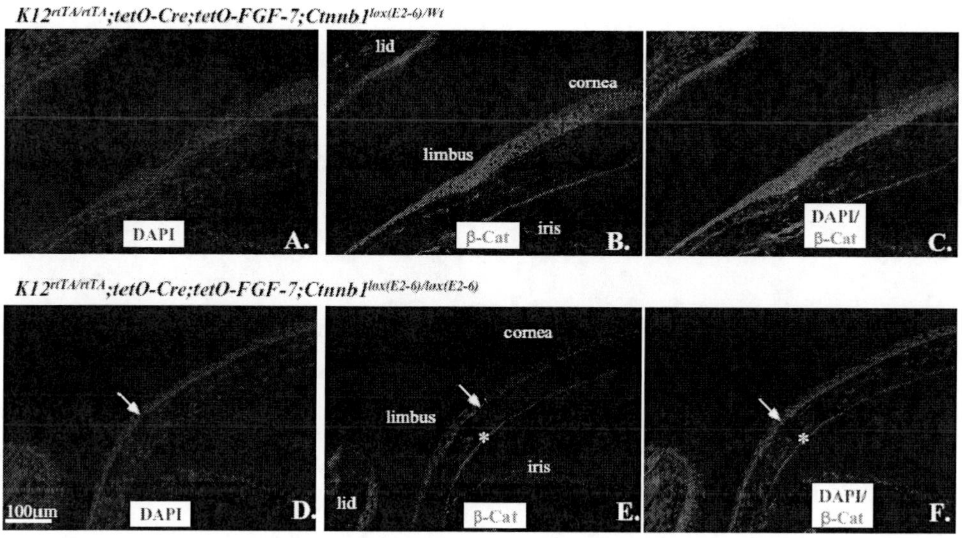

Figure 4. Dox-dependent Deletion of *Ctnnb1* in Corneal Epithelium. Paraffin sections (5 μm) were incubated with mouse monoclonal anti-β-catenin antibody (red). Nuclear counsterstaining was performed with DAPI (blue). Noted that *Krt12$^{rtTA/rtTA}$/tetO-FGF-7/tetO-Cre/Ctnnb1$^{lox(E2-6)Wt}$* (A-C) showed positive staining for β-catenin in ocular surface epithelia including eyelid and corneal epithelium and endothelium. Interestingly, *Krt12$^{rtTA/rtTA}$/tetO-FGF-7/tetO-Cre/Ctnnb1$^{lox(E2-6)/lox(E2-6)}$* offspring (D-F) showed negative staining for β-catenin only in the central corneal epithelium but positive staining in eyelid epithelium and limbal basal epithelial cells (arrow) , and endothelial cells (asterisk) in which K12 is not expressed.

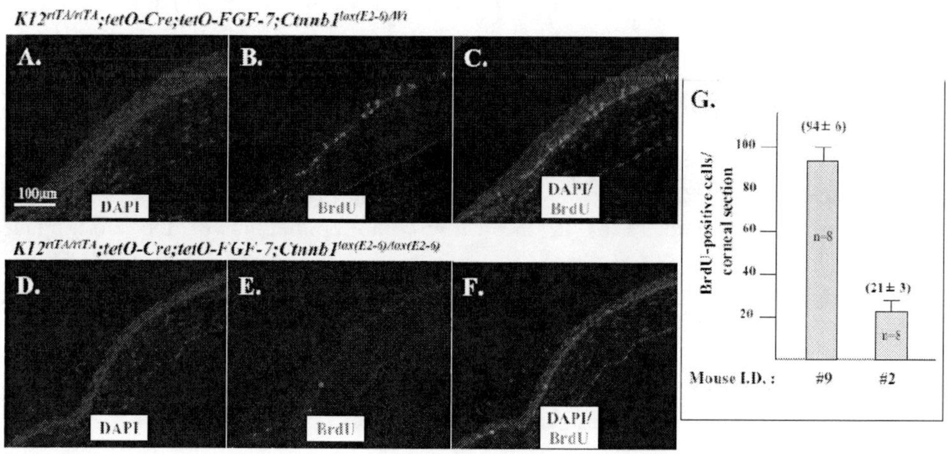

Figure 5. *Ctnnb1* is responsible for the corneal epithelial hyperproliferation induced by FGF-7. Paraffin sections (5 μm) were incubated with mouse monoclonal anti-BrdU antibody (red). Nuclear counterstaining was performed with DAPI (blue). Noted that BrdU positive cells in *Krt12$^{rtTA/rtTA}$/tetO-FGF-7/tetO-Cre/ Ctnnb1$^{lox(E2-6)/Wt}$* (B) appeared in the basal and suprabasal layers but were only found in the limbal basal cells of the *Krt12$^{rtTA/rtTA}$/tetO-FGF-7/tetO-Cre/ Ctnnb1$^{lox(E2-6)/lox(E2-6)}$* offspring (E). Quantitative analysis (G) showed that BrdU positive cells in *Krt12$^{rtTA/rtTA}$/tetO-FGF-7/tetO-Cre/ Ctnnb1$^{lox(E2-6)/Wt}$* increased about 4.5 folds as compared with *Krt12$^{rtTA/rtTA}$/tetO-FGF-7/tetO-Cre/ Ctnnb1$^{lox(E2-6)/lox(E2-6)}$* littermate.

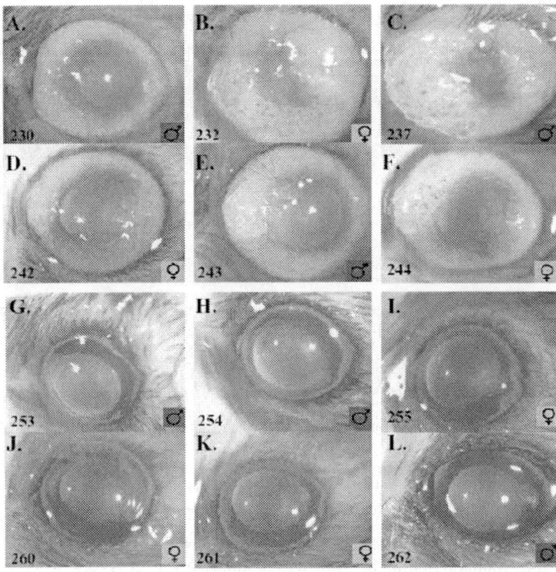

Genotypes	Duration of Dox induction
Krt12$^{rtTA/rtTA}$;*tetO-FGF-7;* *tetO-Cre;Ctnnb1*$^{lox(E2-6)/Wt}$	P1-P60
	P7-P60
Krt12$^{rtTA/rtTA}$;*tetO-FGF-7;* *tetO-Cre;Ctnnb1*$^{lox(E2-6)/lox(E2-6)}$	P1-P60
	P7-P60

Figure 6. Mice lacking β-Catenin were resistant to corneal epithelial tumor formation induced by excess FGF-7. Mouse eyes from *Krt12*$^{rtTA/rtTA}$*/tetO-FGF-7/tetO-Cre/Ctnnb1*$^{lox(E2-6)/Wt}$ (A-F) or *Krt12*$^{rtTA/rtTA}$*/tetO-FGF-7/tetO-Cre/Ctnnb1*$^{lox(E2-6)/lox(E2-6)}$ (G-L) littermates that have been administered with Dox from P1 to P60 (A-C and G-I) or from P7 to P60 D-F and J-L) were examined under a stereomicroscopy. It was noted that tumors appeared more prominent in the corneal/limbal area than the central cornea region. Tumor was never noticed in *Krt12*$^{rtTA/rtTA}$*/tetO-FGF-7/tetO-Cre/Ctnnb1*$^{lox(E2-6)/lox(E2-6)}$ mice.

Immunohistochemical staining showed that β-catenin expression was completely undetectable in the corneal epithelium but was expressed in other ocular tissues of the *Krt12*$^{rtTA/rtTA}$*/tetO-FGF-7/tetO-Cre/Ctnnb1*$^{lox(E2-6)/Wt}$ quadruple transgenic mice (Figure 4). Likewise, BrdU positive-corneal epithelial cells in *Ctnnb1*$^{lox(E2-6)/Wt}$ background were 4.5-fold less than those in the *Ctnnb1*$^{lox(E2-6)/lox(E2-6)}$ background. (Figure 5). Most importantly, neoplasia was never found (n=6) in the *Ctnnb1* null-epithelium of the Dox-induced *Krt12*$^{rtTA/rtTA}$*/tetO-FGF-7/tetO-Cre/Ctnnb1*$^{lox(E2-6)/lox(E2-6)}$ quadruple transgenic mice up to 2 months, whereas prominent tumors were found in all (n=6) Dox-induced *Krt12*$^{rtTA/rtTA}$*/tetO-FGF-7/tetO-Cre/Ctnnb1*$^{lox(E2-6)/Wt}$ mice. (Figure 6).

Discussion

The mammalian ocular surface ectoderm undergoes very dynamic cellular events to give rise to epithelia of the lens, eyelids, conjunctiva, limbus, cornea, lacrimal gland, and meibomian glands. Ectopic over-expression of FGF-7- or FGF-10-driven by the α-crystalline promoter (*α-Crypr*), which is activated at mouse E11.5, converted corneal fate into lacrimal gland [34, 35]. However, over-expression of FGF-7 using the K12 promoter, which is activated at mouse E14.5, did not induce an ectopic lacrimal gland but resulted in squamous cell carcinoma [30]. Although it was not documented in the *α-Crypr*-FGF-7 or in the *α-Crypr*-FGF-10 mice, we found that β-catenin was aberrantly activated in the corneal neoplasia of Dox-induced *Krt12$^{rtTA/rtTA}$/tetO-FGF-7* mice. Similarly, pannus surgically removed from human OSSN patients also exhibited excess FGF-7 and nuclear translocation of β-catenin (Figure 1). These findings implicate that up-regulation of the FGF-7 signaling pathway(s) may be a common mechanism in tumorigenesis resulting from limbal stem cells undergoing oncogenic transformation as a result of insults such as infection with the human papilloma virus 16 (HPV16) and/or long-term UVB exposure, which are two major risk factors associated with human OSSN [36, 37]. Interestingly, it has been demonstrated that FGF-7 level within the cancer lesion was elevated throughout the progression of multi-stage epidermal carcinogenesis in *K14-HPV16* transgenic mice [38, 39]. Similarly, the exposure to UVB irradiation can induce a rapid intracellular production of ROS (reactive oxidative stress), which in turn is capable of triggering phosphorylation and activation of the FGF-7 receptor, FGFR2-IIIb, resembling to those induced by FGF-7 [40]. Therefore, the FGF-7/FGFR2-IIIb signaling is likely the hub that integrates the input through UVB and HPV with the genesis and formation of OSSN. This may explain why excess FGF-7 caused OSSN phenotype in the Dox-treated *Krt12$^{rtTA/rtTA}$/tet-O-FGF-7* double transgenic mouse model [30].

In HTCE cells, FGF-7 was able to induce up-regulation of phospho-ERK1/2, phospo-AKT-1, and phospho-GSK-3β (Ser-9) as well as increasing the cellular level of β-catenin and its nuclear translocation. In *Krt12$^{rtTA/rtTA}$/tetO-FGF-7* transgenic mouse, over-expression of FGF-7 in corneal epithelium had no impact on Wnt genes expression (Figure S1), suggesting that β-catenin activation by FGF-7 signaling was transmitted via Wnt independent pathway. As a result, the expression level of cyclin D was up-regulated (Figure 1F, 1H) and cell proliferation activity was enhanced by FGF-7 treatment (Figure. 1I, 1J). Our observation showed that VEGF-A, known to be downstream targets of β-catenin, and VEGFR1 (FLT1) was increased whereas the soluble VEGFR1/FLT1 expression was decreased in FGF-7-treated HTCE cells (Figure 1H). Since *FLT1* and s*FLT1* mRNA are derived from the splicing variant of the same gene, up-regulation of FLT1 by FGF-7 may be at the transcriptional and/or post-transcriptional level, but the down-regulation of s*FLT1* mRNA has to be regulated in the post-transcriptional level. The molecular mechanism underlying the differential regulation of *FLT1* versus s*FLT1* in corneal epithelial cells treated with FGF-7 requires further investigation. Nevertheless, our data suggest that β-catenin may be a key player in mediating FGF-7 signaling to promote corneal epithelial cell proliferation and to induce neovascularization by stimulation of pro-angiogenic factor synthesis and inhibition of anti-angiogenic factor production, which can lead to corneal neoplasia formation (Figure 7).

In the present study, we demonstrated that one-functional allele of the *Ctnnb*1 gene was sufficient to mediate FGF-7 triggered tumorigenesis in the corneal epithelium, however, loss of both alleles suppressed FGF-7-induced neoplasia (Figures 5). These results strongly argue for the fact that FGF-7-triggered OSSN is dependent on β-catenin mediation. Our data also suggest that although β-catenin is not essential for the maintenance of corneal epithelial homeostasis, it is required for corneal neoplasia. In support of our data, a recent report shows that ablation of the β-catenin gene results in complete tumor regression in mouse skin [41].

Repression of the Wnt/β-catenin signaling pathway has been implicated as an essential prerequisite for the differentiation of corneal epithelial progenitor cells into a non-keratinizing stratified epithelium during corneal morphogenesis [25].

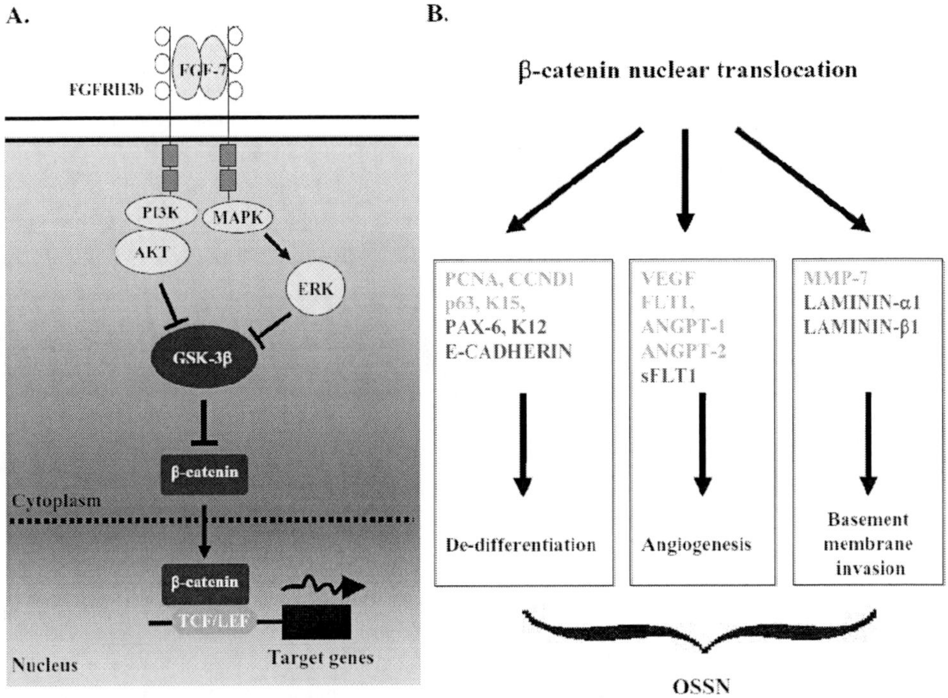

Figure 7. Model of activation of FGF-7/β-catenin signaling pathways leading to OSSN. (A) Depicted is how activation of the FGF-7/FGFRII3b on the cell membrane can lead to activation of β-catenin through different mechanisms. In corneal epithelial cell, activation by FGF-7 of either PI3 kinase (PI3K)/AKT or MAP kinase/ERK can promote GSK-3β phosphorylation, which is associated with translocation of β-catenin to the nucleus where it associates with TCF/LEF to bind and regulate the target genes. (B) Nuclear translocation of β-catenin resulted directly or indirectly in up-(green) and down- (red) regulation of wide variety of genes that lead to de-differentiation of corneal epithelial cells, promoting cell proliferation, increasing pro-angiogenic factors, decreasing anti-angiogenic factor, up-regulation of MMP-7 and disruption of basement membrane components, by which eventually resulted in the development of OSSN [32] .

It has been previously documented that dominant active β-catenin driven by *Le-Cre* suppresses surface ectoderm invagination at E9.5~E10.5. These data suggest that Wnt/β-catenin signaling is unfavorable during the early stages of ocular surface morphogenesis [42]. Therefore, tight regulation of β-catenin signaling to control these processes is essential for the maintenance of corneal epithelium homeostasis, with any compromise leading to tumor formation (Figure 7).

Materials and Methods

Animals: Experimental animals were housed under pathogen-free conditions in accordance with institutional guidelines. Animal care and use conformed to the ARVO Statement for the Use of Animals in Ophthalmic and Vision Research. All animal protocols were approved by the Institutional Animal Care and Use Committee (IACUC) of the

University of Cincinnati. Generation of Compound Transgenic Mice: Genetic modified mouse lines such as *Krt12*^{rtTA} [43], *tetO-FGF-7* [44], *tetO-Cre* [45], and *Ctnnb1*$^{lox(E2-6)}$ [31] mice have been previously described. Compound transgenic mice were generated via natural breeding of individual mouse lines and the genotypes were identified by polymerase chain reaction (PCR) using oligonucleotide primers specific for each transgene. In all experiments the time when the vaginal plug was detected was taken as E0.5 of embryonic development. To activate *tetO-Cre* and/or *tetO-FGF-7* expression, mice were injected once with Dox (Sigma, 80 µg/g body weights) [46] and fed Dox-chow (1g/Kg chow, Bioserv). Control animals were fed regular chow.

Cell Culture

The culture condition for HTCE was described previously [33]. When growing to 70% confluent in KSFM with supplements, HTCE cells were starved for 24h in KSFM without supplements and treated without or with FGF7 (50ng/ml) in KSFM without supplements for 48hr. Immunocytochemical and western blotting analyses were performed as described below.

Immunocytochemical, Immunohistochemical, and Western Blotting Analyses: The following antibodies were used: goat anti-human FGF-7 (AF-251-NA, 1:50) (RandD systems, Inc.); rabbit anti-cytokeratin K12 antibody [47], rabbit antiAKT-1 (1:1000), rabbit anti-pAKT-1 (Ser-473) (1:1000) (Cell signaling); rabbit anti-ERK2 (1:500), anti-pERK1/2 (1:500) (Neuromics, Inc), mouse anti-BrdU (1:100); mouse PC10 anti-PCNA (1:100) (Lab Vision, Corp.); mouse anti-β-catenin (1:50, BD Biosciences, San Jose, CA); rabbit anti-β-catenin (1:2000, Sigma). Alkaline phosphatase-conjugated donkey anti-goat or anti-rabbit IgG (1:250, Novagen, Inc) was used for detecting FGF-7 and β-catenin in human specimens. Alex488 or Alexa555- conjugated donkey anti-goat-IgG, goat anti-rabbit IgG, or rabbit anti-mouse IgG secondary antibody (1:500) (Molecular Probes, Inc.) was used for immunofluorescent staining. For immunocytochemical staining, HTCE cells subcultured in slide chambers were treated with or without FGF-7 (50 µg/ml) for 30 min, fixed with 4% paraformaldehyde in PBS at 4 °C for 10min and equilibrated with PBS. For immunohistological analysis, excised eyes were fixed in 4% paraformaldehyde in PBS and paraffin embedded. Tissue sections (5µm) were de-paraffinized and hydrated in a graded ethanol series. The primary antibody was incubated at the dilutions mentioned above in TBST-3% BSA at 4 °C overnight. Non-specific antibody binding was removed by washing with TBST three times. Alexa 488 or 555-conjugated secondary antibody was then incubated at RT for 1hr. Sections for immunofluorescent analysis were mounted (SlowFade Light Antifade Kit; Molecular Probes, Eugene, OR) in the presence of 4,6-diamidino-2-phenylindole (DAPI), observed with an Nikon Eclipse E800 epifluorescent microscope and were photographed with a SPOT RT digital camera system (Diagnostic Instruments, Inc.). For western blotting analysis, total cellular lysates from mouse corneas or HTCE cells were prepared in lysis buffer [Tris-HCl (50 mM, pH7.4), NaCl (250 mM), EDTA (5mM), NP-40 (0.1%), NaF (25mM), and 1X protease inhibitor cocktail (Sigma P8340)] and subjected to sodium dodecyl sulfate-polyacrylamide gel electrophoresis (SDS-PAGE)/western blotting. First antibody was incubated in dilutions as mentioned above in TBST-3% BSA at 4 °C overnight.

Table 2. DNA sequences of primer sets and their PCR products for the qRT-PCR

Primer name	Primer sequence	PCR size (bp)
hsFLT1-F	5' AGGGGAAGAAATCCTCCAGA	81
hsFLT1-R	5' GAGATCCGAGAGAAAACAGC	
hVEGF-F	5' GCACCCATGGCAGAAGGAGG	90
hVEGF-R	5' TCCACACGTGGAACCGGATT	
hCCND1-F	5' CCGTCCATGCGGAAGATC	168
hCCND1-R	5' GAAGACCTCCTCCTCGCACT	
hGAPDH-F	5' CTCTGACTTCAACAGCGACACC	161
hGAPDH-R	5' CTTACTCCTTGGAGGCCATGTG	

Alexa 680-conjugated or alkaline phosphatase (AP)-conjugated secondary antibody was then incubated at RT for 1hr. Positive signals were visualized using Odyssey Imager (LI-COR® Bioscience, Inc) or by Western blue® substrate for AP. (Promega, Inc.)

Real-Time Reverse Transcription-Polymerase Chain Reaction: HTCE cells growing to 70% confluent in KSFM with supplements, treated with FGF7 (50ng/ml) for 48hr after 24h starvation in KSFM without supplements, 10μg total RNA was annealed to random primer and reverse transcribed with avian reverse transcriptase kits from Promega Inc. (Madison, WI) according to manufacture's instruction. The single strand cDNA was subjected to PCR with primer pairs as shown in Table 2. Real Time RT-PCR was performed by using the CFX96™ real-time system equipped on a C1000™ Thermal Cycler (BioRad). After initial step at 95°C for 10min, 40 cycles at 95°C 30sec, 62°C 30sec, 72°C 30sec. The cycle threshold values were used to calculate the normalized expression of human cyclin D1, FLT1, and VEGF-A against GAPDH using Q-Gene software.

Statistical Analysis: Two-tailed Student's t-test (Excel, Microsoft, Redmond, WA) was used in analyzing the number of BrdU-positive cells. All quantification data are presented as mean+/-sd.

Acknowledgment

This work is supported by grants from NIH/NEI RO1 EY12486 (CYL), EY13755 (WWK), Research Prevent Blindness, Ohio Lions Foundation for Eye Research.

References

[1] Grigoryan T, Wend P, Klaus A, Birchmeier W (2008) Deciphering the function of canonical Wnt signals in development and disease: conditional loss- and gain-of-function mutations of beta-catenin in mice. *Genes Dev.* 22: 2308–2341.

[2] Huang H, He X (2008) Wnt/□-catenin signaling: New (and old) players and new. *Curr. Opi. Cell Bio.* 20:119–125.

[3] Logan CY, Nusse R (2004) The Wnt signaling pathway in development and disease. *Annu. Rev. Cell Dev. Biol.* 20:781-810.

[4] Amerongen R, Berns A (2006) Knockout mouse models to study Wnt signal transduction. *Trends in Genetics* 22:678-689.

[5] Orsulic S, Huber O, Aberle H, Arnold S, Kemler R (1999) E-cadherin binding prevents β-catenin nuclear localization and beta-catenin/LEF-1-mediated transactivation. *J. Cell Sci.* 112:1237-1245.

[6] Eblaghie MC, Song SJ, Kim JY, Akita K, Tickle C, Jung HS (2004) Interactions between FGF and Wnt signals and Tbx3 gene expression in mammary gland initiation in mouse embryos. *J. Anatomy,* 205:1-13.

[7] Wahl MB, Deng C, Lewandoski M, Pourquié O (2007) FGF signaling acts upstream of the NOTCH and WNT signaling pathways to control segmentation clock oscillations in mouse somitogenesis. *Development* 134:4033-4041.

[8] Musgrove EA (2004) Wnt signalling via the epidermal growth factor receptor: a role in breast cancer? *Breast Cancer Res,* 6:65-68.

[9] Richard-Parpaillon L, Héligon C, Chesnel F, Boujard D, Philpott A (2002) The IGF Pathway Regulates Head Formation by Inhibiting Wnt Signaling in Xenopus. *Developmental Biology* 244:407-417.

[10] Zhu W, Shiojima I, Ito Y, Li Z, Ikeda H, Yoshida M, Naito AT, Nishi J, Ueno H, Umezawa A, Minamino T, Nagai T, Kikuchi A, Asashima M, Komuro I (2008) IGFBP-4 is an inhibitor of canonical Wnt signalling required for cardiogenesis. *Nature* 454:345-349.

[11] Novak A, Hsu SC, Leung-Hagesteijn C, Radeva G, Papkoff J, Montesano R, Roskelley C, Grosschedl R, DedharS (1998) Cell adhesion and the integrin-linked kinase regulate the LEF-1 and beta-catenin signaling pathways. *Proc. Natl. Acad. Sci. U S A* 95:4374-4379.

[12] Katoh M, Katoh M (2007) WNT Signaling Pathway and Stem Cell Signaling Network. *Clinical Cancer Research* 13:4042-4045.

[13] Zhao H, Cui Y, Dupont J, Sun H, Hennighausen L, Yakar S (2005) Overexpression of the tumor suppressor gene phosphatase and tensin homologue partially inhibits wnt-1-induced mammary tumorigenesis. *Cancer Res.* 65:6864-6873.

[14] Fujino H, Regan JW (2001) FP prostanoid receptor activation of a Tcf/beta-catenin signaling pathway. *J. Biol. Chem.* 276:12489-12492.

[15] 15.Mulholland DJ, Cheng H, Reid K, Rennie PS, Nelson CC (2002) The Androgen Receptor Can Promote beta-Catenin Nuclear Translocation Independently of Adenomatous Polyposis Coli. *J. Biol. Chem.* 277:17933-17943.

[16] Saez E, Rosenfeld J, Livolsi A, Olson P, Lombardo E, Nelson M, Banayo E, Cardiff RD, Izpisua-Belmonte JC, Evans RM (2004) PPARgamma signaling exacerbates mammary gland tumor development. *Genes Dev.* 18:528–540.

[17] Mulholland DJ, Dedhar S, Coetzee GA, Nelson CC (2005) Interaction of nuclear receptors with the Wnt/beta-catenin/Tcf signaling axis: Wnt you like to know? *Endocr. Rev.* 26:898-915.

[18] Schweizer L, Rizzo CA, Spires TE, Platero JS, Wu Q, Lin TA, Gottardis MM, Attar RM (2008) The androgen receptor can signal through Wnt/beta-Catenin in prostate

cancer cells as an adaptation mechanism to castration levels of androgens. *BMC Cell Biol.* 9:4.

[19] Almeida M, Han L, Martin-Millan M, O'Brien CA, Manolagas SC (2007) Oxidative stress antagonizes Wnt signaling in osteoblast precursors by diverting beta-catenin from T cell factor- to forkhead box O-mediated transcription. *J. Biol. Chem.* 282:27298-27305.

[20] He TC, Sparks AB, Rago C, Hermeking H, Zawel L, da Costa LT, Morin PJ, Vogelstein B, Kinzler KW (1998) Identification of c-MYC as a target of the APC pathway. *Science* 281:1509-1512.

[21] Tetsu O, McCormick F (1999) Beta-catenin regulates expression of cyclin D1 in colon carcinoma cells. *Nature* 398:422-426.

[22] Lin SY, Xia W, Wang JC, Kwong KY, Spohn B, Wen Y, Pestell RG, Hung MC (2000) □-catenin, a novel prognostic marker for breast cancer: its roles in cyclin D1 expression and cancer progression. *Proc. Natl. Acad. Sci. USA,* 97:4262-4266.

[23] Nelson WJ, Nusse R (2004) Convergence of Wnt, beta-Catenin, and Cadherin Pathways. *Science* 303:1483-1487.

[24] Kao WW, Liu, CY (2003) The use of transgenic and knock-out mice in the investigation of ocular surface cell biology. *Ocul. Surf.* 1:5-19.

[25] Mukhopadhyay M, Gorivodsky M, Shtrom S, Grinberg A, Niehrs C, Morasso MI, Westphal H (2006) DKK2 plays an essential role in the corneal fate of the ocular surface epithelium. *Development*, 33:2149-2154.

[26] Pearton DJ, Yang Y, Dhouailly D (2005) Trans-differentiation of corneal epithelium into epidermis occurs by means of a multistep process triggered by dermal developmental signals. *Proc. Natl. Acad. Sci. U S A* 102:3714–3719.

[27] Lyu J, Joo CK (2005) Wnt-7a up-regulates matrix metalloproteinase-12 expression and promotes cell proliferation in corneal epithelial cells during wound healing. *J. Biol. Chem.* 280:21653-21660.

[28] Lee GA, Hirst LW (1995) Ocular surface squamous neoplasia. *Surv. Ophthalmol.* 39:429-450.

[29] Tseng SCG (1989) Concept and application of limbal stem cells. *Eye,* 3:141-157.

[30] Chikama T, Liu CY, Meij JT, Hayashi Y, Wang IJ, Yang L, Nishida T, Kao WW (2008). Excess FGF-7 in corneal epithelium causes corneal intraepithelial neoplasia in young mice and epithelium hyperplasia in adult mice. *Am. J. Pathol.* 172:638-649.

[31] Brault V, Moore R, Kutsch S, Ishibashi M, Rowitch DH, McMahon AP, Sommer L, Boussadia O, Kemler R. (2001) Inactivation of the beta-catenin gene by Wnt1-Cre-mediated deletion results in dramatic brain malformation and failure of craniofacial development. *Development,* 128:1253-1264.

[32] Zhang Y, Call MK, Yeh LK, Liu H, Kochel T, Wang IJ, Chu PH, Taketo MM, Jester JV, Kao WW, Chia-Yang Liu. (2010) Aberrant expression of a □-catenin gain-of-function mutant induces hyperplastic transformation in the mouse cornea. *J. Cell Sci.* 123: 1285-1294.

[33] Robertson DM, Li L, Fisher S, Pearce VP, Shay JW, Wright WE, Cavanagh HD, Jester JV (2005) Characterization of growth and differentiation in a telomerase-immortalized human corneal epithelial cell line. *Invest. Ophthalmol. Vis. Sci.* 46:470-478.

[34] Lovicu FJ, Kao WW, Overbeek PA (1999) Ectopic gland induction by lens-specific expression of keratinocyte growth factor (FGF-7) in transgenic mice. *Mech. Dev.* 88:43-53.

[35] Govindarajan V, Ito M, Makarenkova HP, Lang RA, Overbeek PA (2000) Endogenous and ectopic gland induction by FGF-10. *Dev. Biol.* 225:188-200.

[36] Scott IU, Karp CL, Nuove GJ (2002) Human papillomavirus 16 and 18 expression in conjunctival intraepithelial neoplasia. *Ophthalmology*, 109:542-547.

[37] Kiire CA, Dhillon B (2006) The aetiology and associations of conjunctival intraepithelial neoplasia. *Br. J. Ophthalmol.* 90:109-113.

[38] Arbeit JM, Olson DC, Hanahan D. Upregulation of fibroblast growth factors and their receptors during multi-stage epidermal carcinogenesis in K14-HPV16 transgenic mice. *Oncogene*.1996;13:1847-1857

[39] Pietras K, Pahler J, Bergers G, Hanahan D. Functions of paracrine PDGF signaling in the proangiogenic tumor stroma revealed by pharmacological targeting. *PLoS Med.* 2008;5:e19.

[40] Marchese C, Maresca V, Cardinali G, Belleudi F, Ceccarelli S, Bellocci M, Frati L, Torrisi MR, Picardo M. UVB-induced activation and internalization of keratinocyte growth factor receptor. *Oncogene*. 2003;22:2422-24231.

[41] Malanchi I, Peinado H, Kassen D, Hussenet T, Metzger D, Chambon P, Huber M, Hohl D, Cano A, Birchmeier W, Huelsken J (2008) Cutaneous cancer stem cell maintenance is dependent on beta-catenin signaling. *Nature* 452:650-654.

[42] Smith AN, Miller LA, Song N, Taketo MM, Lang RA (2005) The duality of beta-catenin function: a requirement in lens morphogenesis and signaling suppression of lens fate in periocular ectoderm. *Dev. Biol.* 285:477-489.

[43] Chikama T, Hayashi Y, Liu CY, Terai N, Terai K, Kao CW, Wang L, Hayashi M, Nishida T, Sanford P, Doestchman T, Kao WW (2005) Characterization of tetracycline-inducible bitransgenic Krt12rtTA/+/tet-O-LacZ mice. *Invest. Ophthalmol. Vis. Sci.* 46:1966-1972.

[44] Lowry WE, Blanpain C, Nowak JA, Guasch G, Lewis L, Fuchs E (2005) Defining the impact of b-catenin/Tcf transactivation on epithelial stem cells. *Genes. Dev.* 19:1596-1611.

[45] Tichelaar JW, Lu W, Whitsett JA (2000) Conditional expression of fibroblast growth factor-7 in the developing and mature lung. *J. Biol. Chem.* 275:11858-11864.

[46] Utomo AR, Nikitin AY, Lee WH (1999) Temporal, spatial, and cell type-specific control of Cre-mediated DNA recombination in transgenic mice. *Nat. Biotechnol.* 17:1091-1096.

[47] Liu CY, Zhu G, Westerhausen-Larson A, Converse RL, Kao CW, Su TT, Kao WW (1993) Cornea-specific expression of K12 keratin during mouse development. *Curr. Eye Res.* 12:963-974.

In: Endothelium and Epithelium
Editors: J. Carrasco and M. Mota, pp. 247-284
ISBN 978-1-61470-874-2

Chapter XII

Mammary Gland Involution: Events, Regulation and Influences on Breast Disease

***Vaibhav P. Pai*[1,2,3] *and Nelson D. Horseman*[1]**
[1]Department of Molecular and Cellular Physiology,
University of Cincinnati, Cincinnati, OH, USA
[2]Systems Biology and Physiology Program,
University of Cincinnati, Cincinnati, OH, USA
[3]Current address: Tufts Center for Regenerative and Developmental Biology,
Biology Department, Tufts University, Medford, MA, USA

Abstract

Lactation is under homeostatic control, and sustained milk stasis or weaning leads to mammary gland involution. This transition is under the control of systemic and local factors. Of all the phases of mammary gland development, involution is the least well understood. Recent epidemiological observations suggest that involution events promote breast diseases, including breast cancer. This has renewed interest in understanding involution, and its regulation. In this chapter we discuss the events occurring during mammary gland involution and the role of various systemic and local factors in regulating these events. We also briefly review about how these processes can influence the progression of breast diseases.

1. Introduction

Postnatal mammary gland development occurs in a cyclical fashion in association with each pregnancy. Lactation is believed to be the functionally climactic stage of the gland. However, it is now evident that proper execution of involution, the resolution and breakdown of the lactating gland is equally important for pristine functioning of the gland during subsequent lactation cycles.

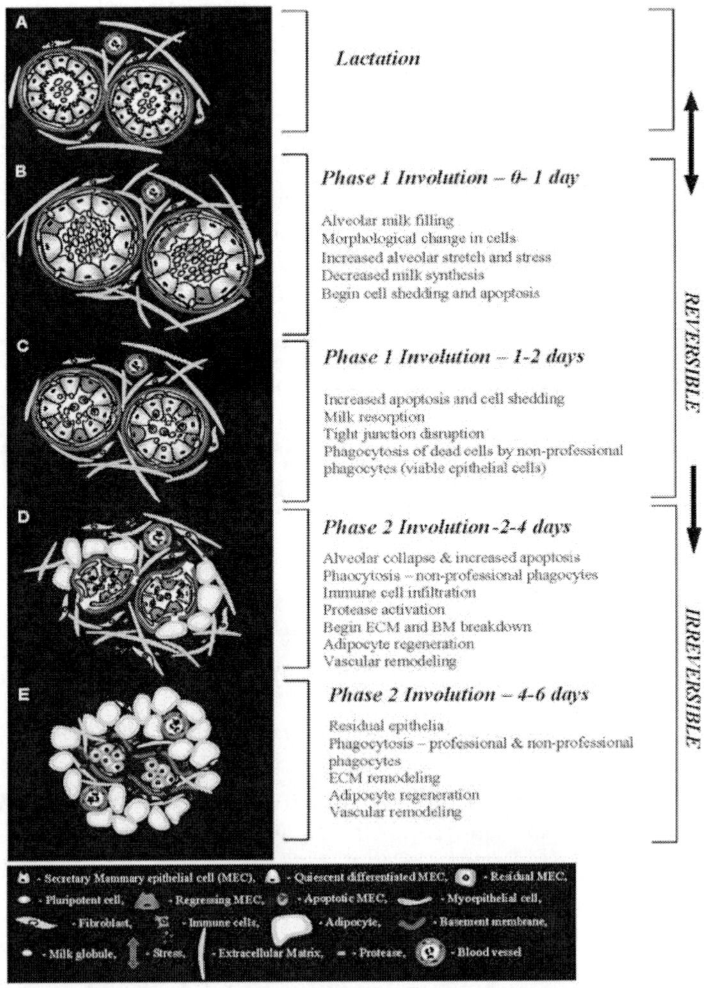

Figure 1. Diagrammatic representation of mouse mammary gland involution arbitrarily divided into stages (Reversible and Irreversible). The mouse mammary alveolus during various times of involution is diagrammatically represented. Events occurring during that time frame within the gland are indicated on the right. (A) Lactation is characterized by a well-formed alveolar architecture lined by secretory mammary epithelial cells. (B) Induction of involution is characterized by lack of suckling and milk removal from the gland. This results in alveolar engorgement, which exerts mechanical stress (red arrows) on the alveoli. This is associated with a change in cellular morphology, inhibition of milk secretion and induction of local factors involved in involution. (C) Cell shedding (red rounded cells) and cell death (dark cells) begins by 12 hrs after milk stasis and is highly increased by day 1 -2. Viable epithelial cells function as non-professional phagocytes and engulf the dead cells and residual milk. Reintroduction of suckling and removal of milk by this point returns the gland to lactation. (D) Upon continued milk stasis, epithelial tight junctions are disrupted followed by induction of ECM and BM remodeling. This makes the involution process irreversible. ECM remodeling is associated with alveolar collapse and high epithelial cell death. Concomitant vascular remodeling and adipogenesis also occurs. (E) During the final stages of involution professional phagocytes (macrophages and neutrophils) efficiently clear the cellular debris and milk. Adipose tissue (white cells) occupies a majority of the gland and residual epithelial cells (maily ductal) are maintained within the gland.

A) Morphology

Lactation consists of the establishment of a high energy-requiring milk secreting epithelium. At the culmination of lactation mammary glands primarily consist of epithelial tissue (90% of the mass in the rodent gland), of which a large portion is arranged in secretory alveoli containing a distended lumen into which milk is secreted [1] (Figure 1A). Epithelial cells also line the ducts through which milk is discharged to the suckling offspring. The epithelial structures are encapsulated on the basolateral side by star-shaped myoepithelial cells (Figure 1A). Both myoepithelial and secretory epithelial cells rest on a layer of basement membrane, which distinguishes the epithelial tissue from the stromal tissue [1] (Figure 1A). The stroma consists of extracellular matrix (ECM), connective tissue cells, dedifferentiated adipose cells, immune cells, and vasculature, and serves to support the epithelium[2](Figure 1A).

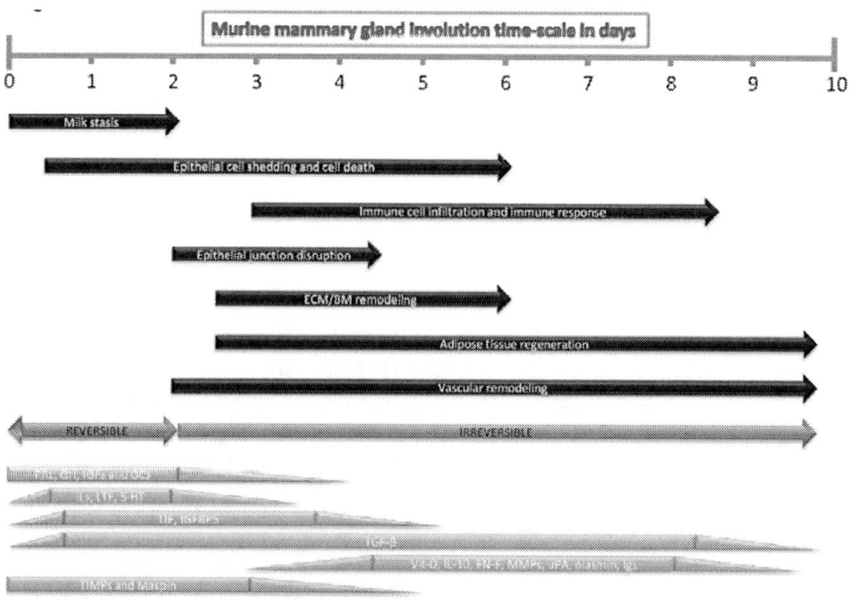

Figure 2. Time-scale of murine mammary gland involution events and the pattern of expression of systemic and local factors. This is a diagramatic representation of the time-scale (in days) of events occurring during murine mammary gland involution. The black arrows represent each event. The blue arrows show the reversibility of involution on a timescale. The green tabs represent systemic hormone and their levels on involution timescale. The orange tabs represent different patterns of expression of local factors involved in involution with examples mentioned in the tab. Involution begins by absence of suckling resulting in milk stasis and in turn inhibition of milk synthesis, which is observed until day 2 involution. This is mediated by rapid induction of local factors like 5-HT. Milk stasis is associated with the beginning of epithelial cell shedding and death along with vascular remodeling all induced by locally released factors. All these events are reversible upon reintroduction of suckling and hence form the reversible phase of involution. On the other hand, continued absence of suckling results in disruption of epithelial junctions by day 3-4. This is associated with a drop in systemic hormone levels and induction of another set of local factors. Concomitantly there is induction of ECM/BM remodeling along with adipose tissue regeneration which go on till the end of involution. These events make the involution process irreversible. For details about which factors are involved in which involution events please see table 1.

Involution is a remarkable process during which the lactating gland is restored to a virgin-like state in which the epithelial cells are returned to a rudimentary state. This is accomplished through several events that occur in a harmonious manner. For clarity we have segregated these events into the following catagories: a) milk stasis and inhibition of milk secretion, b) epithelial regression via cell death, c) immune response and clearance of debris, d) tight junction complex and barrier regulation, e) extracellular matrix (ECM) remodeling, f) adipose tissue remodeling and g) vascular remodeling (Figure 2 – black arrows). In the body of this chapter we attempt to discuss in detail each of these events and their regulation with respect to involution.

All of the involution events as mentioned above occur in a time-span of 8-10 days during a synchronized mouse mammary involution (Figures 1 and 2). Each of the involution events is executed at a different pace within the time frame of involution. The involution time-frame as a whole is divided into two phases based on the reversibility of the involution process [3] (Figures 1 and 2). During the "reversible phase" (Phase 1: 2 days in mice) the gland can revert to a state of milk production and secretion if the suckling stimulus is reintroduced (Figure 1B and C, and Figure 2 blue arrows). In the "irreversible phase" (Phase 2: days 3-10 in mice) the gland is unable to return to lactation without being re-stimulated by pregnancy levels of hormones (Figure 1D and E, and Figure 2 blue arrows).

B) Models of Study

Although the mammary glands are a definitive feature of all mammals there are notable differences in mammary gland biology across species. Species in which mammary gland biology has the most direct practical impact (humans and dairy cattle) are challenging research subjects. As in many other areas of biomedical research, laboratory rodents are the primary research models, and the mouse serves as the archetype because of the extraordinary advances in our understanding of this species. However, comparative biology provides crucial insights into the functional intricacies of the mammary glands, and we will focus attention on various other mammals in which mammary gland involution differs substantially from the mouse.

Mammary gland development has been studied largely using murine models (rats and mice). Natural involution begins in an asynchronous manner in each alveolar cluster as the demand for milk gradually wanes because the pups begin eating other foods. The asynchronous nature of natural involution is not conducive to investigation, and hence different manipulations are used to synchronize involution. These manipulations include: a) pup removal at peak lactation (around day 10 in mice), and b) unilateral teat sealing. The teat sealing model is useful because the effects of milk accumulation can be studied in the presence of a stimulatory systemic hormonal milieu, which is maintained by the continued presence of suckling pups.

Bovine mammary glands typically undergo regenerative involution, which involves an overall turnover of epithelial cells without significant tissue remodeling. A similar regenerative involution process can be observed in mice with concurrent pregnancy (Figure 3). In fur seals, there is uncoupling of suppressed lactation from involution, which allows the mothers to leave their calves for long foraging trips at sea [4]. For *in vitro* studies, two models of differentiated mammary cultures are available: "3D" cultures in a colloidal medium such as reconstituted extracellular matrix or collagen [5, 6][5, 6][5], or culturing on a permeable membrane support, (*e.g.*, "Transwell®" dishes) [7, 8].

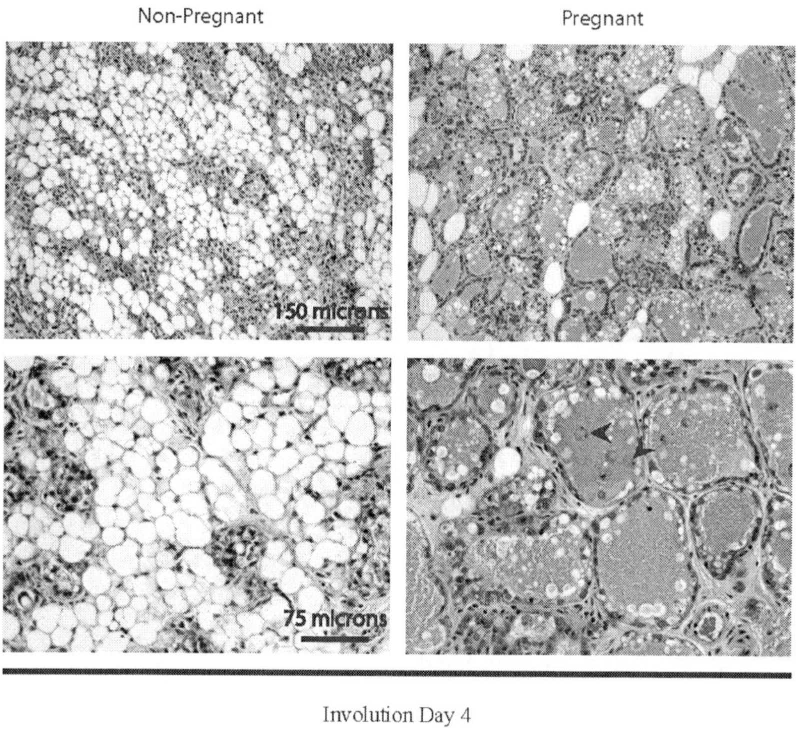

Figure 3. Pregnancy inhibits the second phase of involution. Micrographs of mouse mammary gland at day 4 involution are represented. Involution was induced by removal of pups at peak lactation (day 10). One group of mice was mated to be concomitantly pregnant at the time of involution. Comparison is made between a non-pregnant involuting mouse and a mouse with concomitant pregnancy at the time of involution. Mammary glands of pregnant mice are reminiscen of the first phase of involution, as seen by the change in epithelial cell morphology, cell shedding and death (arrowheads) and accumulation of milk within the alveoli. Pregnancy and associated hormonal changes are able to prevent the second phase of involution, as seen by maintenance of highly organized alveoli in comparison to remodeled tissue and adipocyte repopulation seen in the non-pregnant mice.

2. Milk Stasis and Inhibition of Milk Secretion

Milk accumulation and alveolar engorgement occurs in the absence of suckling, and is the physiological trigger for inhibition of milk secretion [4, 9][4, 9] (Figure 1B). During milk stasis the mammary epithelial cells (MECs) transition from secretory columnar epithelium to a non-secretory, quasi-squamous morphology. Unilateral teat sealing experiments in mice have shown that inhibition of milk secretion occurs locally, even in the continued presence of high levels of lactogenic hormones [*i.e.,* prolactin (PRL)] [10, 11][10, 11]. Hence, local signals within the alveoli inhibit milk synthesis in response to milk stasis. It is important to note here that very brief periods of milk accumulation also occur during lactation between bouts of nursing. These brief milk accumulation periods may trigger some responses that occur at the beginning of involution (extended milk stasis), but they are stopped short of inducing full scale involution by timely reintroduction of the suckling stimulus.

The physiological mechanisms regulating inhibition of milk secretion are poorly understood. Mechanical stretch may play a role, but experiments in goats suggested that simple distension of the glands is not sufficient to inhibit milk secretion[12]. Some studies have suggested that a secreted factor is responsible for milk stasis-induced inhibition of milk secretion [12]. The identity of such a factor has been the subject of intense debate for several decades. Most likely, the inhibition of milk secretion is the manifestation of multiple factors acting simultaneously. Our laboratory has identified serotonin (5-hydroxytryptamine, 5-HT), which is synthesized and secreted by MECs, as a potent inhibitor of milk secretion [13] and a potential signal that is responsible for several responses to milk stasis.

A) Inhibitors of Milk Secretion

Serotonin (5-HT) is a biogenic monoamine synthesized locally by MECs in the mouse, bovine and human mammary glands (Figure 4A) [13, 14]. It has classically been studied as a neurotransmitter, and its synthesis is catalyzed by two separate genes, one of which is exclusively neuronal [tryptophan hydroxylase (TPH) 2], and another which is expressed in non-neuronal cells (TPH2) [15, 16][15, 16]. The 5-HT system is complex, consisting of more than 15 different receptors, encoded by 14 genes, which are classified into 7 families (5-HT_{1-7}). In addition 5-HT exposure is controlled by a serotonin-specific reuptake transporter (SERT) and multiple catabolic enzymes, and non-receptor mechanisms of action have been documented [17-19][17-19]. Normal human mammary epithelium expresses 5-HT_{1D}, 5-HT_{2B}, 5-HT_{3A}, and 5-HT_7 receptors [7, 8]. Bovine mammary epithelium has a slightly different 5-HT receptors profile (1B, 2A, 2B, 4 and 7). In addition to the presence of various 5-HT signaling receptors, SERT expression has been verified in human, bovine and mouse mammary epithelial cells[20-22] (Figure 4A).

Milk stasis is a potent inducer of 5-HT synthesis and secretion by MECs in mouse mammary gland [13, 20]. In turn, 5-HT potently inhibits milk protein gene expression in cell and organ cultures of mouse mammary glands. Additionally, $TPH1^{-/-}$ mice (devoid of peripheral 5-HT) fail to inhibit milk secretion upon stasis, which results in engorged glands that leak milk. Analogous to the effect of 5-HT on milk secretion in mouse, 5-HT secreted in the bovine mammary gland inhibits milk protein secretion and increasing the bioavailability of 5-HT (by blocking reuptake transporter-SERT) suppresses bovine lactation [22]. In addition, 5-HT_{2B}, 5-HT_4, and 5-HT_7 receptors have been implicated in inhibiting milk protein expression in 3D collagen cultures of primary bovine mammary epithelial cells [8][8]. Due to the function of 5-HT in the nervous system, SERT inhibitors (SSRIs) are a commonly prescribed medication for humans. This has allowed the study of changes in 5-HT bioavailability on lactation and milk secretion in women. Women taking SSRI antidepressants exhibited a delayed onset of copious milk secretion [21], further supporting a role for 5-HT in the regulation of milk secretion in humans.

It is not known whether 5-HT directly regulates milk protein gene expression through its receptor signals. However, one mechanism employed by 5-HT (in human, mouse and bovine) is the disruption of epithelial tight junctions (TJs) [20, 22]. TJs are crucial for maintaining the columnar secretory epithelial cells and directing vectorial secretion [23][23]. This effect of 5-

HT is mediated by the 5-HT_7 receptor [20, 24]. The effect of 5-HT on MEC TJs is discussed in detail in the later sections of this chapter. To date, 5-HT is the most exhaustively studied inhibitor of milk secretion in multiple mammalian species (*e.g.,* mouse, bovine and human).

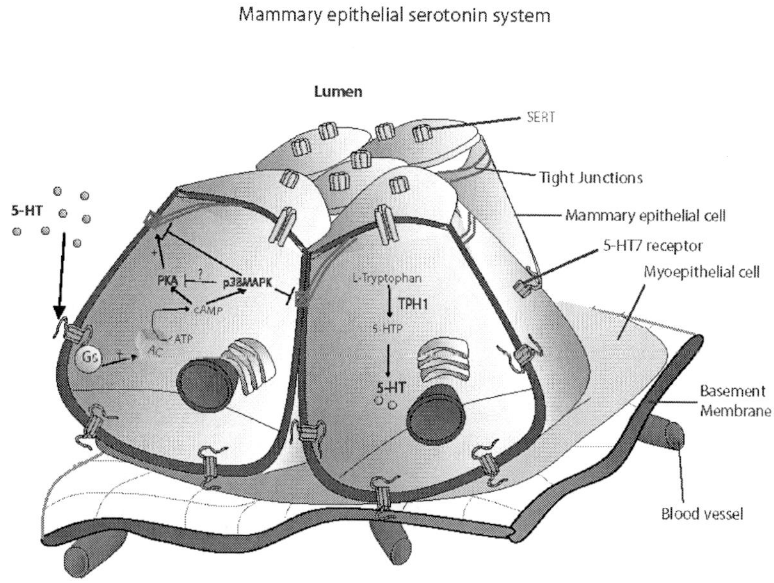

(A)

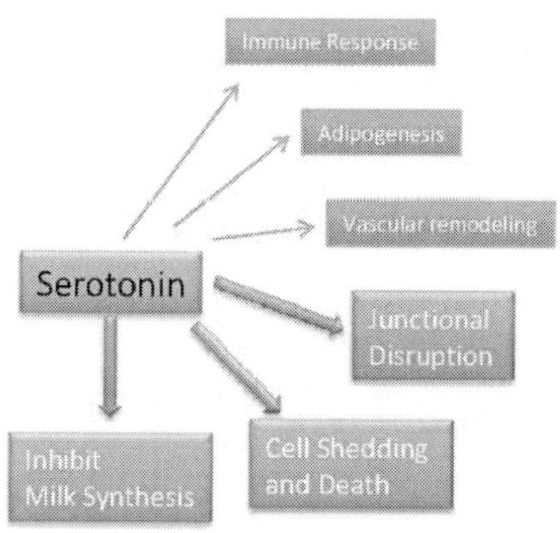

(B)

Figure 4. Mammary epithelial serotonin system and its mechanism of action (A) Diagramatic representation of mammary epithelial serotonin system. Mammary epithelial cells synthesize serotonin and secrete it into their surroundings. This serotonin acts in an autocrine-paracrine manner through its receptors. Among these 5-HT_7 receptor has been localized to the basolateral aspect of the mammary epithelial cells. 5-HT through 5-HT_7 receptor generates two signals: a cAMP-PKA signal and a cAMP-p38MAPK signal. Serotonin reuptake transporter (SERT) is present on the apical membrane of the mammary epithelial cells and is involved in recycling and metabolism of mammary serotonin. (B) Schematics representing serotonin function during involution. Block arrows show functions of serotonin that have been experimentally detected. Line arrows show possible unexplored functions of serotonin in mammary gland involution.

Another locally secreted factor involved in inhibition of milk secretion is the iron-binding glycol-protein of the transferrin family, lactoferrin (LTF). LTF is synthesized by the guinea pig, mouse, pig and human mammary epithelial cells, and its expression is dramatically increased (~ 100 fold) upon milk stasis [25-27]. LTF has been shown to suppress casein expression in bovine 3D mammosphere cultures [28]. Analogously, LTF transgenic female mice fail to sustain their pups due to lack of milk [29]. Histological evaluation of LTF transgenic mouse mammary glands show extensive secretory products held within apical membrane outgrowths of alveolar epithelial cells. This suggests an impaired cellular secretory process. Thus, 5-HT and LTF may be working on different aspects of the milk secretion process (gene expression, TJs, exocytosis, etc.) to ultimately inhibit milk secretion in response to stasis.

Stasis-induced local inhibition of milk secretion occurs in spite of normal levels of systemic lactogenic hormone. It has been shown that other locally induced factors, such as the Interleukin (IL)-6 family of cytokines, decrease the sensitivity of the epithelial cells to lactogenic hormones by suppressing their STAT5 mediated signaling (Figure 5) [30].

In summary, a combination of direct inhibition of milk protein synthesis and secretion and desensitization of epithelial cells to lactogenic hormones is employed by local factors to inhibit milk secretion in response to milk stasis (Table 1). It is important to note that these local factors (*i.e.,* 5-HT, LTF, cytokines) may also be induced during lactation where temporary inhibition of milk secretion might become necessary between bouts of nursing. The effects of 5-HT and LTF on milk secretion are reversible based upon the duration of milk stasis. Extended milk stasis results in a cascade of secondary events that induce irreversible changes in the gland.

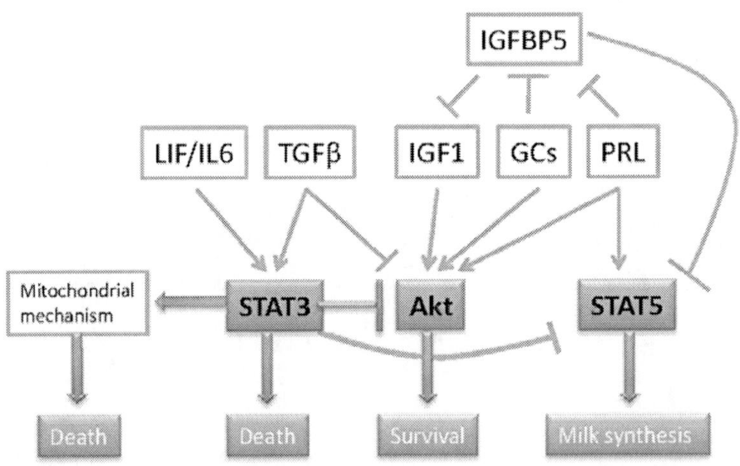

Figure 5. Schematics of important local and systemic signals involved in involution associated cell death. All the orange arrows and tabs are signals inducing involution-associated cell death. All green arrows and tabs are signals associated with preventing involution-associated cell death. Systemic signals like PRL, GC and IGF1 induce survival of mammary epithelial cells mainly through Akt and Stat5 signals. However, local factors induced during involution, not only induce Stat3-mediated cell death but also suppress the survival signal mediated by Akt and Stat5. In addition, Stat3 inhibits Stat5 and induces IGFBP5 (not shown).

Table 1. Lists important systemic and local players for each involution event. It also specifies the time of upregulation of these factors and their mode of action during involution

	Important genes and signaling molecules	Direction of change during involution	peak change	Induced by	major signaling mediator	Other important actions	Refs
Milk stasis							
	TGFb3	Expression and activity increased	6hrs		Akt inactivation and mitochanodrial mecahnism and Stat3 activation	supress stat5 and inturn milk protein synthesis	{{1727 Flanders,K.C. 2009}}
	5-HT	increased	24hrs		Exact mechanism unknown possible via one of its GPCR receptors		{{144 Matsuda,M. 2004; }}
	lactoferrin (LTF)	Highly increased	day 1	IL-1β, cell shape change and cytoskeleton reorganization	IL-1b	suppress milk synthesis and secretion (possibly indirect)	{{827 Baumrucker,C.R. 2006; }}
	LIF	highly increased	12 hrs-day3		stat3 activation	inactivate stat5 and thus suppress milk synthesis and secretion	{{1049 Tiffen,P.G. 2008; }
	Oncostatin-M(OSM)	increased	day 2	LIF/stat3	stat3 activation	inactivate stat5 and thus suppress milk synthesis and secretion	{{1049 Tiffen,P.G. 2008; }
Cell death							
First phase	TGFb3	Expression and activity increased	6hrs	??	Akt inactivation and mitochanodrial mecahnism and Stat3 activation	cell death	{{1727 Flanders,K.C. 2009; }}
	LIF	highly increased	12 hrs-day3		stat3 activation		{{862 Kritikou,E.A. 2003; }}
	IL-6	increased	24hrs		stat3 activation		{{843 Zhao,L. 2002; }}
	IGFBP5	Highly increased	48hrs	stat3 loss of PRL and GC sensitivity	block IGF1 survival signal	decrease anti-apoptotic Bcl2 family genes	{{223 Tonner,E. 2002; }}
	α-lactalbumin	increased	12hrs-???		unknown		
	5-HT	increased	24hrs		via GsPCR -cAMP		{{144 Matsuda,M. 2004; 1733 Pai,V.P. 2009}}
	Epithelial Apical Ca2+ ATPase	rapidly declined	24hrs		Cytosolic Ca2+ accumulation and mitochondrial Ca2+ overload		{{1668 Reinhardt,T.A. 2009}}
	lactoferrin (LTF)	Highly increased	day 1	IL-1β, cell shape change and cytoskeletal reorganization	indirect via IL-1b and natural killer cells and direct inhibition of cell cycle		{{ 818 Son,K.N. 2002; 817 Baratta,M. 2005; }},{{810 Shau,H. 1992;804 Birgens,H.S. 1984; }}, {{811 Damiens,E. 1999; }}

Table 1. Continued

	Important genes and signaling molecules	Direction of change during involution	peak change	Induced by	major signaling mediator	Other important actions	Refs
Second phase	ECM/BM breakdown	increased	day4?		anoikis - loss of attachment		{{1117 Lund,L.R. 2000; }}Science. 1995 Feb 10;267(5199):891-3
	Vit-D sysnthesis enzyme and VDR						
	Vitamin-D receptor	increased	day 4		indirect via TGFb1		{{944 Zinser,G.M. 2004; }}
	Oncostatin-M (OSM)	increased	day 2	LIF-Stat3	Stat3	inactivate stat5, anoikis (via MMP induction)	{{1049 Tiffen,P.G. 2008; }}
	IL-10	increased	day5	?	DR4 and TRAIL		{{838 Sohn,B.H. 2001; }}
	Fibronectin fragments (FN-F)	increased	day 4-5	??	??	induce cel death in mouse mammospheres	{{1109 Schedin,P. 2004; }}
	PRL and PRL signal	decreased	48 hrs	LIF/OSM - Stat3, TGFb, 5-HT Absence of suckling stimulus	stat5		{{31 Flint,D.J. 1994; 394 Travers,M.T. 1996; }}
	GC and GR signal	decrased	72hrs	??	GSK		Development 1996;122(1):181–93.) {{468 Feng,Z. 1995; }}
	IGF1 signal	decreased			decreased IRS-1, IRS 2 and Akt		{{205 Hadsell,D.L. 2001; }}
	Cathepsin	increased	day 3-4	??	autophagy by their proteolytic activity and anoikis by ECM breakdown		
Immune system and celarence							
First phase	GRO1, IL-1α, and IL-13,	Increased	12hrs	LIF-Stat3?? And 5-HT??		pro-inflamatory - Leucocyte and neutrophil attaction and activation	{{959 Stein,T. 2004; 967 Clarkson,R.W. 2004; }}
	IL-1β,	increased	12 hrs	LIF-Stat3?? And 5-HT??		pro-inflamatory - increase neutrophil, macrophage and plasma cells, induce LTF	{{845 Wedlock,D.N. 2004; 834 Nickerson,S.C. 1992; }}
	OSM, uterocalin, SGP2(clustrin)	increased	day 2			anti-inflammatory - prevent leucocyte and neutrophil extravasation	{{967 Clarkson,R.W. 2004; }}
	Acute phase response (APR) genes	increased	day 1	LIF - Stat3		Phagocyte (mostly non-professional) activation and prevent infections	{{959 Stein,T. 2004; 967 Clarkson,R.W. 2004; }}
	Milk fat globule EGF factor 8 (Mfge8)	increased	day 3	??		phogocytosis via binding to phosphotidyl serine on apoptotic cells	{{987 Hanayama,R. 2005; 967 Clarkson,R.W. 2004; }}

	Lactoferrin (LTF)	Highly increased	day 1	IL-1β		antimicrobial and anti-inflammatory	{{804 Birgens,H.S. 1984;805 Sanchez,L. 1992; }}
Second phase	CXCL14, CD68, CSF1, galectin3 and cathepsin-S	Increased	day 4	LIF-Stat3?? And 5-HT??		Macrophage and B-cell attraction	{{959 Stein,T. 2004; 967 Clarkson,R.W. 2004; }}
	Immunoglobulins	increased					{{959 Stein,T. 2004; 967 Clarkson,R.W. 2004; }}
	IL-10	increased	day5			potent immuno suppresent and prevent inflammatory reaction	{{839 Garofalo,R. 1995; }}
Junctional complex regulation							
First phase	5-HT	increased	24 hrs		5-HT7-cAMP-p38MAPK	Disruption of tight junctions	{{145 Stull,M.A. 2007;}}{{1263 Pai,V.P. 2008; }} J Clin Endocrinol Metab. 2010 Feb;95(2):837-46 and Hernandez et al unpublished)
Second phase	PRL and PRL signal	decreased	48 hrs	LIF/OSM - Stat3, TGFb, 5-HT absence of suckling stimulus	stat5	facilitate TJ disruption	{{31 Flint,D.J. 1994; 394 Travers,M.T. 1996; }}
	GC and GR signal	decrased	72hrs	??	GSK	facilitate TJ disruption	{{294 Stelwagen,K. 1999; 551 Thompson,G.E. 1996; }}
ECM remodeling							
First phase	Maspin	High	day 4			inhibit serine protease like uPA and prevent plasmin generation	Dev Biol. 1999 Nov 15;215(2):278-87
	TIMPs - Tissue inhibitors of MMPs	High	day 3			inhibit MMPs	J Cell Biol 118:1271–1282
	IGFBP5	Highly increased	48hrs	stat3		induces tPA, inhibits PAI-1 and thus facilitates plasmin generation	{{224 Sorrell,A.M. 2006; }}
Second phase	uPA - Urokinase plasminogen activator	Highly increased	day 4			Plamsin generation	Development. 1996 Jan;122(1):181-93
	Plasmin	Highly increased	day 4-5	uPA, tPA and Pkal		ECM and BM degradation, induction of MMPs	{{1117 Lund,L.R. 2000; 1110 Lijnen,H.R. 2001}}
	MMPs - Matrix metalloproteases	Highly increased	day 3-4	Plasmin and MMPs		Proteolytic degradation of ECM and BM	{{1109 Schedin,P. 2004; }}
	Fibronectin (FN) and its fragments (FN-F)	Highly increased	day 4-5	MMPs and Plasmin		cell death and integrin disruption	{{1109 Schedin,P. 2004; }}
	laminin fragment-DIII	increased	day 4-5	MMPs and Plasmin		Induce MMP2 and 9 and block integrin signal	{{1105 Schenk,S. 2003;}}
	Cathepsins	increased	day3-4			BM degradation	J Mammary Gland Biol Neoplasia (2009) 14:171–179{{1128 Burke,M.A. 2003; 1127 Guenette,R.S. 1994; }}

Table 1. Continued

	Important genes and signaling molecules	Direction of change during involution	peak change	Induced by	major signaling mediator	Other important actions	Refs
	Oncostatin-M(OSM)	increased	day 4	LIF/stat3	stat3	???	{{1049 Tiffen,P.G. 2008; }}
	GC and GR signal	decrased	72hrs	??		Inhibit uPA , MMPs and cathepsins	{{396 Lund,L.R. 1996; }}
	PRL and PRL signal	decreased	48 hrs	LIF/OSM - Stat3, TGFb, 5-HT	stat5	inhibit MMPs and plasmin generation, induce TIMPs	{{1036 Talhouk,R.S. 1992; }}{{395 Tonner,E. 2000; 106 Allan,G.J. 2002}}
	GH	decreased	day 2			inhibit MMPs and plasmin generation, induce TIMPs	{{1036 Talhouk,R.S. 1992; }}{{395 Tonner,E. 2000; 106 Allan,G.J. 2002}}
Adipose tissue remodeling							
First phase	IL-1β,	increased	12 hrs			inhibits aurocine PRL secretion by pre-adipocytes and hence prevents pre-adipocyte differentiation	{{842 Path,G. 2001; }}
	IL-6	increased	24hrs			Induces lipolysis (loss of lipids) and prevents pre-adipocyte differentiation	{{842 Path,G. 2001; }}
	5-HT	increased	24 hrs		5-HT1 and 5-HT2A	5-HT1 mediated inhibition of preadipocyte differentiation and 5-HT2A mediated induction of preadipocyte differentiation	{{1192 Uchida-Kitajima,S. 2008; }}
second phase	Pkal mediated plasmin generation	increased		Factor XII	Plasmin	breakdown of fibronectin ECM surrounding pre-adipocytes	{{1150 Selvarajan,S. 2001; }}
	TIMPs	locally increased (preadipocytes)	?			Induce PPAR-γ and c-ebp mediated pre-adipocyte differentiation	{{1161 Alexander,C.M. 2001; }}
	MMPs - Matrix metalloproteases	Highly increased	day 3-4	Plasmin and MMPs		inhibit the action of TIMPs on pre-adipocytes	{{1161 Alexander,C.M. 2001; }}
	Cathepsins S and K	increased	day 4			Induce adipocyte differentiation by proteolytic action on fibronectin and collagen	{{1194 Taleb,S. 2006; }}Endocr J. 2009 Mar;56(1):55-63)
	PRL and PRL signal	decreased	48 hrs	LIF/OSM - Stat3, TGFb, 5-HT	stat5	suppresses lipogenesis and triglyceride storage in adipocytes	{{24 Barber,M.C. 1992; }}

GH	decreased	day 2			suppresses lipogenesis and triglyceride storage in adipocytes	{{24 Barber,M.C. 1992; }}
Leptin and its receptor (OB-Rb)	increased	day 10	drop in PRL		induces preadipocyte differentiation	{{923 Lin,Y. 2007; 924 Aoki,N. 1999; }}
Autocine PRL	increased	?	?	?	induces preadipocyte differentiation	{{1189 Brandebourg,T. 2007; }}
Vascular remodeling						
VEGF and VEGFR	epithelia and surrounding stroma - decreased	?	phagocytosis		vascular capillary regression	{{1184 Pepper,M.S. 2000;1185 Hovey,R.C. 2001; }}
	adipose and surrounding stroma - increased		adipocytes		angiogenesis and vascular development	{{1184 Pepper,M.S. 2000;1185 Hovey,R.C. 2001; }}
PRL and PRL signal (systemic)	decreased	48 hrs	LIF/OSM - Stat3, TGFb, 5-HT		induces angiogenesis	{{456 Gaytan,F. 1997; 425 Struman,I. 1999; }}
PRL fragment 16K	increased in epithelia				anti-angiogenic via inhibition of VEGF and FGF	{{456 Gaytan,F. 1997; 425 Struman,I. 1999; }}
5-HT	increased	24 hrs			mitogenic for endothelial and vascular smooth muscle cells	{{1446 Eddahibi,S. 2006; 1445 Pakala,R. 1994; }}

3. Epithelial Regression via Cell Death

During lactation the mammary gland consists of an elaborate epithelial machinery for milk synthesis and secretion, which occupies majority of the gland. Involution involves massive epithelial tissue regression (~80%), mainly via epithelial cell death [31, 32]. The dying cells can be visualized as caspase-positive cells being shed into the lumen [33]. This wave of epithelial cell death transcends both phases of involution (Figure 1 and 2) [34]. In mouse mammary glands, epithelial cell death is seen within 12 hours of weaning and peaks between day 2 and day 3 (Figure 1) [32-34]. The timing of this peak in cell death (between day 2 and 3) suggests that local factors induced during Phase 1 of involution are likely to be major inducers of epithelial cell death. Analogously, unilateral teat sealing (milk stasis) in mice shows only local induction of cell death in the sealed gland [11]. Interestingly, this cell death occurs even in presence of high levels of systemic lactogenic survival factors like PRL, glucocorticoids (GCs) and insulin-like growth factors (IGF-I).

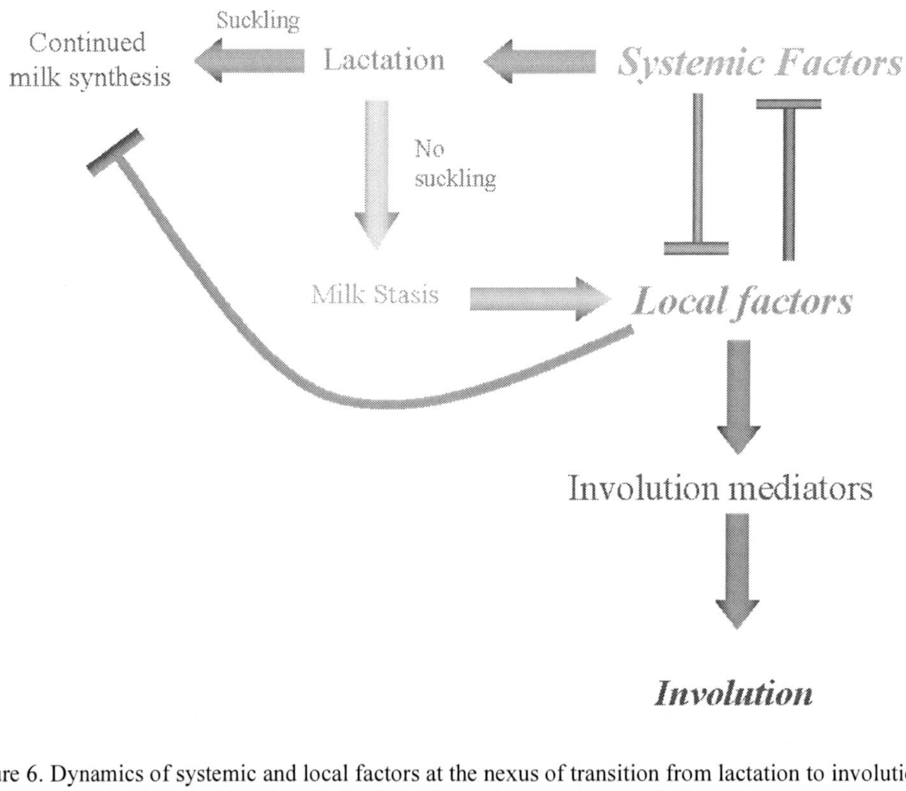

Figure 6. Dynamics of systemic and local factors at the nexus of transition from lactation to involution. The green arrows represent signals inducing lactation and preventing involution. Orange arrows represent signals inducing involution. Systemic factors like PRL and GCs drive lactation and milk synthesis. The systemic factors also keep the induction of local mammary gland factors in check. This status quo is maintained as long as milk is regularly removed from the gland via suckling. Absence of suckling results in milk stasis which is a potent inducer of local mammary gland factors. These local factors inhibit further milk secretion by direct inhibition of milk synthesis as well as indirectly making the gland insensitive to systemic factors. Continued absence of suckling leads to induction of other local involution mediators driving the gland into involution.

A fine balance between survival factors (mostly systemic hormones) and cell death factors (local factors like 5-HT, ILs, transforming growth factor- β; TGFβ) regulates epithelial regression (Table 1). Epithelial regression, initially reversible during early involution (2 days), becomes irreversible as involution progresses. Akt serves as a master sentinal in regulating survival signals while STAT3 and the Bcl2 family of proteins are the major intracellular regulators of cell death [32] (Figure 6). It is intriguing to note that cell death occurs in a heterogeneous fashion and not all mammary epithelial undergo apoptosis. The mechanisms employed by the surviving cells to evade the involution death signals are not yet known.

A) Inducers of Cell Death during Phase 1 of Involution

Studies over the past decade have established 5-HT as a crucial local regulator of multiple involution events. The presence of multiple 5-HT receptors in the mammary epithelium facilitates simultaneous regulation of different functions by 5-HT [7, 8, 35][7, 8, 20, 35]. *In vitro*, 5-HT inhibits cell growth of primary human MECs and induces cell death on prolonged exposure (3 days) [7, 8, 35][7]. This growth inhibitory effect of 5-HT is mediated via 5-HT_7 receptor activation of p38MAPK. Analogously, exposure of mouse mammary explants to 5-HT results in alveolar collapse, with cells having pychnotic, fragmented nuclei (a sign of cell death) even in the presence of PRL [13][13]. Interestingly, in bovine mammary epithelial cultures 5-HT_{1B} and 5-HT_{2A} act as anti-apoptotic signals [7, 8, 35][8]. This difference might be a manifestation of specific selective pressure on the bovine mammary gland for high milk production. Whether 5-HT induces epithelial cell death in the bovine mammary gland remains to be thoroughly tested.

Mammary epithelium mainly consists of 2 layers of cells; apical secretory epithelial cells and basal myoepithelial cells. In addition, suprabasal multipotent cells are interspersed between the two layers and can replenish both apical or basal cells as needed [7, 8, 35, 36][35, 36]. Using *in vivo* studies in mouse and a Transwell® model of differentiated human mammary epithelial cells, we have shown that 5-HT induces epithelial cell shedding and cell death [7, 8, 35][35]. Cell shedding is seen as early as 4 hours after 5-HT exposure and progressively increases with the time of exposure. Analogously, $TPH1^{-/-}$ mice (devoid of peripheral 5-HT) show a complete absence of cell shedding in response to milk stasis. However, sustained exposure to 5-HT (3 days) induces apoptosis in the suprabasal cells causing regressive and irreversible changes to the epithelium.

In addition to 5-HT, TGFβ3 and ILs are also highly induced during early involution at 6 hours and 24 hours, respectively [37-39] (Figure 2). Targeted expression of TGFβ3 to the mouse mammary epithelium results in cell death [37]. In contrast, mice expressing dominant-negative TGFβ3 receptors (TβRII) have significantly impaired cell death upon milk stasis [40]. Transplantation of $TGF\beta3^{-/-}$ mouse mammary epithelium into a cleared fat-pad of wild-type host inhibited cell death (by ~70%) upon milk stasis [37]. suggesting that TGFβ3 acts locally in an autocrine manner in inducing epithelial cell death. Because TGFβ is stored as a latent pro-hormone in the extracellular matrix [39, 41-43][41-43], it is likely that its release in response to early signals that result in protease activation is a critical step in driving early involution forward.

IL-6 expression peaks at day 1 after pup removal and drops to baseline by day 2 in mice (Figure 2). IL-6$^{-/-}$ mice show delayed involution due to decreased epithelial cell death [44]. Unlike other IL-6 family cytokines, leukemia inhibitory factor (LIF) expression remains high until day 3 (Figure 2). LIF induces epithelial cell death in mouse mammary glands as seen via pellet implantation studies [45]. Analogously, LIF$^{-/-}$ mice show impaired mammary epithelial cell death and delayed involution [46].

A potential role for the insulin-like growth factors (IGF) in involution is suggested by the expression of IGF-binding proteins (IGFBPs) during involution [47, 48][47, 48]. IGFBP-5 binds to and sequesters IGF-I. In mouse, IGFBP5 levels are highly upregulated (50 fold) and peak by day 2 after pup removal (Figure 2) [49]. IGFBP5 transgenic mice show accelerated involution with decreased IGF-I signal and increased cell death [50]. Conversely, IGFBP5 -/- mice have delayed involution due to decreased epithelial cell death [47].

Fur seals provide a unique example where milk stasis is uncoupled from epithelial regression [4]. Lactating fur seals participate in long foraging trips which mandate the suspension of suckling. Interestingly, the composition of the milk in these mammals differs from that of other mammals studied [51, 52]. Specifically, α-lactalbumin, which is present in the milk of most species, was completely absent in fur seals. This suggested an important role of α-lactalbumin in epithelial regression. Accordingly, α-lactalbumin pellets induce mammary epithelial apoptosis and α-lactalbumin$^{-/-}$ mice retain lactation morphology even 4 days after pup removal [53, 54]. Biochemical analysis has shown that accumulation of α-lactalbumin results in multimeric structures that can induce cell death [55]. A complicating fact is the dual roles of α-lactalbumin, which is both a secreted milk protein, and a subunit, along with β-galatosyltransferase, of the multimeric enzyme lactose synthase [55, 56][55, 56]. Because lactose is the most important osmolyte in the milk of most species, effects of α-lactalbumin deficiency are likely to differ in mice and other species, compared with the specialized system in fur seals.

B) Inducers of Cell Death during Phase 2 of Involution

Phase 2 of involution consists of the breakdown of extracellular matrix (ECM) and basement membrane (BM). Loss of attachment-dependent survival signal primes the mammary epithelial cells for cell death by anoikis [32, 57, 58]. In addition, breakdown products of ECM proteins (*e.g.*, fibronectin and laminin fragments) are also pro-apoptotic [59]. Other local factors induced during the second phase of involution promote epithelial cells death. Vitamin-D (Vit-D) receptor and oncostatin-M (OSM) (IL-6 family cytokine) are induced locally 3 days after pup removal in mice (Figure 2) [30, 60]. Vit-D receptor$^{-/-}$ and OSM$^{-/-}$ mice show decreased epithelial regression only after 2 days [30, 60].

C) Influence of Systemic Hormones on Epithelial Cell Death

Systemic hormones like PRL, GC and IGF-I are critical survival (anti-apoptotic) factors for mammary epithelial cells (Figure 6) [61-63]. GCs also prevent ECM and BM breakdown (discussed later in this chapter). During early involution PRL and GC treatment suppresses cell death [3, 10, 40, 49, 64]. However, as involution progresses, local factors, such as TGFβ,

5-HT, ILs, IGFBP5, cause MECs to become refractory to systemic hormone action by inhibition of PRL and IGF-I signaling (Figure 2 and 6) [63-65]. Finally, continued absence of the suckling stimulus results in a sustained lower levels of systemic PRL and GC (Figure 4) [10].

Taken together, a progressive increase in local pro-apoptotic factors is accompanied by a gradual decrease in survival factor signals (Table 1). Early in the process, the epithelial architecture is maintained by the recent exposure to elevated systemic hormones (Figure 6). In the absence of suckling stimuli the gland also becomes refractory to the systemic hormones (Figure 2 and 5). These combined effects create permissive conditions for epithelial regression, which is also facilitated by ECM and BM remodeling. By the end of involution the epithelium regresses, resulting in a rudimentary ductal tree morphologically similar to a virgin gland. How the ductal epithelial cells survive this intense pro-apoptotic environment remains unknown.

4. Role of Epithelium in Immune Response and Clearance of Cell Debris

A) Involution-Associated Immune Response

Mammary gland involution results in massive amounts of residual milk accumulation and cellular debris. Efficient clearing of these waste products is crucial for the health of the gland, which otherwise faces impaired lactation following subsequent pregnancies, and increased risk of diseased states like ductal ectasia, mastitis, inflammation, and breast cancer [66, 67]. Immune cells are present in the mammary gland at all stages of development, including involution [68]. The precise role of the immune system during involution has only recently begun to be elucidated. Gene expression patterns and physiological evaluations indicate a distinct involution-associated immune response, which resembles a wound healing process [38]. The involution-associated immune response includes: a) a primary neutrophil activation, b) a secondary macrophage and local acute-phase response (APR) activation, and c) a late B-lymphocyte response, without any signs typical of infection.

Neutrophil and Leukocyte activation: With the onset of involution there is a burst in the expression of pro-inflammatory cytokines and their receptors (e.g., IL-1α, IL-1β, and IL-13) within 12 hours of weaning (Figure 2 and table 1) [38, 69]. Analogously, an early increase in neutrophils (by day 1) is detected in the gland by immunohistochemistry. However, it is of interest to note that although neutrophils are attracted early (day 1) into the involuting gland, their extravasation into the regressing tissue is not seen until day 3-4 of involution. A concomitant release of anti-inflammatory factors by the mammary epithelium possibly prevents neutrophil extravasation, thus attenuating the inflammatory response [69]. Similar observations have been made during mammary involution of ruminants (ewes/sheep and cows) and monogastric mammals (sows and guinea pigs) [66, 70-72]. These observations put into question the involvement of leukocytes/neutrophils in clearance of cellular debris and residual milk during early involution.

Regulators of the epithelial expression of pro- and anti-inflammatory cytokines are not fully understood. Infusion of cell and fat-free involution secretion into non-lactating glands of ewes/sheep elicited an immune reaction, as measured by leukocytes count [73]. This suggests the presence of a secreted immune mediator during involution. Serotonin A role for 5-HT as an immune mediator is well established [74], and our lab has shown 5-HT to be highly induced during early involution [13]. Hence, 5-HT may be one factor that elicits immune reactions directly by activation of immune cells, or indirectly by inducing release of cytokines (Figure 2 and 5).

Phagocyte and acute phase response activation: Gene expression analysis during mouse involution revealed that 12 acute phase response (APR) genes were highly upregulated by day 1 and remained high until day 3 [38, 69]. Contrarily, macrophage and B-cell chemokine (e.g., CXCL14) and macrophage activators (e.g., CD68, colony stimulating factor-1 (CSF1)) were induced later (day 3-4) [38, 69]. Analogously, a delayed increase in macrophage infiltration (day 4) was observed immunohistochemically. Interestingly, phagocyte and APR marker (CD14) showed potent upregulation exclusively in the luminal epithelial cells (see 4b for details) [75, 76][75, 76]. These results suggest two things: a) an APR is induced in the absence of infiltrating immune cells like monocytes and macrophages (classical inducers of APR), and b) induction of APR and possible phagocytic activity by MECs during early involution (discussed below in clearance section).

B-lymphocyte activation: Out of 145 genes upregulated during involution, 49 encoded immunoglobulin (Ig) genes, which were induced by day 3 in mouse involution (Figure 2) [38, 69]. This was accompanied by a large increase in plasma cells between day 2 and 4 of involution. Similar observations have been made during ruminant involution [77, 78]. Most of the Igs (IgA, IgM and IgG) present during involution are synthesized locally by the plasma cells underlying the epithelium. The precise nature of Ig regulation and its function during mammary involution is not yet known.

B) Clearance of Cell Debris

Detecting phagocyte/macrophage involvement in involution is complicated due to presence of macrophage markers on the surface of non-professional phagocytes (viable epithelial cells) [67, 79, 80]. The professional phagocytes (macrophages) appear late (day 3) in an involuting mouse mammary gland. The sheer number and mass of dead epithelial cells, the large extent of residual milk and the short timescale over which majority of clearance occurs (by day 3-4 in mouse [76], strongly suggest a major role of non-professional phagocytes in clearance (Figure 1). The contribution of professional phagocytes (macrophages) in clearance appears to be rather limited. By involution day 3, a majority of viable mammary epithelial cells contain ingested apoptotic cells, casein micelles and milk fat globules within their cytoplasmic vaculoles [75, 81]. The engulfment of these materials by mammary epithelial cells has been shown to be mediated through phosphatidyl serine receptor (PSR) and milk fat globule EGF factor 8 (Mfge8 - lactadherin in humans) [67, 69]. Mfge8$^{-/-}$ mice show accumulated milk fat globules and apoptotic cells in the involuting mammary gland [67, 82]. As involution progresses the Mfge8$^{-/-}$ mice develop inflammation and show mammary gland ectasia (dilation of the ductual structure) with impaired mammary gland development during subsequent lactation cycles. These studies further emphasize the importance of proper clearance as executed by the non-professional phagocytes.

As involution progresses, the number of epithelial cells functioning as non-professional phagocytes drastically decreases, along with a precipitous increase in apoptotic cells that need to be cleared. Thus, non–professional phagocytes, which are adjacent to apoptotic cells and residual milk, initiate the clearing process and prevent an inflammatory reaction by release of apoptotic cell contents. The professional phagocytes appear to function later and serve to complete clearance of cell debris restoring the gland to a virgin-like state.

Mammary gland involution is associated with a carefully orchestrated immune response devoid of a robust inflammatory reaction. This involution-associated immune response is three tiered. First, leukocytes and neutrophils are recruited into the gland. However, their involvement during early involution is limited by the release of immunosuppressive/anti-inflammatory cytokines by non-professional phagocytes (viable mammary epithelial cells). With the progression of involution and induction of an acute phase response the involvement of professional phagocytes (macrophages and neutrophils) increases. This is likely essential for complete clearing of the remaining cell debris and residual milk from the gland. Finally, a late B-cell response is induced the role of which is as yet unknown. The involution immune response thus ensures a safe clearance of cellular debris and residual milk.

5. Epithelial Tight Junctions and Barrier Regulation during Involution

Tight junctions (TJs) are intimately associated with defining the functional polarity of epithelial cells and guide vectorial secretion of substances, including milk within the mammary gland, to the apical surface [83, 84]. TJs form an apical seal between the mammary epithelial cells and thus function as a barrier for paracellular transport of fluids and ions and effectively compartmentalize the lumen and the interstitial space. Mammary epithelial TJs play a criticial role in mammary gland function and their disruption facilitates crucial events in the progression of mammary involution.

When involution is initiated by withdrawal of suckling, TJs are maintained for a brief period of time (18-24 hrs in bovine and goats) followed by their gradual breakdown [85, 86]. However, this TJ opening is reversible if the suckling stimulus (milk removal) is reintroduced early enough. In mice TJ opening has been shown to be reversible only through day 2 involution, beyond which it marks a transition to irreversible phase of involution [86]. The reversibility of TJs is attributed to a) the potent positive actions of systemic hormones (PRL and GC) [87, 88], and b) removal of accumulated milk (Table 1). However, upon continued absence of suckling, there is a drop in these systemic hormone levels and a decreased sensitivity to them (Figure 4). This in combination with local TJ disruptive factors makes the process irreversible [20, 89, 90]. It has been proposed that TJ disruption followed by leakage of milk components into the interstitium acts as one of the triggers for inducing other irreversible changes (e.g.,ECM breakdown) during involution [90-92]. 5-HT is the only locally-secreted factor that has been extensively studied for its actions on regulation of mammary epithelial TJs.

A) Factors Influencing Tight Junction Complexes during Involution

One possibility is that the compromised TJs observed after milk stasis are due to epithelial cell death. However, this theory is called into question by studies where shedding of dead epithelial cells occurs without a compromise in the TJ integrity {{[35];1214 Bement,W.M. 2002; 1100 Rosenblatt,J. 2001; }}.

Locally synthesized and secreted 5-HT has been found to regulate epithelial TJs in bovine, murine and human mammary gland (Figure 2 and 5) [20, 22, 35][20, 22, 35]. Studies in Transwell® cultures show that 5-HT regulates mammary epithelial TJs in a biphasic manner [24]. At low concentrations and shorter time points, 5-HT promotes TJ integrity, whereas sustained exposure to higher concentrations of 5-HT resulted in TJ disruption. The physiological basis for the early effect of 5-HT is not clear, but is believed to be a compensatory response to increased intraluminal pressure (by milk stasis) to maintain the compartmentalization. This biphasic effect of 5-HT on mammary epithelial TJs was confirmed *in vitro* and *in vivo* (in mice) through manipulation of endogenous mammary 5-HT levels and activity [21]. Similarly in bovine, 5-HT has been shown to disrupt mammary epithelial TJs both *in vitro* (3D cultures) and *in vivo* [22][22]. Both actions of 5-HT (potentiating and disruptive) on TJs are mediated though 5-HT_7 receptor in human mammary epithelial cells. This occurs via a switch in the downstream signal from conventional Gs-cAMP-PKA pathway to a Gs-p38MAPK pathway in response to increased exposure to 5-HT (Figure 5) [24]. Finally, the reversibility of TJ disruption upon removal of milk is due to replenishment of the apical cells. However, prolonged exposure to 5-HT induces apoptosis in these replenishing cells in addition to disrupting TJs making the process irreversible [20, 22, 35][35].

6. Extracellular Matrix (ECM) and Basement Membrane (BM) Remodeling

A) Mammary ECM Composition and Function

The critical role of ECM signals in mammary gland development and homeostasis has only recently began to be investigated [93]. Mammary gland stroma mainly consists of adipose tissue, connective tissue, fibrocytes and ECM, as well as vascular components [93]. A specialized form of ECM called basement membrane (BM) serves as an anchor for epithelial and myoepithelial cells (Figure 1 and 7). Stromal fibroblasts and adipocytes induce the myoepithelial cells to synthesize and secrete BM components [93]. These components, which include laminin, collagen, and fibronectin, engage integrins (α5, α6, β1 and β4) on the basal epithelial cell surface [93-95]. The integrins then transduce important biochemical (survival, differentiation and functional) and biophysical (changes in cell shape and morphology through cytoskeletal changes) signals in the mammary epithelial cells [95-97]. Integrin signals are the most studied aspect of ECM signaling. Loss of integrin signaling during ECM remodeling has been proposed to drive epithelial cell death during Phase 2 of involution [57, 59, 98, 99].

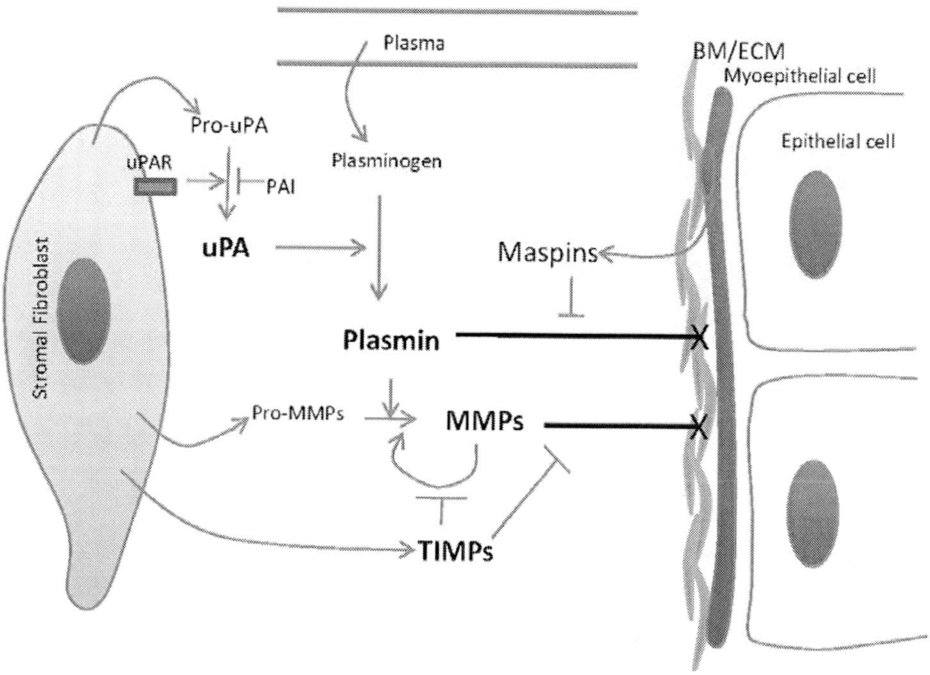

Figure 7. Plasminogen and MMP protease systems involved in matrix remodeling during mammary gland involution. Stromal fibroblasts are the key players in ECM/BM remodeling during involution. During the first phase of involution high expression of maspin by myoepithelial cells and TIMPs by fibroblasts prevents ECM/BM breakdown. This is further facilitated by low levels of pro-uPA and pro-MMP expression by the fibroblasts. Upon receiving appropriate signals during the second phase there is down regulation of TIMPs and maspin and a concomitant upregulation of pro-uPA and pro-MMPs. Active uPA is generated by its receptor uPAR. This results in uPA mediated generation of plasmin from circulating plasminogen. Plasmin directly breaks down ECM/BM as well as activates MMPs. Once activated MMPs can undergo a self activation loop and further induce ECM/BM breakdown.

B) ECM Changes during Mammary Involution

ECM remodeling during murine mammary involution is characterized by breakdown and removal of BM, which becomes thick, folded and discontinuous (Figure 1D) [100, 101]. Removal of BM starts after day 2 involution and peaks by day 4 (Figure 1D). This has been shown to occur via proteolytic hydrolysis and not by phagocytosis [100]. The proteolytic breakdown of ECM is irreversible and marks the Phase 2 of involution. Interestingly, unlike the mammary epithelial cells, myoepithelial cells remain relatively well-organized during involution, with remnants of collapsed alveoli within irregularly-defined rings (Figure 1D) [102]. The myoepithelial cells continue to surround the residual ductal buds at the end of murine involution [100, 101].

The influence of mammary ECM on epithelium during involution has been elegantly demonstrated by a series of experiments using 3D mammosphere cultures of mouse mammary epithelial cells using ECM extracted from mid-involuting (day 4-6) and late-involuting (day 8-10) rat mammary glands [103]. Mid-involuting ECM induced cell death of mammary epithelial cells, whereas late-involuting ECM induced formation of duct-like structures that were highly elongated with bifurcations. A detailed analysis of the ECM components at different stages of involution in rats showed that proteolytic enzymes (*i.e.,* matrix metalloproteases, MMPs), were dramatically increased (as high as 50 fold) with peaks on day 3 involution (Figure 2 brown arrows) [103]. Analogously, ECM and BM fragments further stimulate MMP secretion. (Figure 2 brown arrows) [103, 104].

C) Executors of ECM Remodeling

Two main ECM remodeling systems have been demonstrated in the mammary gland (Figure 7): a) the plaminogen (Plg) system, including Plg activators [urokinase Plg activator (uPA), tissue Plg activator (tPA) and plasma kallikrein (PKal)] and Plg inhibitors [Plg activator inhibitors (PAIs) and α2-anti-plasmin (aAP)], and b) the MMP system, which includes several MMPs and their inhibitors [Tissue inhibitor of MMPs (TIMPs)] [1].

In the mammary gland, the Plg system is present on the stromal cells (fibrocytes), except for Plg (zymogen), which is synthesized in the liver and released into circulation (Figure 7) [3, 58]. Plasmin is generated from Plg via proteolytic cleavage by serine protease Plg activators. uPA is the prominent Plg activator in mammary gland, with minor contributions by PKal [1]. Interestingly, serine protease inhibitors like maspin (inhibit uPA and Plg activation) are highly expressed in the involuting mouse mammary gland until day 4 followed by a gradual decrease in expression (Figure 2) [105]. Maspin is synthesized and secreted by the myoepithelial cells, juxtaposed to the BM (Figure 7). Such geographical proximity to the BM may be effective in preventing premature BM breakdown and maintaining the reversible nature of first phase of involution. Contrarily, uPA levels are low for the first 3 days of mouse mammary involution followed by a 30-fold increase by day 4 (Figure 2) [3]. In addition to directly degrading BM, plasmin also activates MMPs (Figure 7) [106, 107] The importance of the Plg system was demonstrated in $Plg^{-/-}$ mice which showed an absence of ECM/BM breakdown and alveolar regression even in presence of uPA and MMPs [58]. $Plg^{+/-}$ mice showed haplosufficient phenotype suggesting the importance of plasmin levels in executing involution ECM/BM remodeling. Overall, epithelial secretions during involution induce the stromal cells (mainly fibroblasts) to indirectly generate plasmin which in turn affects the epithelial structure and function through their actions on the ECM.

MMPs are a large family of matrix degrading proteases that are secreted as zymogens. MMP activity is regulated at three levels, a) transcription, b) cleavage dependent activation, and c) activity regulation by Tissue Inhibitor of MMP (TIMP) (Figure 7) [1]. In mouse mammary glands, MMPs are mainly synthesized by the stromal fibrocytes Figure 7) [3]. TIMPs have also been shown to be expressed in the stromal fibrocytes. Low expression of MMPs and elevated levels of TIMP1 during the first 2 days of mouse mammary involution prevent MMP action and maintain the reversibility of gland into lactation (Figure 2) [3, 108]. Analogously, $TIMP3^{-/-}$ mice show loss of reversibility to lactation during the first 2 days of involution [109]. Day 4 involution is marked by a dramatic increase in MMPs (~ 50 fold) (Figure 2). These patterns indicate the importance of timing of MMPs and TIMPs expression to maintain the reversible nature of Phase 1 of involution and the ECM/BM breakdown during the second phase. Plasmin-mediated activation of MMPs is one of the several modes of activation [58] which promotes entry of MMPs into a self activation loop (Figure 7). This is demonstrated in MMP3 (stromelysin-1) epithelial transgenic mice which show increased activation of endogenous MMPs, resulting in precocious involution [110, 111].

In summary, it is the balance between proteolytic enzyme activators and inhibitors that determines the timing and rate of ECM/BM degradation (Table 1). During the involution Phase 1 proteolytic activity is prevented by low expression of MMPs and uPA, and by the presence of protease inhibitors such as TIMP and PAIs and maspin. This is essential in maintaining the reversibility of the Phase 1 involution. The start of the second phase (day 3-4 in mice) is marked by a significant shift in this balance. TIMPs PAI and maspin levels progressively decrease where as plasmin and MMP levels increase and peak by day 6 of involution in the mouse (Figure 7) [69, 102, 108].

D) Regulation of ECM/BM Remodeling Systems

The murine teat-sealing model showed that the balance between proteolytic enzymes and their inhibitors is under endocrine regulation [108]. During Phase 1 of involution, the presence of high levels of systemic hormones like PRL and GH maintains high TIMP expression and suppresses MMP expression. This prevents plasmin generation and ECM/BM breakdown [112, 113]. Similarly, GC administration during mouse involution suppressed uPA induction and inhibited uPA and MMP activity [3]. At the transition into the phase 2 of involution the endocrine influence wanes due to a drop in systemic hormone levels [63-65, 114] resulting in a decrease in protease inhibitors and induction of uPA and MMPs resulting in ECM/BM breakdown.

7. Adipose Tissue Remodeling

As the systemic hormonal milieu changes with the progression of involution, the pre-adipocytes redifferentiate and accumulate lipids to once again occupy a major portion of the glandular space (Figure 1E). The adipogenesis process can be broken down into four aspects; a) hormonal regulation of pre-adipocyte differentiation, b) transcriptional and metabolic changes for preadipocytes to differentiate into adipocytes, c) remodeling of pre-adipocyte ECM for expansion of adipocytes, and d) vascularization of adipose tissue aiding lipid accumulation

Hormonal regulation of preadipocyte differentiation; Systemic hormones such as GH and PRL have opposing actions in the adipose and glandular tissues of the mammary gland. PRL supplementation during murine involution prevents adipogenesis where as PRL withdrawal during lactation induces premature adipogenesis [115, 116]. Hormones such as leptin are known to induce pre-adipocyte differentiation with leptin and its long form receptor (OB-Rb) are found in mouse mammary adipocytes [117, 118]. Moreover, mammary leptin has been shown to be inhibited by PRL. Leptin and OB-Rb mRNA levels increase and peak by day 10 of mouse involution, suggesting a role in adipogenesis.

Human and mouse mammary glands, especially mammary adipocytes, have been shown to synthesize PRL [119-121]. This local secretion is under autoregulatory feedback control. Surprisingly, autocrine PRL, unlike systemic PRL, has been suggested to facilitate adipocyte differentiation. For detailed information on autocrine PRL influences of adipogenesis see review [119]. The mechanism for the opposing actions of systemic PRL and autocrine PRL in the same tissue is not yet understood. One possible explanation is the change in the PRL receptor expression as seen during adipocyte differentiation.

Serotonin (5-HT) is an autocrine-paracine factor that facilitates transition of the mammary gland from lactation to involution [13, 14]. Interestingly, pre-adipocyte differentiation is associated with a shift in 5-HT receptors from 5-HT_1 (inhibitory) to 5-HT_{2A} (stimulatory) [122]. The possible role of 5-HT in mammary adipose differentiation is uninvestigated.

Transcriptional and metabolic changes; Pre-adipocyte differentiation into adipose tissue begins after 2 days of mouse mammary gland involution (Figure 2) [3]. A decrease in systemic GH and PRL levels, as observed during the Phase 2 of involution, results in changes in pre-adipocyte cell morphology and regulation of transcription factors, such as peroxisome proliferator-activator receptor– γ (PPARγ) and CCAAT-enhancer binding protein – β (C/EBP-β) which are critical for adipogenesis [123]. Along with transcriptional changes, a rapid shift in the activities of several enzymes such as Acyl-CoA cholesterol acyltransferase (ACAT) and Acetyl Co-A Carboxylase (ACC) is seen. This represents a coordinated reduction of lipid synthesis in the glandular tissue and elevated lipogenesis and triglyceride storage in the adipose tissue [124]. Details of transcriptional and metabolic changes during adipogenesis are yet to be elucidated.

Adipocyte ECM remodeling; The differentiation of pre-adipocytes coincides with the breakdown of the ECM during the Phase 2 of involution. $Plg^{+/+}$, $Plg^{+/-}$ and $Plg^{-/-}$ mice showed a significant and progressive decrease in mammary adipose tissue regeneration, suggesting an important role of plasmin [58]. Surprisingly, mice lacking both uPA and tPA (Plg activators) did not show any effect on adipocyte differentiation during involution [123]. Serum serine protease PKal was found to be critical in adipogenesis associated plasmin generation and adipocyte differentiation (Figure 8) [123]. This may be due to the composition of the surrounding ECM. PKal-mediated, but not uPA-mediated, plasmin generation is favored in presence of fibronectin (FN). FN is found to be associated with pre-adipocytes and down regulated around differentiated adipocytes [123]. Such localized plasmin generation may also activate and release local growth and differentiation factors sequestered in the surrounding ECM. This may alter the bioavailability of factors like IGF-I (known inducer of adipogenesis [119]).

Counterintutively, MMPs have been shown to inhibit the process of adipocyte differentiation. Both $MMP3^{-/-}$ mice and TIMP transgenic mice show adipocyte hypertrophy during mammary involution [125]. Interestingly, it was observed that TIMPs directly stimulate pre-adipocyte differentiation and adipogenesis as seen via induction of transcription factors PPARγ and C/EBP-β (Figure 8) [125]. Thus, in case of adipocyte differentiation the MMPs bind TIMPs and block/regulate their action. This is further supported by the expression of MMPs by differentiated adipocytes whereas TIMPs are expressed by pre-adipocytes undergoing differentiation [125]. If TIMPs directly induce adipocyte differentiation then what prevents adipocyte differentiation during the first phase of involution when the TIMP expression is elevated? The answer to this question may again lie in the structural organization of the gland where local synthesis of TIMPs by pre-adipocytes, and not stromal fibroblasts, is required.

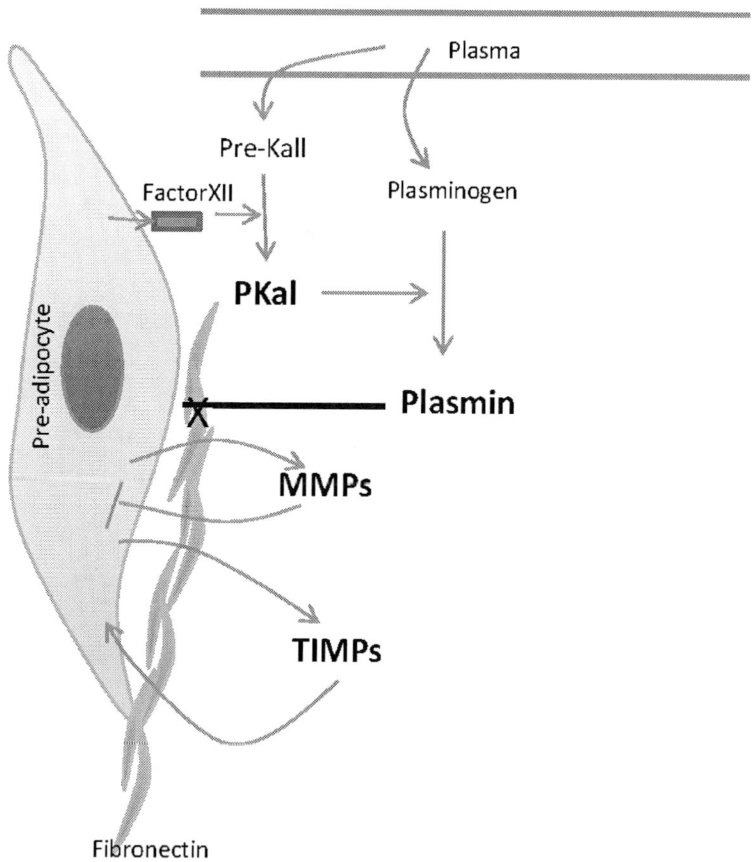

Figure 8. Plasminogen and MMP protease systems involved in adipocyte differentiation during mammary gland involution. Stromal pre-adipocytes secrete factor XII which converts circulating pre-kallikrein (Pre-Kall) to active plasma-kallikrein (PKal). PKal then cleaves circulating plasminogen to generate plasmin. This plasmin then degrades fibronectin deposited around pre-adipocytes. This acts as an important cue for adipocyte differentiation. This is further helped by TIMP secreted by the preadipocytes. In the case of adipocyte differentiation the MMPs act as inhibitors of adipogenesis. They are secreted by adipocytes upon differentiation and hypertrophy.

Vascularization of adipose tissue; Vascular remodeling serves the crucial role of supplying essential nutrients and lipids required for adipogenesis [126, 127]. However, it is the least studied aspect of mammary adipogenesis.

In summary, mammary preadipocyte differentiation during involution is triggered by decreases in systemic hormones which results in breakdown of ECM components that is deposited around the pre-adipocytes. This allows adipocyte differentiation and lipid accumulation under the influence of local (TIMP and fibronectin fragments) and systemic signals. This differentiation is further facilitated by deposition of BM components around the hypertrophic adipocytes and vascular remodleing.

8. Vascular Remodeling

Vascular remodeling consists of two important components; vascular regression and angiogenesis. Vascular regression is relatively less studied in comparison to angiogenesis. During lactation, the vasculature is composed of highly developed capillary networks which form basket-like honeycomb structures enveloping each secretory alveolus [128]. At day 1 of involution, the perivascular capillary basket surrounding each alveolus appears larger than during lactation. This reflects alveolar engorgement due to milk stasis [128]. By day 3 of involution the structure surrounding alveoli exhibits an irregular, collapsed pattern, suggesting local regulation of vascular networks [128]. By day 6 of involution the vascular baskets, similar to the alveoli, are no longer present and are replaced by clusters of capillaries at various stages of regression [128]. By day 10 of involution, the vascular networks in the mouse mammary are similar to those of a virgin gland [129]. Interestingly, although overall the gland undergoes vascular regression during involution, both angiogenesis in adipose tissue and vascular regression in epithelial tissue occur in an overlapping manner. This again suggests a local control of the vasculature by the respective tissue.

Regulators of Involution Vascular Remodeling

Vascular endothelial growth factor (VEGF) is the most studied regulator of the mammary vasculature. Phagocytosis by professional and non-professional phagocytes during mouse involution results in potent VEGF secretion [130]. VEGF is also secreted by murine mammary stromal cells, particularly adipocytes [131, 132]. Involution in mouse mammary gland is characterized by a decline in epithelial VEGF secretion accompanied by a decrease in VEGF receptor (VEGFR) in the adjacent stromal cells. On the other hand, VEGF and VEGFR levels become more prominent in the adipose tissue and their surrounding stroma, respectively [131, 132]. This is in line with the documented epithelial vascular regression and a proposed adipose vascular development during involution [128].

PRL has been shown to have dual actions on angiogenesis. While PRL serves as a potent angiogenic signal, cleaved PRL (16K) is highly anti-angiogenic [133, 134]. The cleaved PRL exerts its anti-angiogenic actions *in vivo* through inhibition of VEGF- and fibroblast growth factor (FGF) -induced endothelial growth, migration and capillary organization [1, 135-137]. Thus, the levels of PRL and proteolytic enzymes determine the mode of action of PRL on mammary vasculature. During involution adipose tissue is characterized by low levels of proteolytic activity and high levels of PRL (autocrine PRL secretion by adipocytes) [119-121]. This would favor the angiogenic actions of PRL concomitant with adipogenesis. Contrarily, the regressing epithelium is characterized by high levels of proteolytic activity and low levels of PRL (systemic) which favors the generation of vaso-inhibitory 16K PRL and facilitates vascular regression.

5-HT synthesized by mammary epithelial cells during mouse involution [13] is vasoactive and mitogenic for the endothelial and vascular smooth muscle cells [138, 139]. The mitogenic action of 5-HT may be mediated through VEGF secretion via sustained activation of p38MAPK as seen in differentiated cultures of human mammary epithelial cells (Figure 5) [24, 140]. Moreover, 5-HT receptors have also been shown to be present in the bovine mammary vascular cells [8] suggesting a possible direct action of 5-HT on mammary vasculature.

9. Involution Associated Breast Diseases

Common breast pathologies include ductal ectasia, and mastitis, inflammation and breast cancer. Although the causal agents of these pathologies vary, the common thread between them is the localized accumulation of secreted fluid due to ductal blockage. Such fluid accumulation is persistent since it occurs at a place and time outside the normal homeostasis. However, such persistent fluid accumulation, in theory, would trigger localized involution-associated responses such as changes in gene expression, cell shedding, apoptosis, ECM and vascular remodeling. This suggests a critical connection between involution-responses and breast pathologies. Occurrence of such involution-responses in places and times where the gland is not equipped with professional and non-professional phagocytes could result in activation of all immune components, resulting in inflammation. Also, such breast pathologies have also been linked to increased breast cancer risk and facilitation of aggressive and metastatic cancers [38, 66, 67, 79]. These inferences suggest crucial roles for local factors regulating involution, such as 5-HT, TGF-β and various cytokines, in mediating breast diseases [7, 141]. This concept is further supported by recent findings demonstrating that lactation and involution are associated with transiently enhanced risk of breast cancers in women [142-144]. In addition, the involution gene signature has recently been shown to resemble that of a wound-healing process and effectively identifies highly metastatic breast cancers [145]. Here, we have kept our focus on discussion of various involution events. Discussing the influence and involvement of each involution event on initiation and progression of breast diseases is out of the scope of this chapter. Nonetheless, it appears that every event occurring during involution (from cell death to vascular remodeling) has at least some aspect promoting the advancement of breast diseases. However, the question that remains is whether involution causes breast lesions or just facilitates the advancement of pre-existing ones.

10. Conclusion

Involution is a multi-step process that is regulated both by systemic hormones and by local factors within the gland (Figure 2). From lactation through the end of involution the mammary gland is transformed from one consisting primarily of secretory epithelium with little stroma and adipose tissue to one that is primarily occupied by adipose tissue and stroma with epithelial ducts interspersed within them (Figure 1A-E).

A substantial amount of information about involution has been collected in last several years, but we have just began to elucidate the complexities of this system. Although we have gained a superficial understanding of what is occurring during involution we do not fully understand the molecular underpinnings of how these processes are regulated. Some basic questions about involution are as yet unanswered. For example, how does milk stasis induce the expression of multiple local factors? Thus far, our understanding of the mechanical forces at play in association with alveolar engorgement and how they affect involution at molecular and cellular level is lacking. Given the extent of cell death signals induced during involution, it is intriguing to note that epithelial cell death occurs in a heterogenous fashion. What is unique about the ductal epithelial cells that survive involution-associated death signals when the rest of the epithelium is undergoing massive regression? These ductal epithelia have been shown to contain pluripotent cells; how are they affected by involution? What induces such a massive induction of plasma cells and immunoglobulin secretions during late involution? And what role do these immunoglobulins play in involution process? The triggers that initiate involution are at least partly studied (Figure 4). However, the factors that stop this intensely dynamic process of tissue remodeling and cell death, to bring the gland into a quasi-dormant state, are yet known. It is debatable whether the processes that occur during involution can result in the transformation of breast cells. However, it is certainly plausible that for a transformed cell existing latently within the mammary gland, involution provides an excellent opportunity for clonal expansion and development into metastatic disease. To the extent that these dynamic involution-like events occur locally in regions of the breast that are isolated by inflammation or ductal occlusion, the same processes may participate in the invasion and metastasis that occurs outside the involution period.

References

[1] Green KA, Lund LR: ECM Degrading Proteases And Tissue Remodelling In The Mammary Gland. *Bioessays* 2005, 27(9):894-903.

[2] Almholt K, Green KA, Juncker-Jensen A, Nielsen BS, Lund LR, Romer J: Extracellular Proteolysis In Transgenic Mouse Models Of Breast Cancer *J Mammary Gland Biol Neoplasia* 2007, 12(1):83-97.

[3] Lund LR, Romer J, Thomasset N, Solberg H, Pyke C, Bissell MJ, Dano K, Werb Z: Two Distinct Phases Of Apoptosis In *Mammary Gland Involution: Proteinase-Independent And -Dependent Pathways. Development* 1996, 122(1):181-193.

[4] Sharp JA, Lefevre C, Brennan AJ, Nicholas KR: The Fur Seal-A Model Lactation Phenotype To Explore Molecular Factors Involved In The Initiation Of Apoptosis At Involution. *J Mammary Gland Biol Neoplasia* 2007, 12(1):47-58.

[5] Enami J, Nandi S: Secretion Of Casein In Cultures Of Mouse Mammary Epithelial Cells On Floating Collagen Gels *J Dairy Sci* 1978, 61(6):729-732.

[6] Debnath J, Mills KR, Collins NL, Reginato MJ, Muthuswamy SK, Brugge JS: The Role Of Apoptosis In Creating And Maintaining Luminal Space Within Normal And Oncogene-Expressing Mammary Acini. *Cell* 2002, 111(1):29-40.

[7] Pai VP, Marshall AM, Hernandez LL, Buckley AR, Horseman ND: Altered Serotonin Physiology In Human Breast Cancers Favors Paradoxical Growth And Cell *Survival Breast Cancer Res* 2009, 11(6):R81.

[8] Hernandez LL, Limesand SW, Collier JL, Horseman ND, Collier RJ: The Bovine Mammary Gland Expresses Multiple Functional Isoforms Of Serotonin Receptors *J Endocrinol* 2009, 203(1):123-131.

[9] Bruckmaier RM, Blum JW: Simultaneous Recording Of Oxytocin Release, Milk Ejection And Milk Flow During Milking Of Dairy Cows With And Without Prestimulation *J Dairy Res* 1996, 63(2):201-208.

[10] Feng Z, Marti A, Jehn B, Altermatt HJ, Chicaiza G, Jaggi R: Glucocorticoid And Progesterone Inhibit Involution And Programmed Cell Death In The Mouse Mammary Gland. *J Cell Biol* 1995, 131(4):1095-1103.

[11] Li M, Liu X, Robinson G, Bar-Peled U, Wagner KU, Young WS, Hennighausen L, Furth PA: Mammary-Derived Signals Activate Programmed Cell Death During The First Stage Of Mammary Gland Involution. *Proc Natl Acad Sci U S A 1997,* 94(7):3425-3430.

[12] Wilde CJ, Addey CV, Casey MJ, Blatchford DR, Peaker M: Feed-Back Inhibition Of Milk Secretion: The Effect Of A Fraction Of Goat Milk On Milk Yield And Composition. *Q J Exp Physiol* 1988, 73(3):391-397.

[13] Matsuda M, Imaoka T, Vomachka AJ, Gudelsky GA, Hou Z, Mistry M, Bailey JP, Nieport KM, Walther DJ, Bader M, Horseman ND: Serotonin Regulates Mammary Gland Development Via An Autocrine-Paracrine Loop. *Dev Cell* 2004, 6(2):193-203.

[14] Hernandez LL, Stiening CM, Wheelock JB, Baumgard LH, Parkhurst AM, Collier RJ: Evaluation Of Serotonin As A Feedback Inhibitor Of Lactation In The Bovine. *J Dairy Sci* 2008, 91(5):1834-1844.

[15] Walther DJ, Bader M: A Unique Central Tryptophan Hydroxylase Isoform. *Biochem Pharmacol* 2003, 66(9):1673-1680.

[16] Walther DJ, Peter JU, Bashammakh S, Hortnagl H, Voits M, Fink H, Bader M: Synthesis Of Serotonin By A Second Tryptophan Hydroxylase Isoform. *Science* 2003, 299(5603):76.

[17] Walther DJ, Peter JU, Winter S, Holtje M, Paulmann N, Grohmann M, Vowinckel J, Alamo-Bethencourt V, Wilhelm CS, Ahnert-Hilger G, Bader M: Serotonylation Of Small Gtpases Is A Signal Transduction Pathway That Triggers Platelet Alpha-Granule *Release Cell* 2003, 115(7):851-862.

[18] Blakely RD, Defelice LJ, Galli A: Biogenic Amine Neurotransmitter Transporters: Just When You Thought You Knew Them. Physiology (Bethesda) 2005, 20:225-231.

[19] Roth BL: *The Serotonin Receptors : From Molecular Pharmacology To Human Therapeutics.* Totowa, N.J.: Humana Press 2006, :618.

[20] Stull MA, Pai V, Vomachka AJ, Marshall AM, Jacob GA, Horseman ND: Mammary Gland Homeostasis Employs Serotonergic Regulation Of Epithelial Tight Junctions. *Proc Natl Acad Sci U S A 2007*, 104(42):16708-16713.

[21] Marshall AM, Nommsen-Rivers LA, Hernandez LL, Dewey KG, Chantry CJ, Gregerson KA, Horseman ND: Serotonin Transport And Metabolism In The Mammary Gland Modulates Secretory Activation And Involution *J Clin Endocrinol Metab* 2010, 95(2):837-846.

[22] Hernandez LL, Collier JL, Vomachka AJ, Collier RJ, Horseman ND: Suppression Of Lactation And Acceleration Of Involution In The Bovine Mammary Gland By A Selective Serotonin Reuptake Inhibitor *J Endocrinol* 2011, 209(1):45-54.

[23] Zahraoui A, Louvard D, Galli T: Tight Junction, A Platform For Trafficking And Signaling Protein Complexes. *J Cell Biol* 2000, 151(5):F31-6.

[24] Pai VP, Horseman ND: Biphasic Regulation Of Mammary Epithelial Resistance By Serotonin Through Activation Of Multiple Pathways. *J Biol Chem* 2008, 283(45):30901-30910.

[25] Welty FK, Smith KL, Schanbacher FL: Lactoferrin Concentration During Involution Of The Bovine Mammary Gland. *J Dairy Sci* 1976, 59(2):224-231.

[26] Masson PL, Heremans JF: Lactoferrin In Milk From Different Species. *Comp Biochem Physiol B* 1971, 39(1):119-129.

[27] Nakamura M, Tomita A, Nakatani H, Matsuda T, Nadano D: Antioxidant And Antibacterial Genes Are Upregulated In Early Involution Of The Mouse Mammary Gland: Sharp Increase Of Ceruloplasmin And Lactoferrin In Accumulating Breast Milk. *DNA Cell Biol* 2006, 25(9):491-500.

[28] Riley LG, Williamson P, Wynn PC, Sheehy PA: Lactoferrin Decreases Primary Bovine Mammary Epithelial Cell Viability And Casein Expression. *J Dairy Res* 2008, 75(2):135-141.

[29] Baumrucker CR, Saurer S, Blum JW, Jungi T, Friis RR: Lactoferrin (Lf): Retinoid Interactions In The Mammary Glands Of Transgenic Mice Overexpressing Human Lf. *Mol Cell Endocrinol* 2006, 251(1-2):56-66.

[30] Tiffen PG, Omidvar N, Marquez-Almuina N, Croston D, Watson CJ, Clarkson RW: A Dual Role For Oncostatin M Signalling In The Differentiation And Death Of Mammary Epithelial Cells In Vivo. *Mol Endocrinol* 2008, .

[31] Marti A, Jehn B, Costello E, Keon N, Ke G, Martin F, Jaggi R: Protein Kinase A And AP-1 (C-Fos/Jund) Are Induced During Apoptosis Of Mouse Mammary Epithelial Cells. *Oncogene* 1994, 9(4):1213-1223.

[32] Baxter FO, Neoh K, Tevendale MC: The Beginning Of The End: Death Signaling In Early Involution *J Mammary Gland Biol Neoplasia* 2007, 12(1):3-13.

[33] Watson CJ, Kreuzaler PA: The Role Of Cathepsins In Involution And Breast Cancer *J Mammary Gland Biol Neoplasia* 2009, 14(2):171-179.

[34] Marti A, Graber H, Lazar H, Ritter PM, Baltzer A, Srinivasan A, Jaggi R: Caspases: Decoders Of Apoptotic Signals During Mammary Involution. Caspase Activation During Involution. *Adv Exp Med Biol* 2000, 480:195-201.

[35] Pai VP, Horseman ND: Multiple Cellular Responses To Serotonin Contribute To Epithelial Homeostasis *Plos One* 2011, 6(2):E17028.

[36] Marshall AM, Pai VP, Sartor MA, Horseman ND: In Vitro Multipotent Differentiation And Barrier Function Of A Human Mammary Epithelium. *Cell Tissue Res* 2008, .

[37] Nguyen AV, Pollard JW: Transforming Growth Factor Beta3 Induces Cell Death During The First Stage Of Mammary Gland Involution. *Development 2000,* 127(14):3107-3118.

[38] Stein T, Morris JS, Davies CR, Weber-Hall SJ, Duffy MA, Heath VJ, Bell AK, Ferrier RK, Sandilands GP, Gusterson BA: Involution Of The Mouse Mammary Gland Is Associated With An Immune Cascade And An Acute-Phase Response, Involving LBP, CD14 And STAT3. *Breast Cancer Res* 2004, 6(2):R75-91.

[39] Flanders KC, Wakefield LM: Transforming Growth Factor-(Beta)S And Mammary Gland Involution; Functional Roles And Implications For Cancer Progression *J Mammary Gland Biol Neoplasia* 2009, 14(2):131-144.

[40] Bailey JP, Nieport KM, Herbst MP, Srivastava S, Serra RA, Horseman ND: Prolactin And Transforming Growth Factor-Beta Signaling Exert Opposing Effects On

Mammary Gland Morphogenesis, Involution, And The Akt-Forkhead Pathway. *Mol Endocrinol* 2004, 18(5):1171-1184.

[41] Saharinen J, Taipale J, Keski-Oja J: Association Of The Small Latent Transforming Growth Factor-Beta With An Eight Cysteine Repeat Of Its Binding Protein LTBP-1 *EMBO J* 1996, 15(2):245-253.

[42] Gleizes PE, Beavis RC, Mazzieri R, Shen B, Rifkin DB: Identification And Characterization Of An Eight-Cysteine Repeat Of The Latent Transforming Growth Factor-Beta Binding Protein-1 That Mediates Bonding To The Latent Transforming Growth Factor-Beta1 *J Biol Chem* 1996, 271(47):29891-29896.

[43] Miyazono K, Olofsson A, Colosetti P, Heldin CH: A Role Of The Latent TGF-Beta 1-Binding Protein In The Assembly And Secretion Of TGF-Beta 1 *EMBO J* 1991, 10(5):1091-1101.

[44] Zhao L, Melenhorst JJ, Hennighausen L: Loss Of Interleukin 6 Results In Delayed Mammary Gland Involution: A Possible Role For Mitogen-Activated Protein Kinase And Not Signal Transducer And Activator Of Transcription 3. *Mol Endocrinol* 2002, 16(12):2902-2912.

[45] Schere-Levy C, Buggiano V, Quaglino A, Gattelli A, Cirio MC, Piazzon I, Vanzulli S, Kordon EC: Leukemia Inhibitory Factor Induces Apoptosis Of The Mammary Epithelial Cells And Participates In Mouse Mammary Gland Involution. Exp Cell Res 2003, 282(1):35-47.

[46] Kritikou EA, Sharkey A, Abell K, Came PJ, Anderson E, Clarkson RW, Watson CJ: A Dual, Non-Redundant, Role For LIF As A Regulator Of Development And STAT3-Mediated Cell Death In Mammary Gland. *Development 2003*, 130(15):3459-3468.

[47] Ning Y, Hoang B, Schuller AG, Cominski TP, Hsu MS, Wood TL, Pintar JE: Delayed Mammary Gland Involution In Mice With Mutation Of The Insulin-Like Growth Factor Binding Protein 5 Gene. *Endocrinology* 2007, 148(5):2138-2147.

[48] Allar MA, Wood TL: Expression Of The Insulin-Like Growth Factor Binding Proteins During Postnatal Development Of The Murine Mammary Gland. *Endocrinology* 2004, 145(5):2467-2477.

[49] Tonner E, Barber MC, Travers MT, Logan A, Flint DJ: Hormonal Control Of Insulin-Like Growth Factor-Binding Protein-5 Production In The Involuting Mammary Gland Of The Rat. *Endocrinology* 1997, 138(12):5101-5107.

[50] Tonner E, Barber MC, Allan GJ, Beattie J, Webster J, Whitelaw CB, Flint DJ: Insulin-Like Growth Factor Binding Protein-5 (IGFBP-5) Induces Premature Cell Death In The Mammary Glands Of Transgenic Mice. *Development 2002,* 129(19):4547-4557.

[51] Schmidt DV, Walker LE, Ebner KE: Lactose Synthetase Activity In Northern Fur Seal Milk *Biochim Biophys Acta* 1971, 252(3):439-442.

[52] Urashima T, Arita M, Yoshida M, Nakamura T, Arai I, Saito T, Arnould JP, Kovacs KM, Lydersen C: Chemical Characterisation Of The Oligosaccharides In Hooded Seal (Cystophora Cristata) And Australian Fur Seal (Arctocephalus Pusillus Doriferus) Milk *Comp Biochem Physiol B Biochem Mol Biol* 2001, 128(2):307-323.

[53] Baltzer A, Svanborg C, Jaggi R: Apoptotic Cell Death In The Lactating Mammary Gland Is Enhanced By A Folding Variant Of Alpha-Lactalbumin *Cell Mol Life Sci* 2004, 61(10):1221-1228.

[54] Stinnakre MG, Vilotte JL, Soulier S, Mercier JC: Creation And Phenotypic Analysis Of Alpha-Lactalbumin-Deficient Mice *Proc Natl Acad Sci U S A 1994,* 91(14):6544-6548.

[55] Svensson M, Sabharwal H, Hakansson A, Mossberg AK, Lipniunas P, Leffler H, Svanborg C, Linse S: Molecular Characterization Of Alpha-Lactalbumin Folding Variants That Induce Apoptosis In Tumor Cells *J Biol Chem* 1999, 274(10):6388-6396.

[56] Hill RL, Brew K: Lactose Synthetase Adv Enzymol Relat Areas *Mol Biol* 1975, 43:411-490.

[57] Boudreau N, Sympson CJ, Werb Z, Bissell MJ: Suppression Of ICE And Apoptosis In Mammary Epithelial Cells By Extracellular Matrix. *Science* 1995, 267(5199):891-893.

[58] Lund LR, Bjorn SF, Sternlicht MD, Nielsen BS, Solberg H, Usher PA, Osterby R, Christensen IJ, Stephens RW, Bugge TH, Dano K, Werb Z: Lactational Competence And Involution Of The Mouse Mammary Gland Require Plasminogen. *Development 2000,* 127(20):4481-4492.

[59] Schedin P, Strange R, Mitrenga T, Wolfe P, Kaeck M: Fibronectin Fragments Induce MMP Activity In Mouse Mammary Epithelial Cells: Evidence For A Role In Mammary Tissue Remodeling. *J Cell Sci* 2000, 113 (Pt 5)(Pt 5):795-806.

[60] Zinser GM, Welsh J: Accelerated Mammary Gland Development During Pregnancy And Delayed Postlactational Involution In Vitamin D3 Receptor Null Mice. *Mol Endocrinol* 2004, 18(9):2208-2223.

[61] Oakes SR, Rogers RL, Naylor MJ, Ormandy CJ: Prolactin Regulation Of Mammary Gland Development *J Mammary Gland Biol Neoplasia* 2008, 13(1):13-28.

[62] Horseman ND: Prolactin And Mammary Gland Development. *J Mammary Gland Biol Neoplasia* 1999, 4(1):79-88.

[63] Hadsell DL, Alexeenko T, Klimentidis Y, Torres D, Lee AV: Inability Of Overexpressed Des(1-3)Human Insulin-Like Growth Factor I (IGF-I) To Inhibit Forced Mammary Gland Involution Is Associated With Decreased Expression Of IGF Signaling Molecules. *Endocrinology* 2001, 142(4):1479-1488.

[64] Flint DJ, Boutinaud M, Whitelaw CB, Allan GJ, Kolb AF: Prolactin Inhibits Cell Loss And Decreases Matrix Metalloproteinase Expression In The Involuting Mouse Mammary Gland But Fails To Prevent Cell Loss In The Mammary Glands Of Mice Expressing IGFBP-5 As A Mammary Transgene. *J Mol Endocrinol* 2006, 36(3):435-448.

[65] Travers MT, Barber MC, Tonner E, Quarrie L, Wilde CJ, Flint DJ: The Role Of Prolactin And Growth Hormone In The Regulation Of Casein Gene Expression And Mammary Cell Survival: Relationships To Milk Synthesis And Secretion. *Endocrinology* 1996, 137(5):1530-1539.

[66] Atabai K, Sheppard D, Werb Z: Roles Of The Innate Immune System In Mammary Gland Remodeling During Involution. *J Mammary Gland Biol Neoplasia* 2007, 12(1):37-45.

[67] Hanayama R, Nagata S: Impaired Involution Of Mammary Glands In The Absence Of Milk Fat Globule EGF Factor 8. *Proc Natl Acad Sci U S A 2005,* 102(46):16886-16891.

[68] Paape MJ, Shafer-Weaver K, Capuco AV, Van Oostveldt K, Burvenich C: Immune Surveillance Of Mammary Tissue By Phagocytic Cells. *Adv Exp Med Biol* 2000, 480:259-277.

[69] Clarkson RW, Wayland MT, Lee J, Freeman T, Watson CJ: Gene Expression Profiling Of Mammary Gland Development Reveals Putative Roles For Death Receptors And Immune Mediators In Post-Lactational Regression. *Breast Cancer Res* 2004, 6(2):R92-109.

[70] Nordin W, Lee CS: Cytology Of Milk In Guinea Pigs. *Acta Anat (Basel)* 1982, 113(2):135-144.

[71] Lee CS, Mccauley I, Hartmann PE: Light And Electron Microscopy Of Cells In Pig Colostrum, Milk And Involution Secretion. *Acta Anat (Basel)* 1983, 116(2):126-135.

[72] Tatarczuch L, Philip C, Bischof R, Lee CS: Leucocyte Phenotypes In Involuting And Fully Involuted Mammary Glandular Tissues And Secretions Of Sheep. *J Anat* 2000, 196 (Pt 3)(Pt 3):313-326.

[73] Colditz IG: Studies On The Inflammatory Response During Involution Of The Ovine Mammary Gland. *Q J Exp Physiol* 1988, 73(3):363-368.

[74] Meredith EJ, Chamba A, Holder MJ, Barnes NM, Gordon J: Close Encounters Of The Monoamine Kind: Immune Cells Betray Their Nervous Disposition. *Immunology* 2005, 115(3):289-295.

[75] Monks J, Rosner D, Geske FJ, Lehman L, Hanson L, Neville MC, Fadok VA: Epithelial Cells As Phagocytes: Apoptotic Epithelial Cells Are Engulfed By Mammary Alveolar Epithelial Cells And Repress Inflammatory Mediator Release. *Cell Death Differ* 2005, 12(2):107-114.

[76] Monks J, Smith-Steinhart C, Kruk ER, Fadok VA, Henson PM: Epithelial Cells Remove Apoptotic Epithelial Cells During Post-Lactation Involution Of The Mouse Mammary Gland. *Biol Reprod* 2008, 78(4):586-594.

[77] Sordillo LM, Nickerson SC: Quantification And Immunoglobulin Classification Of Plasma Cells In Nonlactating Bovine Mammary Tissue. *J Dairy Sci* 1988, 71(1):84-91.

[78] Watson DL, Brandon MR, Lascelles AK: Concentrations Of Immunoglobulin In Mammary Secretion Of Ruminants During Involution With Particular Reference To Selective Transfer Of Igg. *Aust J Exp Biol Med Sci* 1972, 50(4):535-539.

[79] Fadok VA: Clearance: The Last And Often Forgotten Stage Of Apoptosis. *J Mammary Gland Biol Neoplasia* 1999, 4(2):203-211.

[80] Monks J, Geske FJ, Lehman L, Fadok VA: Do Inflammatory Cells Participate In Mammary Gland Involution? *J Mammary Gland Biol Neoplasia* 2002, 7(2):163-176.

[81] Brooker BE: Pseudopod Formation And Phagocytosis Of Milk Components By Epithelial Cells Of The Bovine Mammary Gland. *Cell Tissue Res* 1983, 229(3):639-650.

[82] Atabai K, Fernandez R, Huang X, Ueki I, Kline A, Li Y, Sadatmansoori S, Smith-Steinhart C, Zhu W, Pytela R, Werb Z, Sheppard D: Mfge8 Is Critical For Mammary Gland Remodeling During Involution. *Mol Biol Cell* 2005, 16(12):5528-5537.

[83] Itoh M, Bissell MJ: The Organization Of Tight Junctions In Epithelia: Implications For Mammary Gland Biology And Breast Tumorigenesis. *J Mammary Gland Biol Neoplasia* 2003, 8(4):449-462.

[84] Kohler K, Zahraoui A: Tight Junction: A Co-Ordinator Of Cell Signalling And Membrane Trafficking. *Biol Cell* 2005, 97(8):659-665.

[85] Stelwagen K, Davis SR, Farr VC, Prosser CG, Sherlock RA: Mammary Epithelial Cell Tight Junction Integrity And Mammary Blood Flow During An Extended Milking Interval In Goats. *J Dairy Sci* 1994, 77(2):426-432.

[86] Stelwagen K, Farr VC, Mcfadden HA, Prosser CG, Davis SR: Time Course Of Milk Accumulation-Induced Opening Of Mammary Tight Junctions, And Blood Clearance Of Milk Components. *Am J Physiol* 1997, 273(1 Pt 2):R379-86.

[87] Thompson GE: Cortisol And Regulation Of Tight Junctions In The Mammary Gland Of The Late-Pregnant Goat. *J Dairy Res* 1996, 63(2):305-308.

[88] Zettl KS, Sjaastad MD, Riskin PM, Parry G, Machen TE, Firestone GL: Glucocorticoid-Induced Formation Of Tight Junctions In Mouse Mammary Epithelial Cells In Vitro. *Proc Natl Acad Sci U S A 1992,* 89(19):9069-9073.

[89] Woo PL, Cha HH, Singer KL, Firestone GL: Antagonistic Regulation Of Tight Junction Dynamics By Glucocorticoids And Transforming Growth Factor-Beta In Mouse Mammary Epithelial Cells. *J Biol Chem* 1996, 271(1):404-412.

[90] Shamay A, Shapiro F, Leitner G, Silanikove N: Infusions Of Casein Hydrolyzates Into The Mammary Gland Disrupt Tight Junction Integrity And Induce Involution In Cows. *J Dairy Sci* 2003, 86(4):1250-1258.

[91] Stelwagen K, Politis I, White JH, Zavizion B, Prosser CG, Davis SR, Farr VC: Effect Of Milking Frequency And Somatotropin On The Activity Of Plasminogen Activator, Plasminogen, And Plasmin In Bovine Milk *J Dairy Sci* 1994, 77(12):3577-3583.

[92] Stelwagen K, Farr VC, Davis SR, Prosser CG: EGTA-Induced Disruption Of Epithelial Cell Tight Junctions In The Lactating Caprine Mammary Gland. *Am J Physiol* 1995, 269(4 Pt 2):R848-55.

[93] Kass L, Erler JT, Dembo M, Weaver VM: Mammary Epithelial Cell: Influence Of Extracellular Matrix Composition And Organization During Development And Tumorigenesis *Int J Biochem Cell Bio*l 2007, 39(11):1987-1994.

[94] Xu R, Boudreau A, Bissell MJ: *Tissue Architecture And Function: Dynamic Reciprocity Via Extra- And Intra-Cellular Matrices Cancer Metastasis Rev* 2009, 28(1-2):167-176.

[95] Hynes RO: *Integrins: Bidirectional, Allosteric Signaling Machines Cell* 2002, 110(6):673-687.

[96] Pullan S, Wilson J, Metcalfe A, Edwards GM, Goberdhan N, Tilly J, Hickman JA, Dive C, Streuli CH: Requirement Of Basement Membrane For The Suppression Of Programmed Cell Death In Mammary Epithelium. *J Cell Sci* 1996, 109 (Pt 3)(Pt 3):631-642.

[97] Faraldo MM, Deugnier MA, Lukashev M, Thiery JP, Glukhova MA: Perturbation Of Beta1-Integrin Function Alters The Development Of Murine Mammary Gland. *EMBO J* 1998, 17(8):2139-2147.

[98] Boudreau N, Werb Z, Bissell MJ: Suppression Of Apoptosis By Basement Membrane Requires Three-Dimensional Tissue Organization And Withdrawal From The Cell Cycle. *Proc Natl Acad Sci U S A* 1996, 93(8):3509-3513.

[99] Buckley CD, Pilling D, Henriquez NV, Parsonage G, Threlfall K, Scheel-Toellner D, Simmons DL, Akbar AN, Lord JM, Salmon M: RGD Peptides Induce Apoptosis By Direct Caspase-3 Activation. *Nature* 1999, 397(6719):534-539.

[100] Martinez-Hernandez A, Fink LM, Pierce GB: Removal Of Basement Membrane In The Involuting Breast. *Lab Invest* 1976, 34(5):455-462.

[101] Warburton MJ, Mitchell D, Ormerod EJ, Rudland P: Distribution Of Myoepithelial Cells And Basement Membrane Proteins In The Resting, Pregnant, Lactating, And Involuting Rat Mammary Gland. *J Histochem Cytochem* 1982, 30(7):667-676.

[102] Strange R, Li F, Saurer S, Burkhardt A, Friis RR: Apoptotic Cell Death And Tissue Remodelling During Mouse Mammary Gland Involution. *Development* 1992, 115(1):49-58.

[103] Schedin P, Mitrenga T, Mcdaniel S, Kaeck M: Mammary ECM Composition And Function Are Altered By Reproductive State. *Mol Carcinog* 2004, 41(4):207-220.

[104] Schenk S, Hintermann E, Bilban M, Koshikawa N, Hojilla C, Khokha R, Quaranta V: Binding To EGF Receptor Of A Laminin-5 EGF-Like Fragment Liberated During MMP-Dependent Mammary Gland Involution. *J Cell Biol* 2003, 161(1):197-209.

[105] Zhang M, Magit D, Botteri F, Shi HY, He K, Li M, Furth P, Sager R: Maspin Plays An Important Role In Mammary Gland Development. *Dev Biol* 1999, 215(2):278-287.

[106] Lijnen HR: Plasmin And Matrix Metalloproteinases In Vascular Remodeling. *Thromb Haemost* 2001, 86(1):324-333.

[107] Ambili M, Sudhakaran PR: 60K Gelatinase In Involuting Rat Mammary Gland Is Produced As A 90K Proenzyme. *Biochem Mol Biol Int* 1998, 45(2):389-399.

[108] Talhouk RS, Bissell MJ, Werb Z: Coordinated Expression Of Extracellular Matrix-Degrading Proteinases And Their Inhibitors Regulates Mammary Epithelial Function During Involution. *J Cell Biol* 1992, 118(5):1271-1282.

[109] Fata JE, Leco KJ, Voura EB, Yu HY, Waterhouse P, Murphy G, Moorehead RA, Khokha R: Accelerated Apoptosis In The Timp-3-Deficient Mammary Gland. *J Clin Invest* 2001, 108(6):831-841.

[110] Alexander CM, Howard EW, Bissell MJ, Werb Z: Rescue Of Mammary Epithelial Cell Apoptosis And Entactin Degradation By A Tissue Inhibitor Of Metalloproteinases-1 Transgene. *J Cell Biol* 1996, 135(6 Pt 1):1669-1677.

[111] Thomasset N, Lochter A, Sympson CJ, Lund LR, Williams DR, Behrendtsen O, Werb Z, Bissell MJ: Expression Of Autoactivated Stromelysin-1 In Mammary Glands Of Transgenic Mice Leads To A Reactive Stroma During Early Development. *Am J Pathol* 1998, 153(2):457-467.

[112] Tonner E, Allan GJ, Flint DJ: Hormonal Control Of Plasmin And Tissue-Type Plasminogen Activator Activity In Rat Milk During Involution Of The Mammary Gland. *J Endocrinol* 2000, 167(2):265-273.

[113] Allan GJ, Tonner E, Barber MC, Travers MT, Shand JH, Vernon RG, Kelly PA, Binart N, Flint DJ: Growth Hormone, Acting In Part Through The Insulin-Like Growth Factor Axis, Rescues Developmental, But Not Metabolic, Activity In The Mammary Gland Of Mice Expressing A Single Allele Of The Prolactin Receptor. *Endocrinology* 2002, 143(11):4310-4319.

[114] Alexandrova M: Glucocorticoid Receptor Of Rat Mammary Gland During Pregnancy And Lactation. *Endocrinol Exp* 1986, 20(2-3):293-300.

[115] Agius L, Robinson AM, Girard JR, Williamson DH: Alterations In The Rate Of Lipogenesis In Vivo In Maternal Liver And Adipose Tissue On Premature Weaning Of Lactating Rats: A Possible Regulatory Role Of Prolactin. *Biochem J* 1979, 180(3):689-692.

[116] Ros M, Lobato MF, Garcia-Ruiz JP, Moreno FJ: Integration Of Lipid Metabolism In The Mammary Gland And Adipose Tissue By Prolactin During Lactation. *Mol Cell Biochem* 1990, 93(2):185-194.

[117] Lin Y, Li Q: Expression And Function Of Leptin And Its Receptor In Mouse Mammary Gland. *Sci China C Life Sci* 2007, 50(5):669-675.

[118] Aoki N, Kawamura M, Matsuda T: Lactation-Dependent Down Regulation Of Leptin Production In Mouse Mammary Gland. *Biochim Biophys Acta* 1999, 1427(2):298-306.

[119] Brandebourg T, Hugo E, Ben-Jonathan N: Adipocyte Prolactin: Regulation Of Release And Putative Functions. *Diabetes Obes Metab* 2007, 9(4):464-476.

[120] Reynolds C, Montone KT, Powell CM, Tomaszewski JE, Clevenger CV: Expression Of Prolactin And Its Receptor In Human Breast Carcinoma. *Endocrinology* 1997, 138(12):5555-5560.

[121] Clevenger CV, Chang WP, Ngo W, Pasha TL, Montone KT, Tomaszewski JE: Expression Of Prolactin And Prolactin Receptor In Human Breast Carcinoma. Evidence For An Autocrine/Paracrine Loop. *Am J Pathol* 1995, 146(3):695-705.

[122] Uchida-Kitajima S, Yamauchi T, Takashina Y, Okada-Iwabu M, Iwabu M, Ueki K, Kadowaki T: 5-Hydroxytryptamine 2A Receptor Signaling Cascade Modulates Adiponectin And Plasminogen Activator Inhibitor 1 Expression In Adipose Tissue. *FEBS Lett* 2008, 582(20):3037-3044.

[123] Selvarajan S, Lund LR, Takeuchi T, Craik CS, Werb Z: A Plasma Kallikrein-Dependent Plasminogen Cascade Required For Adipocyte Differentiation. *Nat Cell Biol* 2001, 3(3):267-275.

[124] Barber MC, Travers MT, Finley E, Flint DJ, Vernon RG: Growth-Hormone-Prolactin Interactions In The Regulation Of Mammary And Adipose-Tissue Acetyl-Coa Carboxylase Activity And Gene Expression In Lactating Rats. *Biochem J* 1992, 285 (Pt 2)(Pt 2):469-475.

[125] Alexander CM, Selvarajan S, Mudgett J, Werb Z: Stromelysin-1 Regulates Adipogenesis During Mammary Gland Involution. J Cell Biol 2001, 152(4):693-703.

[126] Smas CM, Sul HS: Control Of Adipocyte Differentiation. *Biochem J* 1995, 309 (Pt 3)(Pt 3):697-710.

[127] Crandall DL, Hausman GJ, Kral JG: A Review Of The Microcirculation Of Adipose Tissue: Anatomic, Metabolic, And Angiogenic Perspectives. *Microcirculation* 1997, 4(2):211-232.

[128] Djonov V, Andres AC, Ziemiecki A: Vascular Remodelling During The Normal And Malignant Life Cycle Of The Mammary Gland. *Microsc Res Tech* 2001, 52(2):182-189.

[129] Matsumoto M, Nishinakagawa H, Kurohmaru M, Hayashi Y, Otsuka J: Pregnancy And Lactation Affect The Microvasculature Of The Mammary Gland In Mice. *J Vet Med Sci* 1992, 54(5):937-943.

[130] Golpon HA, Fadok VA, Taraseviciene-Stewart L, Scerbavicius R, Sauer C, Welte T, Henson PM, Voelkel NF: Life After Corpse Engulfment: Phagocytosis Of Apoptotic Cells Leads To VEGF Secretion And Cell Growth. *FASEB J* 2004, 18(14):1716-1718.

[131] Pepper MS, Baetens D, Mandriota SJ, Di Sanza C, Oikemus S, Lane TF, Soriano JV, Montesano R, Iruela-Arispe ML: Regulation Of VEGF And VEGF Receptor Expression In The Rodent Mammary Gland During Pregnancy, Lactation, And Involution. *Dev Dyn* 2000, 218(3):507-524.

[132] Hovey RC, Goldhar AS, Baffi J, Vonderhaar BK: Transcriptional Regulation Of Vascular Endothelial Growth Factor Expression In Epithelial And Stromal Cells During Mouse Mammary Gland Development. *Mol Endocrinol* 2001, 15(5):819-831.

[133] Gaytan F, Morales C, Bellido C, Aguilar E, Sanchez-Criado JE: Role Of Prolactin In The Regulation Of Macrophages And In The Proliferative Activity Of Vascular Cells In Newly Formed And Regressing Rat Corpora Lutea. *Biol Reprod* 1997, 57(2):478-486.

[134] Struman I, Bentzien F, Lee H, Mainfroid V, D'Angelo G, Goffin V, Weiner RI, Martial JA: Opposing Actions Of Intact And N-Terminal Fragments Of The Human

Prolactin/Growth Hormone Family Members On Angiogenesis: An Efficient Mechanism For The Regulation Of Angiogenesis. *Proc Natl Acad Sci U S A 1999,* 96(4):1246-1251.

[135] Lee H, Struman I, Clapp C, Martial J, Weiner RI: Inhibition Of Urokinase Activity By The Antiangiogenic Factor 16K Prolactin: Activation Of Plasminogen Activator Inhibitor 1 Expression. *Endocrinology* 1998, 139(9):3696-3703.

[136] Ferrara N, Clapp C, Weiner R: The 16K Fragment Of Prolactin Specifically Inhibits Basal Or Fibroblast Growth Factor Stimulated Growth Of Capillary Endothelial Cells. *Endocrinology* 1991, 129(2):896-900.

[137] D'Angelo G, Struman I, Martial J, Weiner RI: Activation Of Mitogen-Activated Protein Kinases By Vascular Endothelial Growth Factor And Basic Fibroblast Growth Factor In Capillary Endothelial Cells Is Inhibited By The Antiangiogenic Factor 16-Kda N-Terminal Fragment Of Prolactin. *Proc Natl Acad Sci U S A* 1995, 92(14):6374-6378.

[138] Eddahibi S, Guignabert C, Barlier-Mur AM, Dewachter L, Fadel E, Dartevelle P, Humbert M, Simonneau G, Hanoun N, Saurini F, Hamon M, Adnot S: Cross Talk Between Endothelial And Smooth Muscle Cells In Pulmonary Hypertension: Critical Role For Serotonin-Induced Smooth Muscle Hyperplasia. *Circulation* 2006, 113(15):1857-1864.

[139] Pakala R, Willerson JT, Benedict CR: Mitogenic Effect Of Serotonin On Vascular Endothelial Cells. *Circulation* 1994, 90(4):1919-1926.

[140] Palmieri D, Astigiano S, Barbieri O, Ferrari N, Marchisio S, Ulivi V, Volta C, Manduca P: Procollagen I COOH-Terminal Fragment Induces VEGF-A And CXCR4 Expression In Breast Carcinoma Cells. *Exp Cell Res* 2008, 314(11-12):2289-2298.

[141] Roberts AB, Wakefield LM: The Two Faces Of Transforming Growth Factor Beta In Carcinogenesis. *Proc Natl Acad Sci U S A* 2003, 100(15):8621-8623.

[142] Schedin P, O'Brien J, Rudolph M, Stein T, Borges V: Microenvironment Of The Involuting Mammary Gland Mediates Mammary Cancer Progression. *J Mammary Gland Biol Neoplasia* 2007, 12(1):71-82.

[143] Schedin P: Pregnancy-Associated Breast Cancer And Metastasis. *Nat Rev Cancer* 2006, 6(4):281-291.

[144] Polyak K: Pregnancy And Breast Cancer: The Other Side Of The Coin. *Cancer Cell* 2006, 9(3):151-153.

[145] Stein T, Salomonis N, Nuyten DS, Van De Vijver MJ, Gusterson BA: A Mouse Mammary Gland Involution Mrna Signature Identifies Biological Pathways Potentially Associated With Breast Cancer Metastasis *J Mammary Gland Biol Neoplasia* 2009, 14(2):99-116.

Index

B

C

D

F

G

H

I

M

N

O

P

Q

R

S

T

U

V

W

X

Y

Z